Advances in Regenerated Asphalt Mixtures

Advances in Regenerated Asphalt Mixtures

Editors

Yuanyuan Li
Tao Bai
Jizhe Zhang
Yangming Gao

MDPI • Basel • Beijing • Wuhan • Barcelona • Belgrade • Manchester • Tokyo • Cluj • Tianjin

Editors
Yuanyuan Li
School of Civil Engineering and Architecture
Wuhan Institute of Technology
Wuhan
China

Tao Bai
School of Civil Engineering and Architecture
Wuhan Institute of Technology
Wuhan
China

Jizhe Zhang
Qilu School of Transportation
Shandong University
Jinan
China

Yangming Gao
School of Civil Engineering and Built Environment
Liverpool John Moores University
Liverpool
United Kingdom

Editorial Office
MDPI
St. Alban-Anlage 66
4052 Basel, Switzerland

This is a reprint of articles from the Special Issue published online in the open access journal *Materials* (ISSN 1996-1944) (available at: www.mdpi.com/journal/materials/special_issues/Regenerated_Asphalt_Mixtures).

For citation purposes, cite each article independently as indicated on the article page online and as indicated below:

LastName, A.A.; LastName, B.B.; LastName, C.C. Article Title. *Journal Name* **Year**, *Volume Number*, Page Range.

ISBN 978-3-0365-7349-6 (Hbk)
ISBN 978-3-0365-7348-9 (PDF)

Contents

About the Editors

Yuanyuan Li

Yuanyuan Li is an Associate Professor at the School of Civil Engineering and Architecture, Wuhan Institute of Technology. He worked as the Marie Curie early-stage researcher at the University of Nottingham from 2018 to 2020. His work focused on the intelligent detection and fast repairing of bituminous pavement diseases. His current research areas are novel and green bituminous pavement materials, UV ageing mechanisms and anti-UV ageing methods for bitumen, the maintenance of bituminous pavement, etc. His studies were funded by the National Science Foundation of China (NSFC), Science and technology projects of the Department of Transportation of Hubei Province, the Science and Technology Project of Fujian Provincial Department of Science and Technology, and the Fundamental Research Funds for the Central Universities, CHD.

Tao Bai

Tao Bai is a Professor at the School of Civil Engineering and Architecture, Wuhan Institute of Technology. His interest areas include structures and materials for asphalt pavement, solid waste and RAP resource utilization, and eco-friendly maintenance technologies. His research has been funded by the NSFC, Natural Science Foundation Youth Fund, Science and Technology Projects of the Department of Transportation of Hubei Province.

Jizhe Zhang

Jizhe Zhang is an Associate Professor at Shandong University. He is mainly engaged in the research of water damage mechanism of asphalt concrete, composite modification of asphalt materials, and regeneration technology of asphalt materials, amongst others. He has published more than 30 academic papers. He presided over the project of NSFC, Shandong Provincial Natural Science Foundation, and the Postdoctoral Science Foundation of China. He also participated in the project funded by the Engineering and Natural Science Research Council (EPSRC) of the UK, where he conducted in-depth systematic research on the adhesion mechanism of asphalt–aggregate interface and the adhesion failure law after encountering water.

Yangming Gao

Yangming Gao is currently a Senior Lecturer at Liverpool John Moores University, the UK. His research interests include the mechanics of pavement materials, molecular dynamics simulations, multiscale and multiphysics modelling, ageing, fatigue damage, and healing of bituminous materials, as well as durable, sustainable, and smart road materials and technologies. He serves on the RILEM technical committee (278-CHA) and is an active member on the Transportation Research Board (TRB) and American Society of Civil Engineers (ASCE). He was the recipient of the 2020 Marie Skłodowska-Curie Actions (MSCA) Individual Fellowships and 2019 ARMOURERS and BRASIERS'Prize.

materials

MDPI

Editorial

Advances in Regenerated Asphalt Mixtures

Yuanyuan Li [1], Tao Bai [1,*], Jizhe Zhang [2] and Yangming Gao [3,*]

1 School of Civil Engineering and Architecture, Wuhan Institute of Technology, Wuhan 430074, China
2 Qilu School of Transportation, Shandong University, Jinan 250061, China
3 School of Civil Engineering and Built Environment, Liverpool John Moores University, Byrom Street, Liverpool L3 3AF, UK
* Correspondence: baigs08@wit.edu.cn (T.B.); y.gao-3@tudelft.nl (Y.G.)

This Special Issue is devoted to research on asphalt pavement materials, including asphalt binders, asphalt mixtures and recycled asphalt pavement (RAP). It called for outstanding papers on asphalt pavement recycling materials and new asphalt pavement materials to expand the current understanding of asphalt pavement materials.

Asphalt binder plays very important role in the asphalt mixture. It binds the granular aggregate together and provides the material with strength, meaning that the performance of the asphalt mixture largely depends on the technical performance of the asphalt binder. With respect to the asphalt binder, it is difficult for the performance of conventional petroleum asphalt to meet the requirements of new pavements. Therefore, the modification of the asphalt binder is suggested to improve the performance of the asphalt binder and asphalt mixture [1,2]. With respect to the asphalt mixture, the main efforts are directed toward the strength of the mixture, with the primary aim of reducing the waste of resources, improving the performance of asphalt materials and reducing the economic pressure on the RAP. A short review of the papers in this Special Issue follows.

Current asphalt pavements have large and complex axle loads, and the performance of asphalt pavements deteriorates rapidly with age. However, the performance of asphalt binders can be significantly improved with the use of asphalt modifiers. The asphalt modifier can be a solid waste, such as crumb rubber (CR), or a rock asphalt, such as Bourdon rock asphalt (BRA). Currently, the uncertainty of CR's swelling mechanism in asphalt limits the application of crumb rubber modified bitumen (CRMB) in the road field. To overcome this limitation, a comprehensive study was carried out on CRMB, swelling rubber in CRMB (SCR) and recycled asphalt after CR action (CRRB) [3]. It was noted that CR can improve the high-temperature performance, ultra-low-temperature cracking resistance, storage stability and elastic recovery of an asphalt binder. The decrease in the relative content of light components in asphalt binder improved the high-temperature performance of CRRB and worsened its low-temperature performance. The swelling reaction led to an increase in the volume of the CR. Additionally, the CR's microscopic surface became more complex, the small molecule content was significantly higher and the generation of C=C bonds led to an increase in the toughness of the crumb rubber. A Bourdon rock asphalt modified asphalt mixture (BRA-MAM) demonstrates excellent high-temperature performance and moisture damage resistance, reducing the dependence on petroleum asphalt for conventional asphalt pavements. The segregation problem of the BRA in a Bourdon rock asphalt modified asphalt mixture (BRA-MA) has attracted a great deal of scholarly attention. A study of the physical performance and storage stability of BRA-MA by particle size and content of BRA is reported in this Special Issue [4]. Through simulating the separation process of BRA-MA, the viscosity and high-temperature stability of BRA-MA were negatively correlated with the particle size of BRA and positively correlated with its dosage. The degree of separation of the BRA-MA was positively correlated with the particle size, dosage, storage temperature and storage time of the BRA. Less common asphalt modifiers include the mussel-like adhesive L-Dopa methacrylic anhydride (L-DMA), which reacts with

Citation: Li, Y.; Bai, T.; Zhang, J.; Gao, Y. Advances in Regenerated Asphalt Mixtures. *Materials* **2023**, *16*, 2872. https://doi.org/10.3390/ma16072872

Received: 1 April 2023
Accepted: 3 April 2023
Published: 4 April 2023

asphalt through physical reactions. The main physical reactions of L-DMA are hydrogen bonding, aromatic ring conjugation, the conjugation of benzene rings and the conversion of catechol to quinone and semi-quinone at high temperatures [5]. Among these reactions, the adhesion of modified asphalt was improved mainly through a hydrogen bond and aromatic ring conjugation, and its high-temperature performance was improved mainly through a hydrogen bond, benzene ring conjugation and catechol. L-DMA can improve the rutting coefficient, creep recovery rate and compound modulus of an asphalt binder. When the dosage of L-DMA is higher than 10%, the low-temperature rutting performance of the asphalt can be improved to some extent.

In addition, asphalt aging is an important research problem with respect to asphalt performance. Heat, ultraviolet radiation, moisture and other external factors can have effects on an asphalt binder, leading to its aging. An aged asphalt mixture demonstrates poor performance and a short life. Herein, thermal oxygen aging, ultraviolet aging, and still-water erosion tests were conducted to investigate the aging mechanism of asphalt under various aging conditions such as high temperature, ultraviolet radiation, and an aqueous solution, respectively [6]. High temperature and UV composite conditions had a positive impact on the high-temperature stability of asphalt, but their impact on low-temperature cracking resistance was negative. At the same time, ultraviolet rays had a greater impact on the physical and chemical performance of an asphalt binder than water. The mechanism of the thermal-oxidative aging of asphalt was the increase in saturated hydrocarbons and aromatic ring substances and the rapid generation of polar components. The influence of ultraviolet radiation on the physical and chemical performance of asphalt decreases with the deepening of aging. It is exciting that this provides a research idea for asphalt regeneration. On asphalt, when the aging effects of heat, ultraviolet rays and water occur in the positive direction, the performance of the asphalt can be restored from the reverse direction; that is, to supplement the components lost due to aging, achieve asphalt regeneration and achieve the reuse of the asphalt pavement materials. As steel slag powder (SSP) can be used as a substitute for a natural aggregate in asphalt mixtures, the application of steel slag to asphalt roads can alleviate the incongruous situation of the annual production and utilization of steel slag. Research on SSP as a substitute for natural mineral powder in an asphalt mixture has also been carried out in which the anti-aging performance of steel slag asphalt mortar (SSP-MAM) was the focus of the research. The thermal oxidation and UV aging performance of SSP-AM can be achieved through the use of antioxidants and UV absorbers to prepare active composite modified asphalt mortar (SSP-MAM) [7]. In SSP-MAM, antioxidants capture the free radicals generated by light and thermal oxygen while ultraviolet absorbers convert ultraviolet rays into heat, improving their thermal oxidation and ultraviolet aging performance.

The purpose of designing a high-performance asphalt binder is to improve the performance of asphalt mixture and ultimately apply it to asphalt pavement to improve the performance and service life of the asphalt pavement. The research in this Special Issue indicates that improving the interaction between the asphalt masterbatch and aggregate can extend the durability of pavement, thereby improving its overall performance [8]. The current axle load environment and traffic requirements of automobiles require asphalt mixtures to have early strength, rutting resistance, and self-compacting performance. A new type of cement emulsified bitumen mixture (CEBM) demonstrates excellent rutting resistance, water stability, and dynamic stability which meet the corresponding requirements well [9]. The research on the strength formation mechanism, mechanical performance, and road performance of CEBM showed that the presence of water had a small impact on the adhesion between emulsified asphalt and aggregates and had a positive effect on the processability of the CEBM. In CEBM, the hydration products of cement form a skeleton in the aggregate, jointly establishing a spatial network structure with the original adhesion of the asphalt and ensuring the high strength of the CEBM. To ensure that the Marshall stability of the CEBM met the specifications, the curing time of the CEBM was required to be higher than 6 h, and the recommended dosage of emulsified asphalt and cement in CEBM was determined. The performance of asphalt mixtures is affected by various factors, one of which is aggregate gradation (AG). Through the coarse aggregate vibration compaction test, Marshall compaction test, and wheel tracking test of mixtures with different gradings, the effects of AG on the rutting performance (RP) and volume parameters (VPs) of an asphalt mixture were studied [10]. AG has a high impact on the RP and VPs of asphalt mixtures. To achieve a high bone density of asphalt mixtures, the fine and coarse aggregate boundary sieve (BS) should be 2.36 mm, and there should also be an appropriate BS pass rate. This Special Issue also introduces a phosphogypsum-based filler asphalt mixture, which is formed by coupling phosphogypsum with steel slag powder to form phosphogypsum-based filler (PF). When the proportion of phosphogypsum in the steel slag powder was 23%, a PF asphalt mortar (P-AM) achieved the best performance, and the PF had a positive effect on the high-temperature performance and water stability of the asphalt mixture [11].

The research on asphalt mixture cannot remain only at the laboratory level: on-site research is also necessary. The final on-site construction conditions are the last link that determines the performance of the new asphalt mixture pavement. The uncontrollable temperature problem during the construction of asphalt pavement is one of the causes of asphalt pavement damage. Here is a report on the pavement performance and service life issues caused by temperature segregation during paving with asphalt pavement [12]. Through the use of unmanned aerial vehicle infrared thermal imaging technology to collect the temperature distribution of an asphalt mixture pavement construction site and by simulating on-site pavement conditions in the laboratory, it was concluded that the prediction results for the performance of asphalt mixture pavement obtained using unmanned aerial vehicle infrared thermal imaging technology have a high accuracy, and the relationship between the melt temperature, high temperature stability and water stability of asphalt mixture was determined.

When the asphalt mixture reaches its service life, its road performance fails. At the same time, it is necessary to recover the asphalt pavement material. The recycling of recycled asphalt pavement materials is proposed to realize the resource utilization of recycled asphalt pavement materials, mostly including recycled asphalt and binder recycled asphalt mixture (RAM). To overcome the limitation of a single light oil regenerator on the performance of recycled asphalt, a composite regenerator made of tung oil, dioctyl phthalate (DOP), C9 petroleum resin and organic montmorillonite (OMMT) was used to regenerate aged asphalt [13]. The optimum ratio of the tung oil composite regenerate is tung oil: DOP: C9 petroleum resin: OMMT = 25:5:2:3. An asphalt regenerate at this ratio can promote the anti-aging performance of recycled asphalt and the dispersion and dissolution of polar substances, and the structure and morphology of aged asphalt can also be restored. During the process of regenerating aged asphalt with a tung oil composite regenerate, the macromolecules in the asphalt are destroyed and transformed into small molecules. The road performance of steel slag recycled asphalt mixture (SSRAM) and basalt recycled asphalt mixture (BRAM) has received attention [14]. Compared to RAM, SSRAM demonstrates better fatigue resistance, skid resistance, high- and low-temperature performance, and water stability, as well as excellent durability in high-temperature and water environments. SSRAM had better rutting resistance, fatigue performance, and low-temperature cracking resistance than BRAM.

Of course, the outstanding research in this Special Issue is not the end point of asphalt mixture research. There are still some limitations in the research process, such as the high-cost preparation of L-DMA, which limits its application in optimizing asphalt performance. In the future, new SSP-MAM anti-aging agent substitutes are needed to reduce the economic pressure caused by the extensive use of ultraviolet absorbers and antioxidants, and the engineering implementation and energy consumption of SSRAM must be studied in the future. These should not become frustrating issues. In the future, after breakthroughs in key issues in the continuous, pioneering research on asphalt mixtures, perhaps road scholars will find pleasure.

Author Contributions: Writing—original draft preparation, Y.L.; writing—review and editing, Y.L., T.B., J.Z. and Y.G.; funding acquisition, Y.L. and Y.G. All authors have read and agreed to the published version of the manuscript.

Funding: This Special Issue work was founded by the National Natural Science Foundation of China (No. 52108415), the Science and Technology Project of Fujian Provincial Department of Science and Technology (No. 2022H0027), the State Key Laboratory of Silicate Materials for Architectures (Wuhan University of Technology) (No. SYSJJ2022-21), the European Union's Horizon 2020 Research and Innovation Programme under the Marie Skłodowska-Curie grant agreement (No. 101030767), the Open Funding of Engineering Research Center of Ministry of Education for Traffic Pavement Materials, Chang'an University (300102312501), and the Fundamental Research Funds for the Central Universities, CHD.

Acknowledgments: The guest editors would-like to thank all the authors for supporting this Special Issue with their well-organized research work. The help from all the reviewers is also highly appreciated. Special thanks to the editorial assistants of Materials who made this entire Special Issue successful.

Conflicts of Interest: The authors declare no conflict of interest.

References

1. Li, Y.; Feng, J.; Chen, A.; Wu, F.; Wu, S.; Liu, Q.; Gong, R. Effects of Low-Temperature Construction Additives (LCAs) on the Performance of Asphalt Mixtures. *Materials* **2022**, *15*, 677. [CrossRef] [PubMed]
2. Chen, A.; Qiu, Y.; Wang, X.; Li, Y.; Wu, S.; Liu, Q.; Wu, F.; Feng, J.; Lin, Z. Mechanism and Performance of Bituminous Mixture Using 100% Content RAP with Bio-Rejuvenated Additive (BRA). *Materials* **2022**, *15*, 723. [CrossRef] [PubMed]
3. Zhu, H.; Zhang, M.; Li, Y.; Zou, Y.; Chen, A.; Wang, F.; Liu, L.; Gu, D.; Zhou, S. Swelled Mechanism of Crumb Rubber and Technical Properties of Crumb Rubber Modified Bitumen. *Materials* **2022**, *15*, 7987. [CrossRef] [PubMed]
4. Su, Y.; Hu, X.; Wan, J.; Wu, S.; Zhang, Y.; Huang, X.; Liu, Z. Physical Properties and Storage Stability of Buton Rock Asphalt Modified Asphalt. *Materials* **2022**, *15*, 3592. [CrossRef] [PubMed]
5. Wu, J.; Liu, Q.; Wu, S. Bionic Synthesis of Mussel-like Adhesive L-DMA and Its Effects on Asphalt Properties. *Materials* **2022**, *15*, 5351. [CrossRef] [PubMed]
6. Zou, Y.; Pang, L.; Xu, S.; Wu, S.; Yuan, M.; Amirkhanian, S.; Xu, H.; Lv, Y.; Gao, X. Investigation of the Rheological Properties and Chemical Structure of Asphalt under Multiple Aging Conditions of Heat, UV and Aqueous Solution. *Materials* **2022**, *15*, 5711. [CrossRef] [PubMed]
7. Lv, Y.; Wu, S.; Cui, P.; Amirkhanian, S.; Xu, H.; Zou, Y.; Yang, X. Optimization of Asphalt-Mortar-Aging-Resistance-Modifier Dosage Based on Second-Generation Non-Inferior Sorting Genetic Algorithm. *Materials* **2022**, *15*, 3635. [CrossRef] [PubMed]
8. Zhang, J.; Shen, Q.; Ji, P.; Wang, J.; Xiao, Y. Influence of Filler Type and Rheological Properties of Asphalt Mastic on the Asphalt Mastic–Aggregate Interaction. *Materials* **2023**, *16*, 574.
9. Yi, J.; Feng, J.; Li, Y.; Bai, T.; Chen, A.; Gao, Y.; Wu, F.; Wu, S.; Liu, Q.; Li, C. Research on Design and Performance of Self-Compacting Cement Emulsified Bitumen Mixture (CEBM). *Materials* **2022**, *15*, 4840. [CrossRef] [PubMed]
10. Li, W.; Cao, W.; Ren, X.; Lou, S.; Liu, S.; Zhang, J. Impacts of Aggregate Gradation on the Volumetric Parameters and Rutting Performance of Asphalt Concrete Mixtures. *Materials* **2022**, *15*, 4866. [CrossRef] [PubMed]
11. Wan, J.; Han, T.; Li, K.; Shu, S.; Hu, X.; Gan, W.; Chen, Z. Effect of Phosphogypsum Based Filler on the Performance of Asphalt Mortar and Mixture. *Materials* **2023**, *16*, 2486. [CrossRef] [PubMed]
12. Chen, W.; Wei, K.; Wei, J.; Han, W.; Zhang, X.; Hu, G.; Wei, S.; Niu, L.; Chen, K.; Fu, Z. Research on the Road Performance of Asphalt Mixtures Based on Infrared Thermography. *Materials* **2022**, *15*, 4309. [CrossRef] [PubMed]
13. Wang, Q.; Ye, Q.; Luo, J.; Xie, C.; Liu, H.; Liu, J.; Qin, M. Effects of Tung Oil Composite Regenerating Agent on Rheological Properties and Microstructures of Reclaimed Asphalt Binder. *Materials* **2022**, *15*, 3197. [CrossRef] [PubMed]
14. Wang, Z.; Wu, S.; Yang, C.; Xie, J.; Xiao, Y.; Zhao, Z.; Wang, F.; Zhang, L. Quantitative Assessment of Road Performance of Recycled Asphalt Mixtures Incorporated with Steel Slag. *Materials* **2022**, *15*, 5005. [CrossRef] [PubMed]

materials

MDPI

Article

Effects of Tung Oil Composite Regenerating Agent on Rheological Properties and Microstructures of Reclaimed Asphalt Binder

Qimin Wang [1], Qunshan Ye [2,3,*], Junhui Luo [1], Cheng Xie [1], Haobin Liu [1], Jianhua Liu [3] and Mengnan Qin [3]

1 Guangxi Beitou Transportation Maintenance Technology Group Co., Ltd., Nanning 530201, China; wangqm1615@163.com (Q.W.); jhluo85@hotmail.com (J.L.); xiecheng1357@163.com (C.X.); liuhaobin9203@163.com (H.L.)

2 Key Laboratory of Road Structure and Material of Ministry of Transport (Changsha), Changsha University of Science and Technology, Changsha 410114, China

3 Department of Traffic and Transportation Engineering, Changsha University of Science and Technology, Changsha 410114, China; 005848@csust.edu.cn (J.L.); szct@stu.csust.edu.cn (M.Q.)

* Correspondence: yequnshan@csust.edu.cn; Tel.: +86-136-8732-6030

Abstract: The single light oil regenerating agent has certain limitations on the performance recovery of aged asphalt. In this study, tung oil, dioctyl phthalate (DOP), C9 petroleum resin, and organic montmorillonite (OMMT) were used to prepare the composite regenerating agent, and its optimal mix proportion was determined by the orthogonal experimental design. The rheological properties and anti-aging performance of reclaimed asphalt were studied by the dynamic shear rheometer (DSR) and bending beam rheometer (BBR); and the Fourier transform infrared (FTIR) spectrometer, gel permeation chromatography (GPC), and scanning electron microscope (SEM) were adopted to explore its microstructure, morphology, and mechanism of action. The results show that with the addition of tung oil composite regenerating agent, the rheological properties of aged asphalt can be effectively recovered, even better than that of base asphalt. By using the complex modulus aging index (CMAI) and phase angle aging index (PMAI) it is found that the anti-aging performance of reclaimed asphalt is better than that of base asphalt. With the optimal content of the tung oil composite regenerating agent, the contents of characteristic functional groups and macromolecular asphaltenes in the aged asphalt can be reduced, indicating that the composite regenerating agent is beneficial to the dispersion and dissolution of polar substances in the aged asphalt. After aging, a large number of wrinkles appear on the surface of the asphalt. However, the addition of the tung oil composite regenerating agent can make the asphalt surface smooth, which indicates that the tung oil composite regenerating agent can restore the microstructure and morphology of aged asphalt to a certain extent.

Keywords: tung oil composite regenerating agent; aged asphalt; reclaimed asphalt; rheological properties; microstructure and morphology

Citation: Wang, Q.; Ye, Q.; Luo, J.; Xie, C.; Liu, H.; Liu, J.; Qin, M. Effects of Tung Oil Composite Regenerating Agent on Rheological Properties and Microstructures of Reclaimed Asphalt Binder. *Materials* **2022**, *15*, 3197. https://doi.org/10.3390/ma15093197

Academic Editor: Yuanyuan Li

Received: 31 March 2022
Accepted: 26 April 2022
Published: 28 April 2022

1. Introduction

With the rapid development of road construction in China, asphalt pavement is widely used due to its excellent road performance. However, under the comprehensive action of various factors, such as a complex traffic environment and driving load, the aging phenomena of asphalt materials inevitably appear, which weakens the adhesion, aging resistance, low-temperature crack resistance, and other properties of asphalt pavement, eventually producing a large amount reclaimed asphalt pavement (RAP) [1,2]. By rationally treating a large number of old materials, the shortage of raw asphalts can be alleviated and an economical and environmentally friendly society with green transportation can be realized. Therefore, much attention has been paid to the recycling technology of reclaimed asphalt mixtures.

The recycling technology of asphalt is that a certain proportion of a regenerating agent is added to the reclaimed asphalt and effectively achieves its regeneration, thereby prolonging the service life of asphalt pavement [3–5]. The demand for regenerating agents is increasing day by day. However, it is difficult for the traditional single light oil to be used as the regenerating agent to achieve the ideal regeneration effect. In addition, the regeneration cost is growing higher and higher, which restricts the recycling technology. Therefore, it is inevitable to develop a green, environmentally friendly, and economical composite regenerating agent to replace the traditional regenerating agent and single light oil.

At present, regenerants with mineral oil as the main component are widely used. However, its shortcomings such as strong volatility and non-renewable restrict the development of regeneration technology [6]. The reclaimed light oil is usually used as a regenerating agent because of its good performance, low cost, and easily available raw materials. For example, reclaimed vegetable oil, organic oil, aromatic extract, distilled tall oil, and bio-oil can be used as regenerating agents to restore the performance of aged asphalt [7–9]. The reclaimed cooking oil (WCO) is conducive to the physical properties, rheological properties, and other pavement behaviors of asphalt binders [10–12]. Eleyedath et al. [13] believed that the light components contained in such regenerating agents had a small molecular weight and low viscosity, which could quickly restore their performance after mixing with the aged asphalt; however, the difference in molecular weight between these lightweight components and other molecules was too large, which led to poor compatibility, and the rapid loss of small and medium molecules caused the poor durability of asphalt pavement. The use of bio-oil can improve the low-temperature cracking resistance of aged asphalt, and its microstructure is similar to that of base asphalt [14–16]. The residual soybean oil selected as a regenerating agent increases the permeability and reduces the viscosity, which is unfavorable for the high-temperature rutting resistance [17,18]. Moreover, the reclaimed engine oil chosen as a regenerating agent improves the viscoelasticity and flexibility of the asphalt binder, and its anti-stripping performance is comparable to that of base asphalt [19,20]. Zhang et al. [21] used reclaimed wood-derived bio-oil to balance the chemical components of aged asphalt. Tung oil with the main chemical component of fatty acid triglycerides can be used as a natural light oil to supplement and balance the missing components of aged asphalt [22].

From the above-mentioned information, most of the current research focuses on a single oil regenerating agent. However, a single light oil or other aromatic compounds used as a regenerating agent can only improve the fluidity of the asphalt but finds it difficult to restore or improve the overall performance of the reclaimed asphalt. In this paper, tung oil, dioctyl phthalate (DOP), C9 petroleum resin, and organic montmorillonite (OMMT) were compounded to prepare a composite regenerating agent of asphalt. The optimal content of each raw material of the composite regenerating agent was determined by the orthogonal design test, which effectively restored the rheological properties and anti-aging performance of aged asphalt, and combined with the microstructure test, its regeneration mechanism was analyzed as well.

2. Raw Materials and Test Methods

2.1. Raw Materials

The tung oil from a tung oil factory in Mianyang, Sichuan was used as the base oil; DOP from a chemical plant in Yixing, Wuxi was adopted as the plasticizer; C9 petroleum resin from a chemical plant in Dongguan, Guangdong was chosen as the tackifier resin; and OMMT with the advantages of barrier properties, aging resistance, and flame retardancy was selected from a mineral products processing plant in Hebei. The main technical indexes of each raw material are shown in Table 1.

Table 1. Technical indexes of tung oil.

Raw Materials	Appearance	Density/g·cm^{-1}	Flash Point/°C
Tung oil	Yellow liquid	0.943	236
DOP	Colorless oily liquid	0.985	225
C9 petroleum resin	Yellow particle	0.995	260
OMMT	White powder	1.03	-

2.2. Asphalt Preparation

2.2.1. Aged Asphalt

The PG64-22 petroleum asphalt provided by Hunan Baoli International was used as the base asphalt, and the aged asphalt was prepared by the laboratory simulation accelerated aging test method. The specific operation is that the base asphalt is aged by the rolling thin film oven test (RTFOT) for 85 min, and then placed in a pressurized aging vessel (PAV) to accelerate the aging for 20 h. The technical indexes of base asphalt and aged asphalt are shown in Table 2.

Table 2. Technical indexes of substrate asphalt and aged asphalt.

Technical Indexes	Base Asphalt	Aged Asphalt	Test Methods
Penetration (25 °C)/0.1 mm	68.7	21.1	ASTM D5
Ductility (15 °C)/cm	142.0	5.6	ASTM D113
Softening point (ring and ball method)/°C	48.0	65.6	ASTM D36
Viscosity (135 °C)/mPa·s	485	936	ASTM D4402

2.2.2. Reclaimed Asphalt

According to the previous research [22,23], for long-term-aged asphalt with RTFOT aging for 85 min and PAV aging for 20 h, the appropriate content of tung oil is in the range of 2–8%. We must consider that different sources of asphalts have different performances to some extent. The tung oil composite regenerating agent was blended with the aged asphalt (4%, 6%, 8%, 10%, and 12%) to prepare the reclaimed asphalt. The aged asphalt was kept at 135 ± 5 °C with the tung oil composite regenerating agent added. It was sheared for 20 min (3000 r/min) by the high-speed shear, and then continuously stirred for 10 min (500 r/min). The reclaimed asphalt is named according to the content of the regenerating agent, such as R-4% reclaimed asphalt, that is, the content of the tung oil composite regenerating agent in the reclaimed asphalt is 4%.

2.3. Test Methods

2.3.1. Rheological Property Test

The DSR test can measure the complex modulus (G^*) and the phase angle (δ) of the asphalt, and the rutting resistance of asphalt pavement can be characterized by the rutting factor $G^* / \sin\delta$. In this study, high-temperature rheological properties were evaluated by temperature sweep and frequency sweep tests. A rotor with a diameter of 25 mm was selected for base asphalt and reclaimed asphalt, and its setting interval is 1 mm. However, an 8 mm rotor with an interval of 2 mm was chosen for the aged asphalt. The temperature range of the temperature sweep test is 42~72 °C. The temperatures of the frequency sweep are 28 °C, 40 °C, 52 °C, 64 °C, and 76 °C, and its frequency is in the range of 0.1–10 Hz at each temperature. The master curves of the complex modulus and phase angle were constructed by the time-temperature equivalence principle. Based on the time–temperature equivalence principle, the effect of the temperature and loading frequency on the asphalt material was converted into a reduced frequency by using the displacement factor. The displacement factor can be obtained by the WLF empirical equation:

$$Log\alpha_T = \frac{-C_1\left(T - T_{ref}\right)}{C_2 + T - T_{ref}} \tag{1}$$

where α_T is the displacement factor at T; T_R is the reference temperature; and C_1 and C_2 are empirical constants.

In addition, the low temperature cracking resistance of asphalt can be characterized by the BBR test. The BBR test can measure the creep stiffness (*S*) and creep rate (*m*), and the test temperature includes −12 °C, −18 °C, and −24 °C.

2.3.2. Anti-Aging Performance Test

The aging resistance of reclaimed asphalt was analyzed by rheological properties after PAV aging and UV aging. The aging resistance of reclaimed asphalt was analyzed according to the effects of rheological indicators (CMAI, PMAI) on the rheological properties of aged asphalt. The calculation of Formulas (2) and (3) is shown below:

$$\mathrm{CMAI} = \frac{G^*}{G_0^*} \tag{2}$$

$$\mathrm{PMAI} = \frac{\delta}{\delta_0} \tag{3}$$

where G^* is the complex modulus of asphalt after aging; G_0^* is the complex modulus of asphalt before aging; δ is the phase angle of asphalt after aging; and δ_0 is the phase angle of asphalt before aging.

2.3.3. Micro Performance Test

Scanning electron microscopy (SEM) was used to compare the difference between the microstructure and morphology of reclaimed asphalt with the optimum content and base asphalt. The asphalt samples were sprayed with gold prior to SEM.

Fourier transform infrared (FTIR) spectroscopy was adopted to compare the differences in the composition of characteristic functional groups between the reclaimed asphalt with the optimum content and the base asphalt. FTIR spectroscopy studied the physical and chemical changes during the aging and regeneration processes of asphalt, and the information on chemical bonds or functional groups was obtained through the absorption peaks with FTIR spectroscopy. The changes in asphalt functional groups after adding the tung oil composite regenerating agent were analyzed. The FTIR test wavelength range is 500–4000 cm^{-1}, and the number of scans is 32.

Gel permeation chromatography (GPC) analyzed the molecular weight distribution changes of asphalt during the aging and regeneration processes. The molecular weight distribution of asphalt measured by GPC is closely related to the macroscopic properties of asphalt [24]. In this study, the mobile phase is tetrahydrofuran (THF), the concentration of the asphalt sample is 2 mg/mL, and the flow rate is 10 mL/min.

3. Orthogonal Test Design and Analysis

We used orthogonal tables to analyze multi-factor and multi-level experiments [25]. Under the condition that the orthogonal test can ensure the level of each test factor, the same number of tests can simplify the test groups and improve the test efficiency. The components of the tung oil composite regenerating agent mainly composed of tung oil, DOP, C9 petroleum resin, and OMMT were taken as the main factors, and the factor levels of the orthogonal test design are shown in Table 3.

Table 3. Factor levels of orthogonal test design.

Levels	Tung Oil (A)/%	DOP (B)/%	C9 Petroleum Resin (C)/%	OMMT (D)/%
Level 1	75	10	14	1
Level 2	70	15	10	5
Level 3	65	20	6	9

4. Test Results and Analysis

4.1. Orthogonal Test Results of Tung Oil Composite Regenerating Agent

In order to select the primary and secondary effects of each factor on each index, a 25 °C penetration and softening point, a 15 °C ductility, and 135 °C viscosity of reclaimed asphalt were used as evaluation indexes to discuss, analyze, and determine the optimal level of each factor. The orthogonal test results and preferred combinations are shown in Tables 4 and 5, respectively.

Table 4. Orthogonal test results.

No.	Tung Oil (A)/%	DOP(B)/%	C9 Petroleum Resin (C)/%	OMMT(D)/%	Softening Point/°C	Penetration/0.1 mm	Ductility/cm	Viscosity/mPa·s
1	75	10	14	1	48.4	93.5	143.6	480.0
2	75	15	10	5	47.0	100.1	138.2	455.0
3	75	20	6	9	45.1	130.0	142.6	404.0
4	70	10	10	9	46.6	145.2	125.4	447.0
5	70	15	6	1	46.7	103.2	134.3	426.0
6	70	20	14	5	48.4	82.5	121.0	477.5
7	65	10	6	5	46.5	99.6	117.4	451.0
8	65	15	14	9	47.2	98.9	119.3	483.0
9	65	20	10	1	48.2	86.0	97.6	505.0

Table 5. Preferred combinations of orthogonal tests.

Indexes	Preferred Combinations
Softening point/°C	$A_1B_2C_3D_3$
Penetration/(0.1mm)	$A_2B_2C_3D_3$
Ductility/cm	$A_1B_2C_3D_3$
Viscosity/(mPa·s)	$A_1B_2C_3D_3$

It can be seen from Tables 4 and 5 that the performance of aged asphalt can be restored by selecting 70% or 75% of tung oil; however, when the content of tung oil increases from 70% to 75%, the change trend of the penetration is relatively small, while the change trend of the ductility, viscosity, and softening are relatively large. Therefore, the optimal content of tung oil is 75%. With the increase in DOP content, the indexes of the softening point and viscosity first decrease and then increase, while the indexes of the penetration and ductility first increase and then decrease, indicating that the DOP starts to have adverse effects after improving the aging asphalt maximally. As a result, the optimal content of DOP is 15%. To meet the principle whereby the softening point is the minimum while the penetration and ductility are the maximum, the optimal contents of the C9 petroleum resin and OMMT are 6% and 9%, respectively.

According to the comprehensive analysis of the orthogonal test, the best combination of the tung oil composite regenerating agent is $A_1B_2C_3D_3$, namely tung oil: DOP: C9 petroleum resin: OMMT = 25:5:2:3.

4.2. Effects of Composite Regenerating Agent on the Rheological Properties of Reclaimed Asphalt

4.2.1. Complex Modulus

Figures 1 and 2 show the effect of the tung oil composite regenerating agent on the complex modulus and phase angle of aged asphalt. The range of the test temperature is 42–72 °C, and its increase rate is 2 °C/min; the loading frequency ω is 10 rad/s, and the strain control is 12%.

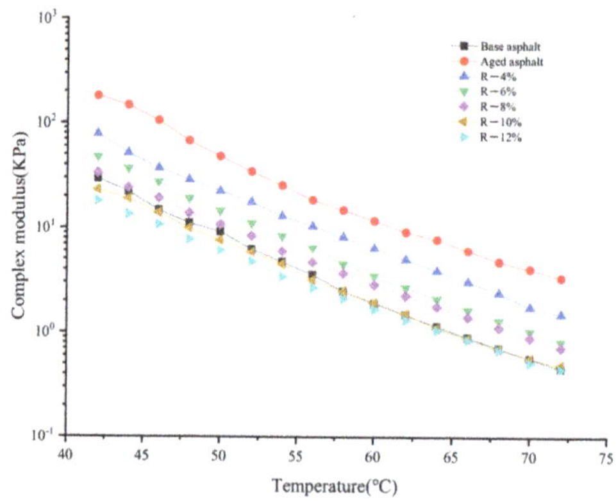

Figure 1. Test results of complex modulus (G^*) of aged asphalt after the regeneration.

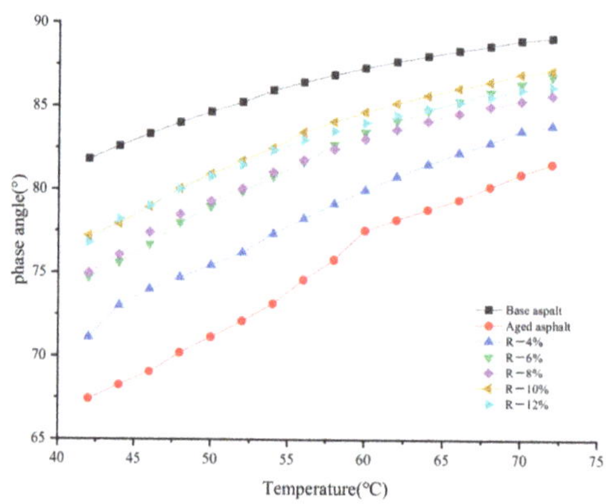

Figure 2. Test results of phase angle (δ) of aged asphalt after the regeneration.

It can be seen from Figure 1 that the complex modulus G^* of all asphalt samples decreases gradually with the increase in temperature. For instance, at the initial test temperature, the G^* of aged asphalt is nearly 6 times higher than that of base asphalt, indicating that the aging makes the asphalt harder. The addition of the tung oil composite regenerating agent can reduce the complex modulus of reclaimed asphalt, because the tung oil can dissolve macromolecular substances and supplement the light components of asphalt, softening the asphalt and reducing the complex modulus. As the content of the tung oil composite regenerating agent increases, the G^* of each reclaimed asphalt gradually decreases, which has a negative influence on the deformation resistance of the asphalt. However, the appropriate content of the tung oil composite regenerating agent can restore the fluidity of the aged asphalt. The G^* of the R-8% reclaimed asphalt is close to or even higher than that of base asphalt, partly because the C9 petroleum resin in the composite regenerating agent is favorable to high temperatures. YAN [22] et al. used tung oil as a regenerating agent to restore the high-temperature rheological properties of aged asphalt only to the level of base asphalt. However, the composite regenerating agent of tung oil in this paper caused the high-temperature performance of R-8% asphalt to be better than that of matrix asphalt. Therefore, the tung oil composite regenerating agent can restore and improve the deformation resistance of aged asphalt.

4.2.2. Phase Angle

In Figure 2, it is shown that after the asphalt is aged, the phase angle δ decreases and the deformation resistance increases. With the addition of the tung oil composite regenerating agent, the phase angle δ of reclaimed asphalt gradually decreases and is still smaller than that of base asphalt, indicating the elastic recovery ability of reclaimed asphalt is better than that of base asphalt.

4.2.3. Rutting Factor

The test results of the rutting factor of reclaimed asphalt are shown in Figure 3. As the temperature increases, the $G^*/\sin\delta$ of all asphalts gradually decreases. Moreover, as the content of the tung oil composite regenerating agent increases, it also declines, indicating that the addition of the regenerating agent and the increase in temperature reduce the deformation resistance of the asphalt. The $G^*/\sin\delta$ of aged asphalt is the largest, indicating its rutting resistance is the best. As the content of the tung oil composite regenerating agent increases, the $G^*/\sin\delta$ of reclaimed asphalt gradually decreases and is close to that of base asphalt. The addition of too much of the tung oil composite regenerating agent can lead to poorer rutting resistance of reclaimed asphalt. Therefore, the proper content of the regenerating agent ensures that the reclaimed asphalt has sufficient rutting resistance. The $G^*/\sin\delta$ of R-8% reclaimed asphalt is very close to or even better than that of base asphalt. Thus, the content of the tung oil composite regenerating agent should not exceed 8%.

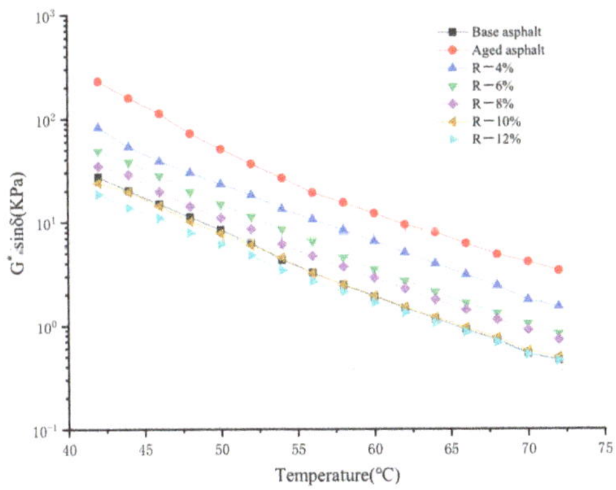

Figure 3. Test results of rutting factor ($G^*/\sin\delta$).

4.2.4. Master Curve

The temperature range of the frequency sweep test is 28–76 °C (the temperature interval is 12 °C), and its sweep frequency is 0.1–10 Hz. According to the WLF equation [26], the master curve of the complex modulus and phase angle constructed at the reference temperature of 20 °C is shown in Figures 4 and 5.

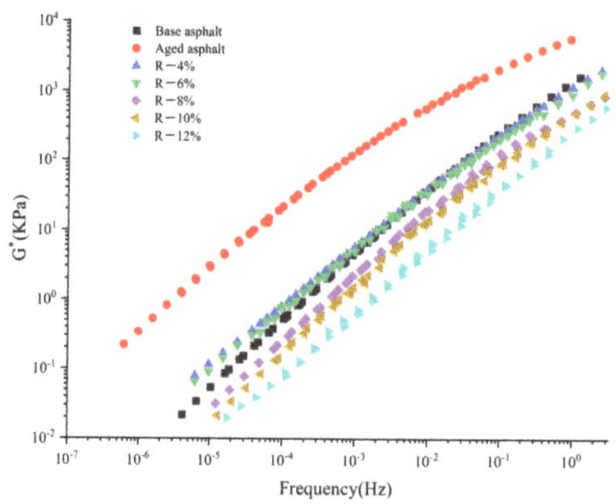

Figure 4. Master curve of complex modulus of reclaimed asphalt.

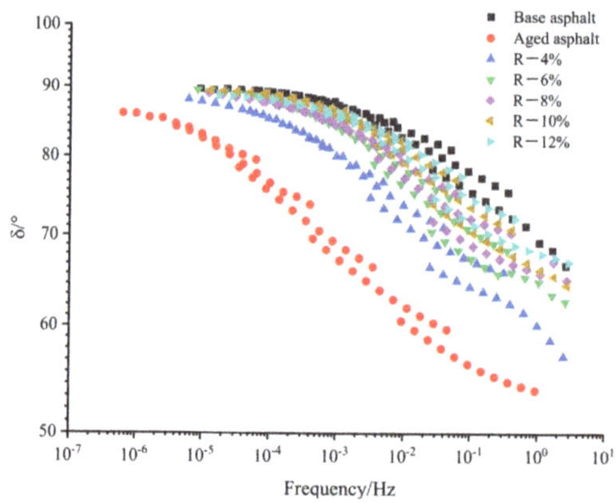

Figure 5. Master curve of phase angle of reclaimed asphalt.

As shown in Figure 4, compared with the base asphalt, the aged asphalt shows a higher complex modulus, which is beneficial to the rutting resistance of RAP at a low frequency and high temperature; G^* has a great linear relationship with the frequency. As the content of the tung oil composite regenerating agent increases, the G^* of the asphalt shifts close to that of base asphalt and increases with the increase in frequency, which means that the asphalt has the advantage of road deformation resistance at a high-frequency state and a low temperature. As the loading frequency decreases and the temperature increases, the G^* of reclaimed asphalt decreases and the G^* of R-8% reclaimed asphalt is the closest to that of base asphalt.

It can be seen from Figure 5 that the aged asphalt has the smallest δ due to the loss of light components, which increases the proportion of elastic components in the asphalt; the addition of the tung oil composite regenerating agent can increase the δ of aged asphalt. As the content of the tung oil composite regenerating agent increases, the δ of the asphalt gradually increases and is close to that of base asphalt, indicating that the tung oil composite regenerating agent can increase the proportion of viscous components in the aged asphalt and improve the viscoelastic properties of aged asphalt; Moreover, when the content of the tung oil composite regenerating agent is 8%, the δ of R-8% asphalt is smaller than that of base asphalt, indicating that the elastic recovery performance of R-8% asphalt is better than that of base asphalt.

The black diagram used to evaluate the viscoelastic properties of asphalt is a diagram of rheological data for asphalt materials in the form of complex modulus and phase angle. Figure 6 shows a black diagram of base asphalt, aged asphalt, and reclaimed asphalt. The curve of aged asphalt is incoherent in the black diagram. However, because the tung oil composite regenerating agent can improve the molecular conformation of aged asphalt to a certain extent, the curve of reclaimed asphalt is smooth, which is basically a coherent curve. The phase angle of reclaimed asphalt is smaller than that of the base asphalt, indicating that the elastic response of reclaimed asphalt is stronger.

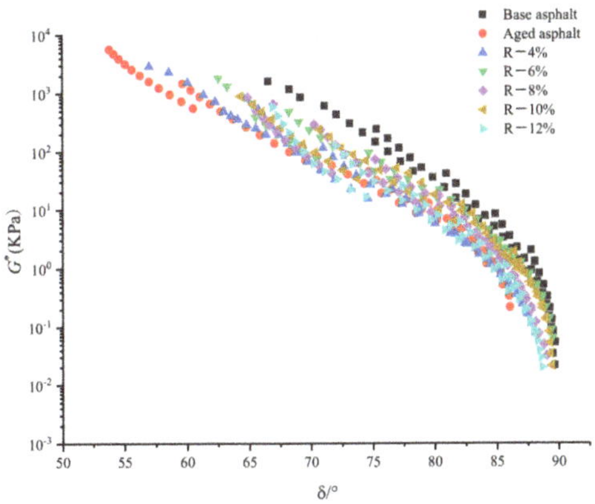

Figure 6. Black diagram of reclaimed asphalt.

4.2.5. Creep Stiffness and Creep Rate

The results of creep stiffness S and creep rate m of different asphalt samples were shown in Figures 7 and 8. The test temperatures are −12 °C, −18 °C, and −24 °C. There is no test result for R-10% and R-12% reclaimed asphalts due to their excessive deformation at −12 °C.

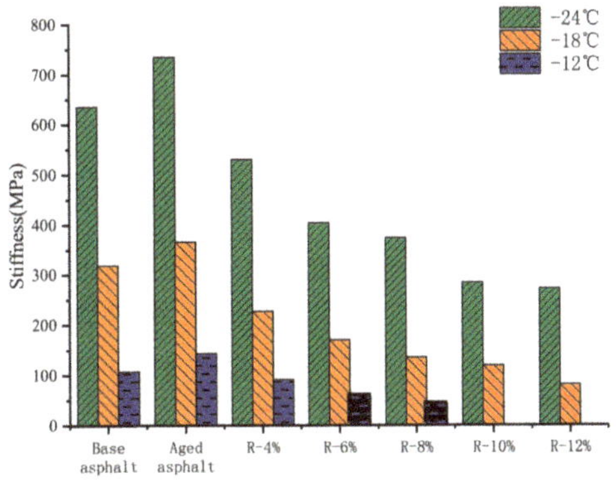

Figure 7. S of different reclaimed asphalt samples.

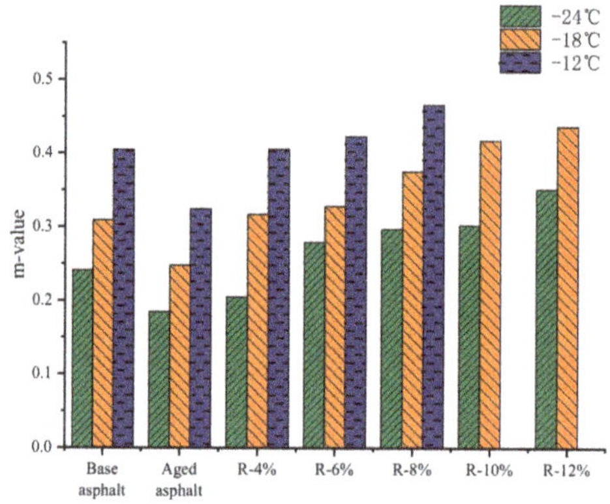

Figure 8. m of different reclaimed asphalt samples.

It can be seen from Figures 7 and 8 that as the content of the tung oil composite regenerating agent increases, the *S* and *m* of reclaimed asphalt gradually decreases and increases, respectively, indicating that with the addition of the tung oil composite regenerating agent, the low-temperature flexibility and cracking resistance of reclaimed asphalt are gradually improved. YAN [22] restored the low-temperature performance of aged asphalt by using tung oil as a regenerating agent. At −18 °C, the S of the regenerated asphalt with 8% tung oil is 170 MPa, and its m is 0.38, while the content of the tung oil composite regenerating agent in this paper is 8%, the S of R-8% asphalt is 135 MPa, and its m is 0.375, indicating that the tung oil composite regenerating agent has better recovery ability compared to the low-temperature performance of aged asphalt. This is because the plasticizer in the tung oil composite regenerating agent can improve the flexibility, low-temperature ductility, and crack resistance of the asphalt. Compared with those of base asphalt, the *S* and *m* of R-8% reclaimed asphalt decrease by 60% and increase by 15.1% at −12 °C, respectively; at −18 °C, the *S* and *m* decrease by 57.7% and increase by 21.4%, respectively; and at −24 °C, the *S* and *m* decrease by 41.1% and increase by 23.2%, indicating that the tung oil composite regenerating agent can not only restore the *S* and *m* of aged asphalt to the level of base asphalt, but also improve the low-temperature crack resistance of reclaimed asphalt with the optimal content, which is better than that of base asphalt.

4.3. Anti-Aging Performance of Reclaimed Asphalt with Composite Regenerating Agent

4.3.1. Thermo-Oxidative Aging Resistance

Figure 9 shows the change trends of CMAI and PMAI of base asphalt and different contents of reclaimed asphalt after the PAV aging. The CMAI of asphalt first increases and then decreases with the increase in temperature. The motion state of asphalt molecules is closely related to the ambient temperature. According to its deformation characteristics under the action of external force, the asphalt can be divided into a glass state, high elastic state, and viscous flow state. Liquids with lower molecular weights usually tend to show lower viscous flow temperatures [27]. Prior to asphalt aging, more light components and fewer heavy components are generated in the asphalt. Under the action of external force, the relaxation time of asphalt molecules is shorter so that the asphalt is more prone to viscous flow, and its viscous flow temperature is relatively low; however, after thermal-oxidative aging, light components decrease and heavy components increase in the asphalt, which leads to an increase in the average relative molecular mass of the asphalt and increases the relaxation time of asphalt molecules; that is, the viscous flow temperature becomes higher. During the temperature scanning process, as the temperature increases, the viscous flow of original asphalt appears earlier than that of aging asphalt at the stage of 42~54 °C so that the complex modulus of original asphalt decays much faster than that of aged asphalt and improves the CMAI of the asphalt. It can be seen from Figure 9a that the CMAI of reclaimed asphalt after PAV aging is smaller than that of base asphalt, indicating that the OMMT in the tung oil composite regenerating agent can effectively block the penetration and propagation of gaseous substances in the asphalt, such as water molecules and oxygen, and delay the aging process of asphalt under thermal-oxidative action. Therefore, the tung oil composite regenerating agent can effectively improve the thermal-oxidative aging resistance of aged asphalt.

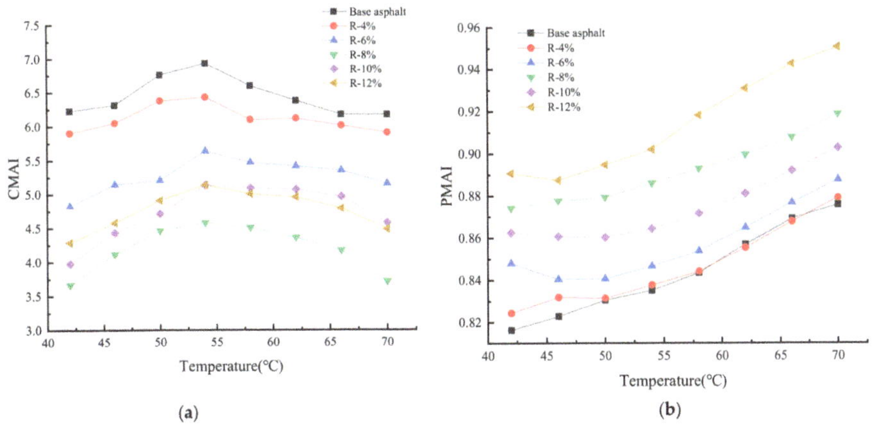

Figure 9. Thermo-oxidative aging resistance of reclaimed asphalt after PAV aging. (**a**) CMAI. (**b**) PMAI.

As the content of the tung oil composite regenerating agent increases, the CMAI of reclaimed asphalt first decreases and then increases, of which the CMAI of R-8% reclaimed asphalt is the smallest, indicating that the reclaimed asphalt has the best thermal-oxidative aging resistance. When the content of the tung oil composite regenerating agent exceeds 8%, the CMAI of reclaimed asphalt gradually increases. Thus, when the content of the tung oil composite regenerating agent is 8%, the reclaimed asphalt can be guaranteed to have better thermal-oxidative aging resistance. It can be seen from Figure 9b that the phase angle of the asphalt decreases after PAV aging, and more viscous components in the asphalt are transformed into elastic components; when the content of the tung oil composite regenerating agent is 4%, the PMAI of reclaimed asphalt is close to that of base asphalt; and when its content exceeds 4%, the PMAI of reclaimed asphalt is larger than that of base asphalt, indicating that the tung oil composite regenerating agent can restore the thermo-oxidative aging resistance of aged asphalt and effectively improve the thermo-oxidative aging resistance of reclaimed asphalt.

4.3.2. UV Aging Resistance

In Figure 10, the change trends of CMAI and PMAI of base asphalt and reclaimed asphalt with different contents after UV aging are shown. It can be seen from Figure 10a that the CMAI of base asphalt ranges from 2.2 to 2.7 and gradually decreases with the increase in temperature; and the CMAI of reclaimed asphalt is lower than that of base asphalt, indicating that the tung oil composite regenerating agent can reflect and absorb the UV light and decrease the damage of UV light to the reclaimed asphalt. When the content of the tung oil composite regenerating agent increases to 8%, the CMAI of reclaimed asphalt is the smallest, and thereby its UV aging resistance is the best. As shown in Figure 10b, the PMAI of base asphalt is 0.92~0.98 and increases linearly with the increase in temperature, while the PMAI of R-8%, R-10%, and R-12% reclaimed asphalts is larger than that of base asphalt, indicating that the tung oil composite regenerating agent can improve the UV aging performance of reclaimed asphalt. However, when the content of the tung oil composite regenerating agent exceeds 8%, the PMAI of reclaimed asphalt tends to decrease. As a result, when the content of the tung oil composite regenerating agent is 8%, it can be ensured that the reclaimed asphalt has better UV aging resistance.

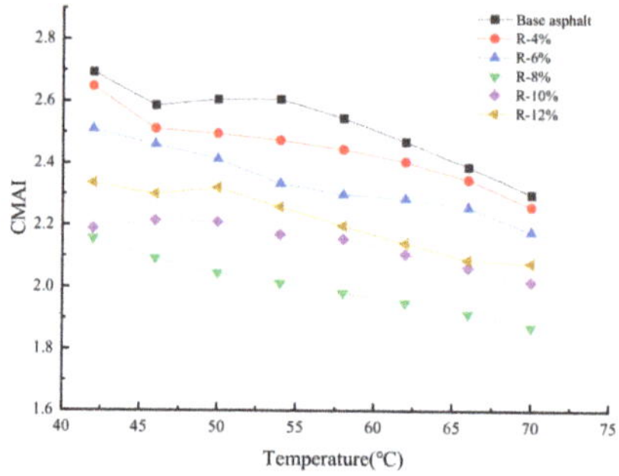

(a)

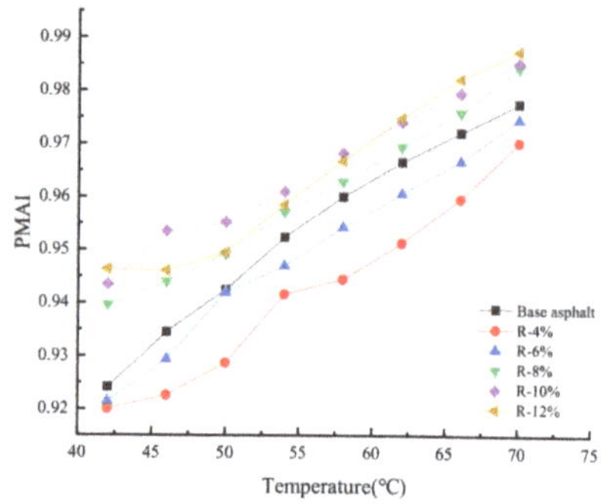

(b)

Figure 10. UV aging resistance of reclaimed asphalt after the UV aging. (**a**) CMAI. (**b**) PMAI.

4.4. Microstructure and Mechanism Analysis of Reclaimed Asphalt with Composite Regenerating Agent

4.4.1. Morphology

A Zeiss sigma 300 SEM was used to collect the surficial micro-morphologies of asphalt samples. The microstructures and morphologies of base asphalt, aged asphalt, and reclaimed asphalt are shown in Figure 11. It can be seen from Figure 11 that the overall surface of base asphalt is in a flat and smooth state, which is basically a homogeneous structure, and a large number of wrinkles appear on the surface of aged asphalt. This is because the light components decrease, the molecular polarity increases, the molecular movement ability is weakened, and the asphalt fluidity deteriorates after aging. With the addition of the tung oil composite regenerating agent, the surface of R-8% reclaimed asphalt tends to be flat and smooth, which is similar to that of base asphalt. In addition, the uneven striped "honeycomb structure" can be seen from the graphs of both aged asphalt and reclaimed asphalt, and the area of a single honeycomb structure of aged asphalt is larger than that of reclaimed asphalt, which may be due to the aggregation of macromolecular asphaltenes of the asphalt [20]. After the tung oil composite regenerating agent is added, the light components increase. Furthermore, the area of the honeycomb structure of R-8% reclaimed asphalt decreases, and its surface tends to be smooth, indicating that the tung oil composite regenerating agent can roughly restore the microstructure and morphology of aged asphalt.

(a)

Figure 11. *Cont.*

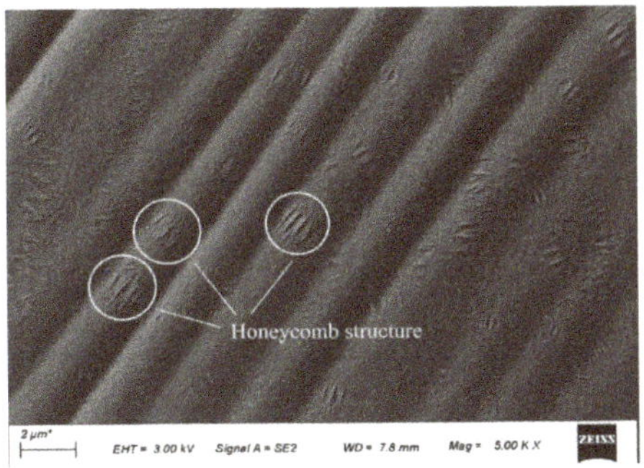

(**b**)

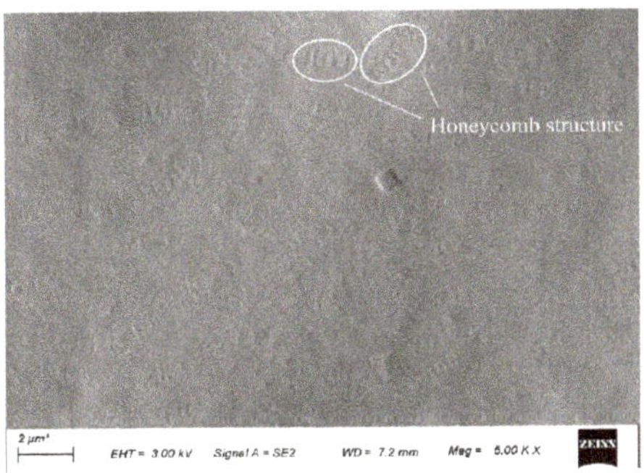

(**c**)

Figure 11. Micromorphologies of different asphalts under SEM (5000 times). (**a**) Base asphalt. (**b**) Aged asphalt. (**c**) R-8%.

4.4.2. Composition of Regenerating Agent and Reclaimed Asphalt

The functional groups of the asphalt samples were determined by the Nicolet iS50 FTIR spectrometer. The wavelength range of the test is 500–4000 cm^{-1}, and the number of scans is 32. The infrared spectra are shown in Figures 12 and 13.

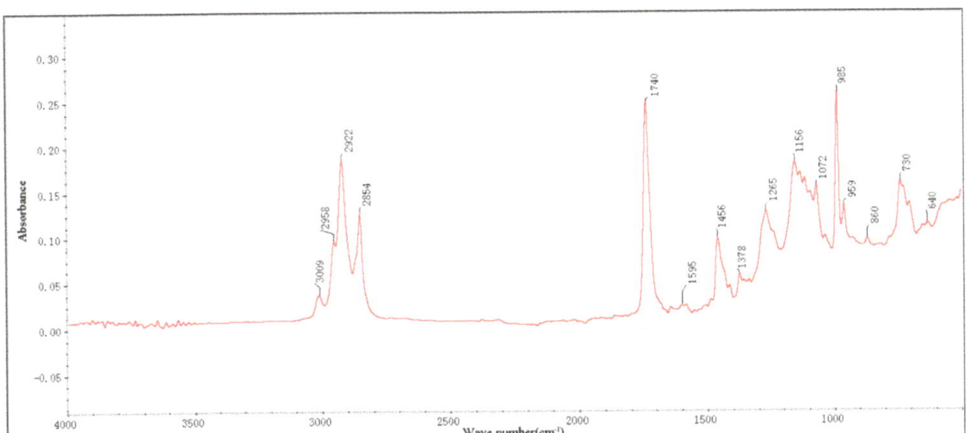

Figure 12. Infrared spectrum of tung oil composite regenerating agent.

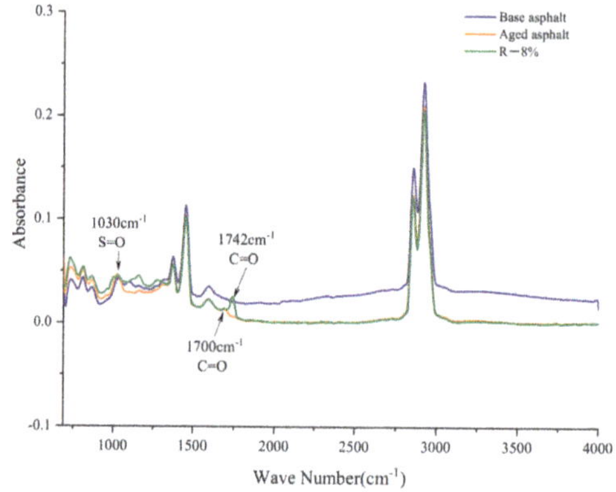

Figure 13. FTIR images of different asphalt samples.

The infrared spectrum of the tung oil composite regenerating agent is shown in Figure 12. Through the analysis of characteristic peaks in the infrared spectrum of the tung oil composite regenerating agent, it is found that the absorption peak near 2958 cm^{-1} is a CH_3 antisymmetric and symmetric stretching vibration, while the absorption peak in the interval of 2922 cm^{-1}~2854 cm^{-1} is a -CH_2 antisymmetric and symmetric stretching vibration, indicating that the tung oil composite regenerating agent contains non-polar methyl and methylene functional groups. The absorption peak near 3009 cm^{-1} is the C-H stretching vibration. The tung oil composite regenerating agent has absorption peaks of C=C stretching vibration of three aromatics near 1595 cm^{-1}, 1456 cm^{-1}, and 1378 cm^{-1}, and C-H bending vibration of the benzene ring in the interval of 900 cm^{-1}~650 cm^{-1}, indicating that the main components of the tung oil composite regenerating agent are light components rich in aromatic hydrocarbons. There is a C=O stretching vibration absorption peak of saturated fatty acid ester near 1740 cm^{-1} and a C-O stretching vibration absorption peak near 1265 cm^{-1}, 1156 cm^{-1}, and 1072 cm^{-1}. Moreover, there is a C-H in-plane bending and stretching characteristic peak in the interval of 900 cm^{-1}~600 cm^{-1}, indicating that the tung oil composite regenerating agent is rich in aromatic compounds and has good compatibility with the asphalt.

Figure 13 shows the infrared spectra of base asphalt, aged asphalt, and reclaimed asphalt. It can be found from Figure 13 that the positions of characteristic peaks of all asphalts are almost the same. Compared with the base asphalt, the aging asphalt has an absorption peak caused by the carbonyl C=O at 1700 cm^{-1}, and the characteristic peak of sulfoxide group S=O at 1030 cm^{-1} increases, which is caused by the oxidation reaction during the aging process of the asphalt. For the reclaimed asphalt, a new characteristic peak appears near 1742 cm^{-1} after the addition of the tung oil composite regenerating agent, which is caused by the C=O stretching vibration of saturated fatty acid ester. In addition, no other characteristic peaks appear in the reclaimed asphalt. Its characteristic peak is almost the same as that of base asphalt, indicating that the tung oil composite regenerating agent has no chemical reaction with the asphalt, rather only physical blending.

4.4.3. Molecular Weight and Distribution of Different Asphalts

The molecular weight and distribution of asphalt were analyzed by Waters 1515 GPC, and tetrahydrofuran (THF) was used as the mobile phase. The concentration and flow rate of the asphalt sample are 2 mg/mL and 10 mL/min, respectively.

In Figure 14, the abscissa of the GPC curve is the molecular weight, and its ordinate is the differential distribution of molecular weight. The GPC is usually divided into 13 blocks, of which blocks 1–5 are macromolecules (LMS), blocks 6–9 are medium molecules (MMS), and blocks 10–13 are small molecules [28]. According to the distribution of molecular weight in Figure 14, the integral areas of LMS, MMS, and SMS of base asphalt, aged asphalt, and reclaimed asphalt were calculated, and the content of each molecule of LMS, MMS, and SMS was obtained, as shown in Figure 15. It can be seen from the figure that compared with those of base asphalt, the LMS content of aged asphalt increases by 12.5%, while the MMS and SMS contents decrease by 5.8% and 6.7%, respectively. This is because, in the process of thermo-oxidative aging, there is a polymerization reaction between aromatic and colloidal components of small molecular weight and asphaltenes of a large molecular weight produced, thereby enhancing the intermolecular force of the asphalt and weakening its molecular movement ability. Compared with those of aged asphalt, the contents of LMS and MMS of R-8% reclaimed asphalt decrease by 3.2% and 1.4%, respectively, and the content of its SMS increases by 4.6%. It is shown that the tung oil composite regenerating agent contains a certain number of medium and small molecules, which can fully supplement small and medium molecules in the components of aging asphalt and dissolve a small part of the macromolecules, thus solving the agglomeration problem of macromolecules.

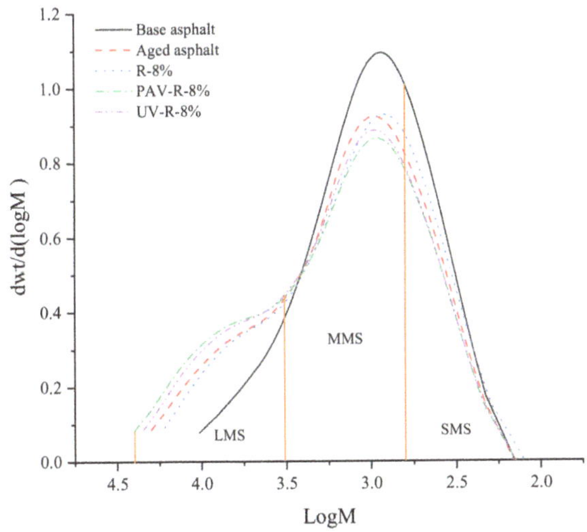

Figure 14. Molecular weight distribution of different asphalts.

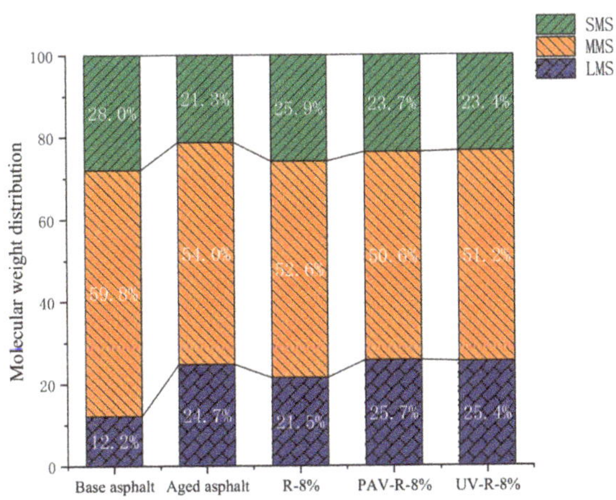

Figure 15. Molecular weight proportion of different asphalts.

In Figure 15, it is shown that after the PAV and UV aging, the LMS and MMS of R-8% reclaimed asphalt have different increasing trends, while its SMS has decreasing trends. Furthermore, the molecular weight of PAV-aged asphalt is larger than that of UV-aged asphalt, indicating that during the UV aging process, the R-8% reclaimed asphalt without thermo-oxidative aging loses few medium and small molecules. According to the change trend of the molecular weight of LMS, MMS, and SMS in Figure 15, the change range of molecular weight of base asphalt and R-8% reclaimed asphalt after the PAV aging was calculated, as shown in Table 6.

Table 6. Change range of molecular weight of asphalt after PAV aging.

Types of Asphalt	LMS (%)	MMS (%)	SMS (%)
Base asphalt	12.5	−5.8	−6.7
R-8% reclaimed asphalt	4.2	−2	−2.2

It can be seen from Table 6 that after PAV aging, the LMS content of base asphalt increases by 12.5%, while its MMS and SMS contents decrease by 5.8% and 6.7%, respectively. Furthermore, the LMS content of R-8% reclaimed asphalt increases by 4.2%, while its MMS and SMS contents decrease by 2% and 2.2%, respectively. It is obvious that the change range of the molecular weight of R-8% reclaimed asphalt is small, indicating that the aging performance of R-8% reclaimed asphalt decays slowly after PAV aging, which is beneficial to the aging resistance of reclaimed asphalt. Moreover, it can be found from Figure 15 that after PAV aging, the SMS content of R-8% reclaimed asphalt is more than that of base asphalt, indicating that the tung oil composite regenerating agent can inhibit the loss of small molecules in the reclaimed asphalt.

5. Conclusions

- The optimal mix proportion of the tung oil composite regenerating agent was determined by the orthogonal design test method; that is, tung oil: DOP: C9 petroleum resin: OMMT = 25:5:2:3.
- As the content of the tung oil composite regenerating agent increases, the rutting factor and creep stiffness gradually decrease and the creep rate increases, indicating that the tung oil composite regenerating agent can restore the rheological properties of aged asphalt, which is even better than that of base asphalt. The CMAI of reclaimed asphalt is smaller than that of base asphalt, while the PMAI of reclaimed asphalt is larger than that of base asphalt. The anti-aging ability of reclaimed asphalt is significantly improved, and the optimal content of the tung oil composite regenerating agent is 8%.
- As the content of macromolecules increases, the fluidity of aged asphalt becomes poor and a wrinkled texture and large honeycomb structure appear on its surface. The addition of the tung oil composite regenerating agent can restore the morphological features of aged asphalt to a certain extent, which makes its surface tend to be flat and smooth, and the size of te honeycomb structure is reduced.
- The FTIR diagram shows that the tung oil composite regenerating agent is mainly composed of light components rich in aromatic hydrocarbons, and the characteristic peaks of reclaimed asphalt are basically consistent with those of base asphalt, indicating that the tung oil composite regenerating agent is beneficial to the dispersion and dissolution of polar substances in the aged asphalt.
- The GPC results of reclaimed asphalt show that the tung oil composite regenerating agent can reduce the content of macromolecule in the aged asphalt, and the change range of molecular weight of reclaimed asphalt after aging is smaller than that of base asphalt, indicating that the aging of reclaimed asphalt decays slowly, which is favorable for the aging resistance of reclaimed asphalt.

Author Contributions: Conceptualization, Q.W. and Q.Y.; methodology, Q.W. and Q.Y.; validation, J.L. (Junhui Luo), J.L. (Jianhua Liu) and C.X.; formal analysis, H.L. and M.Q.; investigation, Q.Y. and M.Q.; resources, J.L. (Junhui Luo), J.L. (Jianhua Liu) and C.X.; data curation, H.L., J.L. (Junhui Luo) and J.L. (Jianhua Liu); writing—original draft preparation, Q.Y., J.L. (Junhui Luo) and J.L. (Jianhua Liu); writing—review and editing, Q.Y. and M.Q.; visualization, Q.W. and Q.Y.; supervision, Q.W. and Q.Y.; project administration, J.L. (Junhui Luo), J.L. (Jianhua Liu) and C.X. All authors have read and agreed to the published version of the manuscript.

Funding: This research was funded by Ministry of Transport of People's Republic of China (No.2020-MS1-005), the Education Department of Hunan Province (No. 20A012), the Department of Science of Changsha City (No. kq2014107), and the Key Laboratory of Road Structure and Material of Ministry of Transport (Changsha University of Science and Technology, No. kfj140301).

Institutional Review Board Statement: Not applicable.

Informed Consent Statement: Not applicable.

Data Availability Statement: The data presented in this study are available on request from the corresponding author.

Conflicts of Interest: The authors declare no conflict of interest.

References

1. Zhang, H.; Liu, H.; Zhang, Z. Study on the mechanism of the repeated asphalt ageing and recycling based on the macro-performance. *Road Mater. Pavement Des.* **2016**, *17*, 920–932. [CrossRef]
2. Zhao, K.; Wang, Y. Influences of aging conditions on the rheological properties of asphalt binders. *Int. J. Pavement Eng.* **2020**, *21*, 653–665. [CrossRef]
3. Zhou, C.L.; Zheng, C.C.; Cheng, D.X. Evaluation of Environmental Pollution from Asphalt Recycling Technology. *Adv. Mater. Res.* **2014**, *898*, 482–485. [CrossRef]
4. Xiao, F.; Yao, S.; Wang, J.; Li, X.; Amirkhanian, S. A literature review on cold recycling technology of asphalt pavement. *Constr. Build. Mater.* **2018**, *180*, 579–604. [CrossRef]
5. Cong, P.; Luo, W.; Xu, P.; Zhao, H. Investigation on recycling of SBS modified asphalt binders containing fresh asphalt and rejuvenating agents. *Constr. Build. Mater.* **2015**, *91*, 225–231. [CrossRef]
6. Suo, Z.; Nie, L.; Xiang, F.; Bao, X. The Effect of Waste Plant Oil on the Composition and Micro-Morphological Properties of Old Asphalt Composition. *Buildings* **2021**, *11*, 407. [CrossRef]
7. Kaseer, F.; Martin, A.E.; Arámbula-Mercado, E. Use of recycling agents in asphalt mixtures with high recycled materials contents in the United States: A literature review. *Constr. Build. Mater.* **2019**, *211*, 974–987. [CrossRef]
8. Zaumanis, M.; Mallick, R.B.; Poulikakos, L.; Frank, R. Influence of six rejuvenators on the performance properties of Reclaimed Asphalt Pavement (RAP) binder and 100% recycled asphalt mixtures. *Constr. Build. Mater.* **2014**, *71*, 538–550. [CrossRef]
9. Ji, J.; Yao, H.; Suo, Z.; You, Z.; Li, H.; Xu, S.; Sun, L. Effectiveness of Vegetable Oils as Rejuvenators for Aged Asphalt Binders. *J. Mater. Civ. Eng.* **2017**, *29*, D4016003. [CrossRef]
10. Zargar, M.; Ahmadinia, E.; Asli, H.; Karim, M.R. Investigation of the possibility of using waste cooking oil as a rejuvenating agent for aged bitumen. *J. Hazard. Mater.* **2012**, *233–234*, 254–258. [CrossRef]
11. Wang, C.; Xue, L.; Xie, W.; You, Z.; Yang, X. Laboratory investigation on chemical and rheological properties of bio-asphalt binders incorporating waste cooking oil. *Constr. Build. Mater.* **2018**, *167*, 348–358. [CrossRef]
12. Jalkh, R.; El-Rassy, H.; Chehab, G.R.; Abiad, M.G. Assessment of the Physico-Chemical Properties of Waste Cooking Oil and Spent Coffee Grounds Oil for Potential Use as Asphalt Binder Rejuvenators. *Waste Biomass Valorization* **2018**, *9*, 2125–2132. [CrossRef]
13. Eleyedath, A.; Swamy, A.K. *Use of Waste Engine Oil in Materials Containing Asphaltic Components//Eco-Efficient Pavement Construction Materials*; Woodhead Publishing: Sawston, UK, 2020; pp. 33–50.
14. Gong, M.; Yang, J.; Zhang, J.; Zhu, H.; Tong, T. Physical–chemical properties of aged asphalt rejuvenated by bio-oil derived from biodiesel residue. *Constr. Build. Mater.* **2016**, *105*, 35–45. [CrossRef]
15. Nizamuddin, S.; Baloch, H.A.; Jamal, M.; Madapusi, S.; Giustozzi, F. Performance of waste plastic bio-oil as a rejuvenator for asphalt binder. *Sci. Total Environ.* **2022**, *828*, 154489. [CrossRef]
16. Lv, S.; Liu, J.; Peng, X.; Liu, H.; Hu, L.; Yuan, J.; Wang, J. Rheological and microscopic characteristics of bio oil recycled asphalt. *J. Clean Prod.* **2021**, *295*, 126449. [CrossRef]
17. Zhang, H.; Wu, J.; Qin, Z.; Luo, Y. The Effect of Bio-Oil on High-Temperature Performance of Bio-Oil Recycled Asphalt Binders. *J. Renew. Mater.* **2022**, *10*, 1025–1037. [CrossRef]
18. Zhang, X.; Ning, Y.; Zhou, X.; Xu, X.; Chen, X. Quantifying the rejuvenation effects of soybean-oil on aged asphalt-binder using molecular dynamics simulations. *J. Clean Prod.* **2021**, *317*, 128375. [CrossRef]

19. Al-Saffar, Z.H.; Yaacob, H.; Mohd Satar, M.K.I.; Saleem, M.K.; Jaya, R.P.; Lai, C.J.; Shaffie, E. Evaluating the Chemical and Rheological Attributes of Aged Asphalt: Synergistic Effects of Maltene and Waste Engine Oil Rejuvenators. *Arab. J. Sci. Eng.* **2020**, *45*, 8685–8697. [CrossRef]
20. Qurashi, I.A.; Swamy, A.K. Viscoelastic properties of recycled asphalt binder containing waste engine oil. *J. Clean Prod.* **2018**, *182*, 992–1000. [CrossRef]
21. Zhang, R.; You, Z.; Wang, H.; Chen, X.; Si, C.; Peng, C. Using bio-based rejuvenator derived from waste wood to recycle old asphalt. *Constr. Build. Mater.* **2018**, *189*, 568–575. [CrossRef]
22. Yan, K.; Peng, Y.; You, L. Use of tung oil as a rejuvenating agent in aged asphalt: Laboratory evaluations. *Constr. Build. Mater.* **2020**, *239*, 117783. [CrossRef]
23. Yan, K.; Lan, H.; Duan, Z.; Liu, W.; You, L.; Wu, S.; Miljković, M. Mechanical performance of asphalt rejuvenated with various vegetable oils. *Constr. Build. Mater.* **2021**, *293*, 123485. [CrossRef]
24. Li, H.; Yang, F.; Zhang, F. Diffusion and Regeneration Mechanism of Waste Composite Oils Rejuvenator in Aged Asphalt. *J. Wuhan Univ. Technol. -Mater. Sci. Ed.* **2021**, *36*, 664–671. [CrossRef]
25. Wu, C.F.J.; Hamada, M.S. *Experiments: Planning, Analysis, and Optimization*; John Wiley and Sons: Hoboken, NJ, USA, 2011.
26. Wang, T.; Wei, X.; Zhang, D.; Shi, H.; Cheng, Z. Evaluation for Low Temperature Performance of SBS Modified Asphalt by Dynamic Shear Rheometer Method. *Buildings* **2021**, *11*, 408. [CrossRef]
27. Rubinstein, M.; Colby, R.H. *Polymer Physics*; Oxford University Press: New York, NY, USA, 2003.
28. Ma, J.; Sun, G.; Sun, D.; Yu, F.; Hu, M.; Lu, T. Application of gel permeation chromatography technology in asphalt materials: A review. *Constr. Build. Mater.* **2021**, *278*, 122386. [CrossRef]

materials

Article

Impacts of Aggregate Gradation on the Volumetric Parameters and Rutting Performance of Asphalt Concrete Mixtures

Weihua Li [1], Weidong Cao [2,*], Xianfu Ren [1], Shurong Lou [1], Shutang Liu [2,*] and Jizhe Zhang [2]

[1] Shandong High Speed Group Weifang Development Co., Ltd., Weifang 262500, China; wfglsj@163.com (W.L.); renxianfu-001@126.com (X.R.); sdgs144@163.com (S.L.)
[2] School of Qilu Transportation, Shandong University, Jinan 250002, China; jizhe.zhang@sdu.edu.cn
* Correspondence: cwd2001@sdu.edu.cn (W.C.); gtreesl@sdu.edu.cn (S.L.)

Abstract: The main objective of this study was to determine the effect of aggregate gradation (AG) on the volumetric parameters (VPs) and rutting performance (RP) of asphalt concrete (AC) mixtures. The boundary sieve (BS) between fine and coarse aggregates was 2.36 mm size, and 15 gradation curves of three nominal maximum aggregate sizes (13.2, 19.0, and 26.5 mm) were designed based on the percentage passing of the BS. A vibrating compaction test of coarse aggregates, Marshall compaction and wheel-tracking tests of AC mixtures with various gradations were conducted. It was found that AG had crucial effects on the VPs and RP of AC mixtures. The AC mixture can be designed as a skeletal dense structure provided that the percentage passing of the BS is appropriate. More notably, AC mixtures with a skeletal dense structure showed the best rutting resistance performance. Therefore, it is important to optimize AG for enhancing the high-temperature RP of AC mixtures.

Keywords: aggregate gradation; asphalt concrete mixture; volumetric parameters; rutting performance; skeletal dense structure

Citation: Li, W.; Cao, W.; Ren, X.; Lou, S.; Liu, S.; Zhang, J. Impacts of Aggregate Gradation on the Volumetric Parameters and Rutting Performance of Asphalt Concrete Mixtures. *Materials* **2022**, *15*, 4866. https://doi.org/10.3390/ma15144866

Academic Editor: Angelo Marcello Tarantino

Received: 20 June 2022
Accepted: 11 July 2022
Published: 13 July 2022

1. Introduction

It is well accepted that the performance of asphalt concrete (AC) pavement is greatly influenced by its aggregate characteristics, since AC mixtures contain approximately 90–95% of mineral aggregate by weight. In particular, aggregate gradation (AG) is a significant factor to be considered in the design of AC mixture. Hence, different methods have been developed and applied on gradation selection [1,2], including Superpave and Marshall mix designs, which meet the requirements for the volumetric properties of AC mixture. The paramount volumetric properties of a mix design include voids in the mineral aggregate (VMA), voids filled with asphalt (VFA), and air voids (AV). Current VMA requirements are built upon the premise that pavement performance and durability reduce when the VMA drops below the minimum value. However, an increase in minimum VMA can attenuate the performance of some mixtures. AG plays an important role in ensuring that an adequate amount of VMA is obtained to achieve the desired performance levels [3]. Therefore, VMA has a crucial effect on the designs of aggregate gradation [4]. According to the Asphalt Institute, the recommendations for optimizing the gradation curves to achieve the desired effects on AV and VMA are largely dependent on the experience of designers in handling different materials [5]. A growing body of research has focused on the association between AG and Maximum Density Line (MDL) and how this connection can affect the VMA of AC mixtures [6].

With regard to volume fraction, the fine and coarse aggregates account for 85–90% of the volume of hot mix asphalt (HMA) [7,8], in which the proportion of coarse aggregates (particle size > 2.36 mm) is also prominent. From a different perspective, coarse aggregates in HMA play dual roles: (i) the internal resistance of HMA is characterized as stone-on-stone skeleton established in the mix design (also for open-graded friction or stone mastic asphalt course), and the coarse aggregate contact of HMA gradation is regarded as the primary source of internal resistance [8,9]; (ii) the percentage voids in the coarse aggregate of asphalt mixture (VCA_{mix}) structured by coarse aggregates can provide a maternal space for VMA and partially restrict the VMA values [4]. Hence, VCA_{mix} is also a critical volume indicator that has no direct relationship with the gradation design.

AG is one of the important properties of HMA, as it has various aspects of mixture performance, including VMA, resistance to permanent deformation, durability and compatibility [10–14]. It has been reported that AG is closely related to the rutting and indirect tensile (IDT) performance. The gradation effect is aggregate specific for rutting. The design binder content can be used to measure the effect of gradation on indirect tensile strength (ITS), which is associated with the VMA values of AG [15]. Sangsefidi and colleagues [16] assessed the moisture susceptibility of warm mix asphalt and the effect of AG on creep. Their findings demonstrated that AG could differently affect the moisture vulnerability and rutting resistance of the studied mixtures. Hafeez and co-workers [17] evaluated the effects of AG with various nominal maximum aggregate sizes on the fatigue, rutting and stiffness performance of stone mastic asphalt (SMA). Their findings showed that stiffer SMA had lower rut values and fatigue life. A mixture of stiffness and rut resistance may enhance with increasing aggregate sizes during AG. Xiao and colleagues [18] found that AG could affect the rutting resistance and moisture susceptibility of open graded friction course (OGFC) mixtures. Kim and co-workers [19] conducted a study to examine the effects of mix gradations related to the Superpave restricted zone on rutting potential, particularly for low traffic volume roadways. They concluded that similar to that for medium to high traffic volume pavements, the restricted zone is not a significant factor influencing the RP of HMA for low traffic volume local pavements. The fineness of AG, rather than the restricted zone, may serve as a promising factor that influences rutting performance. Sun et al. [20] evaluated the skeleton contact stability of a graded aggregate system and analyzed the slip creep properties of asphalt mixture from the geometric characteristics of aggregates. Devulapalli et al. [21] summarized the concerns in the SMA mixtures and gave knowledge about the gradation, stone-on-stone contact, drain down, and stabilizing agents based on a detailed literature review.

According to the relevant literature [7] and our research [22,23], a 2.36 mm size can be employed as the boundary sieve (BS) between fine and coarse aggregates. Hence, the coarse aggregates > 2.36 mm constitute the skeleton structure of asphalt mixtures. The AC mixture has been widely used in the asphalt pavement structure in China, which is often regarded as a dense-suspended gradation structure. However, the dense skeleton structure in AC mixture can also be formed when the suitable AG is designed. The main objective of this study was to assess the quantitative impacts of AG variations (i.e., different percentage passing of 2.36 mm sieve) on the volumetric parameters (VPs) and high-temperature RP of AC mixtures. In this paper, the relations between key volumetric parameters, rutting resistance performance and percent passing of BS for AC-13, AC-20 and AC-25 were established, respectively. In addition, the reference values of the percentage passing of BS were presented for guiding the gradation design of AC mixture with dense skeleton structure.

2. Materials and Experiments

2.1. Materials

2.1.1. Aggregates and Filler

AC mixtures with 3 nominal maximum sizes of 26.5, 19 and 13.2 mm, namely, AC-25, AC-20 and AC-13, were employed in the present study. AC-25, AC-20 and AC-13 were applied in the bottom, middle and upper surface layers of the asphalt pavement, respectively. Considering the mechanical properties and abrasion resistance of crushed basalt stones, they were employed for fine and coarse aggregates in AC-13. Meanwhile, the crushed lime stones were applied for fine and coarse aggregates in AC-25 and AC-20. One basalt fine aggregate (0–3 mm) and three basalt coarse aggregates with different particle sizes (3–5, 5–10 and 10–15 mm) were selected. Table 1 shows the properties of basalt coarse aggregates according to the Chinese specifications [24], while Table 2 shows the properties of lime coarse aggregates with different particle sizes (3–5, 5–10, 10–20 and 20–30 mm). The basic properties of the two kinds of fine aggregates are presented in Table 3. The limestone powder was used as a mineral filler after passing through the #200 sieve. All the materials have met the technical requirements specifications [24].

Table 1. Properties of the basalt coarse aggregates.

Properties	Test Values			Specification
	10–15 mm	5–10 mm	3–5 mm	
Apparent specific gravity	2.835	2.848	2.823	≥2.60
Bulk specific gravity	2.770	2.679	2.703	–
Water absorption (%)	1.0	1.5	1.2	≤2.0
Percent of flat and elongated particles (%)	8.3	9.5	9.8	≤15
Crushed stone value (%)	12.6	12.6	–	≤26
L.A. abrasion (%)	9.3	9.6	9.0	≤28

Table 2. Properties of the lime coarse aggregates.

Properties	Test Values				Specification
	20–30 mm	10–20 mm	5–10 mm	3–5 mm	
Apparent specific gravity	2.737	2.727	2.747	2.673	≥2.50
Bulk specific gravity	2.700	2.687	2.694	2.612	–
Water absorption (%)	0.5	0.5	0.7	0.9	<3.0
Percent of flat and elongated particles (%)	10.3	11.5	9.8	11.0	≤15
Crushed stone value (%)	18.5	18.5	18.5	–	≤28
L.A. abrasion (%)	17.3	17.6	17.2	–	≤30

Table 3. Properties of the two fine aggregates.

Properties	Test Values		Specification
	Basalt	Limestone	
Apparent specific gravity	2.821	2.723	≥2.50
Angularity (s)	38	36	≥30

2.1.2. Asphalt Binder

A 70 penetration grade (Pen 70) asphalt binder was supplied by a commercial petroleum company and was employed in all mixture designs. Table 4 shows the properties of the Pen 70 asphalt binder. The specifications for pavement asphalt binders are met for all data [24].

Table 4. Properties of the Pen 70 asphalt binder.

Index	Softening Point (°C)	Penetration (25 °C, 0.1 mm)	Specific Gravity (15 °C)	Ductility (cm)	
				15 °C	10 °C
Test values	49.6	71	1.021	>100	40

2.2. Mix Design

The size of BS between fine and coarse aggregates was 2.36 mm. Five grading curves of AC-25, AC-20 and AC-13 were designed based on the percent passing (PP) of the BS. The structure and type of the aggregate can be reflected by the PP of the BS. The gradation curves of all aggregates demonstrated S type. Figures 1–3 show the design gradation curves of aggregates with 3 nominal maximum sizes. The gradation curves can be seen in two parts consisting of a coarse section (sieve size is equal or greater than 2.36 mm) and fine section (sieve size is less than or equal to 2.36 mm). Additionally, the coarse sections of five gradation curves for each nominal maximum aggregate size were designed as *S* type curves, and the proportion of the coarse aggregates was constant.

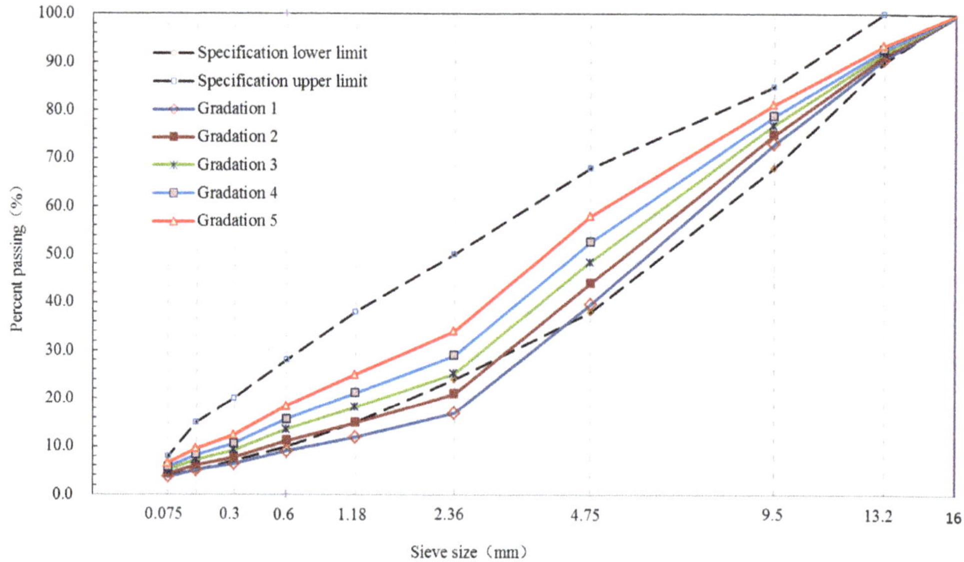

Figure 1. Gradation curves of AC-13.

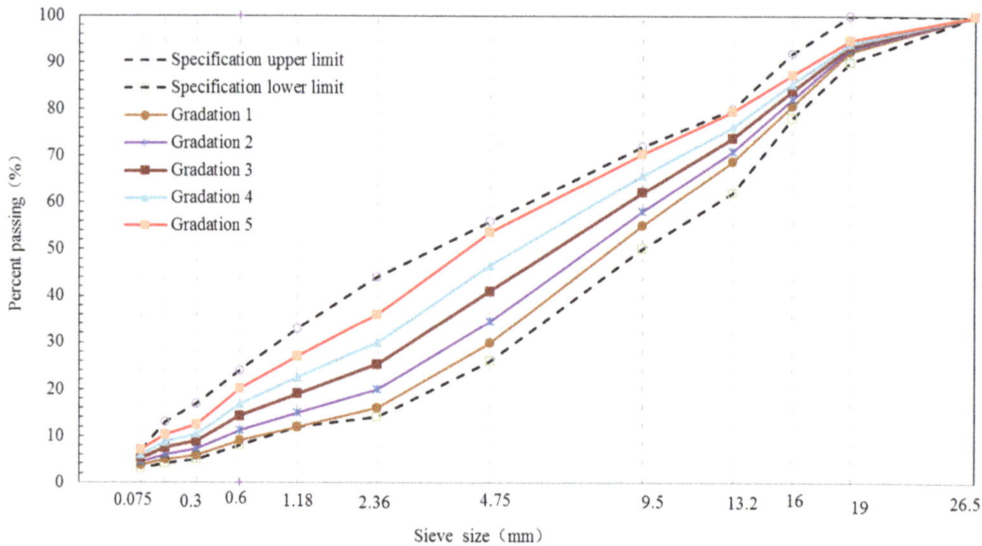

Figure 2. Gradation curves of AC-20.

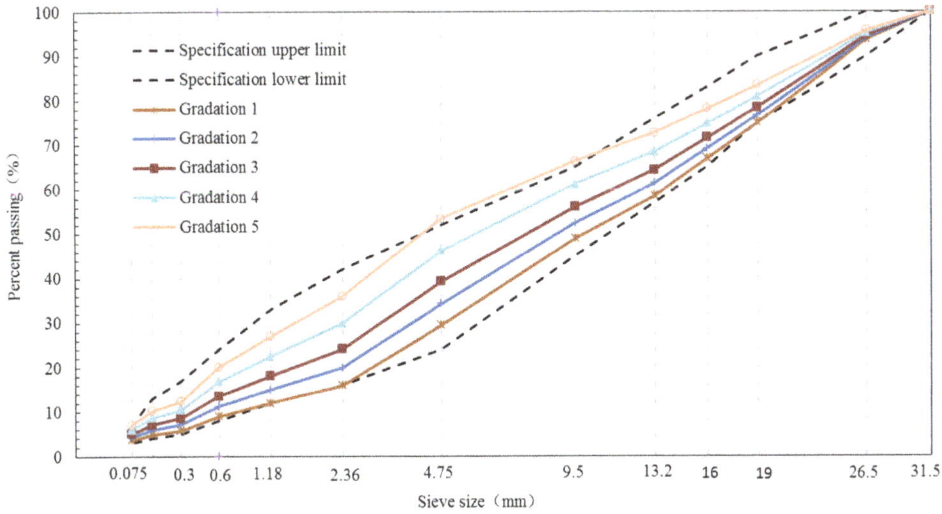

Figure 3. Gradation curves of AC-25.

According to the China specifications [24], our previous research and field experience of AC mixture [25], the asphalt contents of the three AC mixtures were evaluated, which were 3.9%, 4.4% and 5.3% for AC-25, AC-20 and AC-13, respectively. To further assess the individual effect of AG on the VPs and high-temperature RP of AC mixtures, 5 AGs of each AC mixture used the similar asphalt contents.

2.3. Experimental Program and Testing Methods

First of all, VCA formed from the packing of coarse aggregates for three kinds of AC mixtures at the state without any asphalt binders were conducted using vibrating compaction. Then, the VPs and RP of AC mixtures were assessed by determining the effects of 5 levels of AG and 3 levels of nominal maximum aggregate size. A total of 15 AC mixtures were designed with the Marshall method and examined using the two-wheel laboratory tracking devices.

2.3.1. Vibrating Compaction Test of Coarse Aggregates

The vibrating compaction test of coarse aggregates was performed by a vibration table, and the vessel with a weight stack is shown in Figure 4. Firstly, different size coarse aggregates were fully mixed by the proportion according to the coarse sections of designed gradation curves, and the mixed specimen was obtained. Next, the mixed specimen was placed in a 10 L capacity vessel 3 times. After loading the sample into a layer, the vessel was placed onto a vibration table, and the compaction experiment was conducted (the vibration frequency and time were 50 Hz and 90 s, respectively). Lastly, when the final layer was vibrated, the surface of the samples was leveled. To maintain data accuracy, the experimental errors were reduced and the number of parallel experiments was set to 3. The values of VCA can be calculated using Equation (1).

$$\text{VCA} = \left(1 - \frac{\rho}{\rho_b}\right) \times 100 \tag{1}$$

where ρ is the packing density of the mixed coarse aggregates in the vessel; ρ_b is the synthetic bulk density of coarse aggregates with different sizes.

Figure 4. Vibrating compaction test device.

2.3.2. Marshall Compaction Test

To assess the effects of different AGs on the VPs of AC mixtures, the samples were fabricated with 75 blows of Marshall Compactor. The asphalt mixture samples were prepared according to the T072-2011 [26]. Four samples were fabricated for each AG. The VMA, AV and VCA_{mix} of the compacted specimens of various AGs were tested and calculated in terms of the procedures and formulas described in the China Standard Test Methods [26].

2.3.3. Wheel-Tracking Test

To evaluate the high-temperature RP of the AC mixture, the wheel-tracking test was performed by utilizing the wheel-tracking device (Figure 5). Each sample was 50 mm in height and 300 mm × 300 mm in cross-sectional areas. According to the China Standard T0719-2011 [26], the wheel-tracking test was conducted using the 0.7 MPa wheel load at 60 °C temperature under dry conditions. Dynamic stability (DS) is presented by the number of times the wheel passes the sample per rut depth within an interval of 45–60 min, which can be calculated using Equation (2) [26]:

$$DS = \frac{(t_2 - t_1) \times N}{d_2 - d_1} \tag{2}$$

where d_1 and d_2 are rut depth at t_1 (45 min) and t_2 (60 min), respectively; N is the speed of wheel passing over the center of the sample, 42 cycles/min.

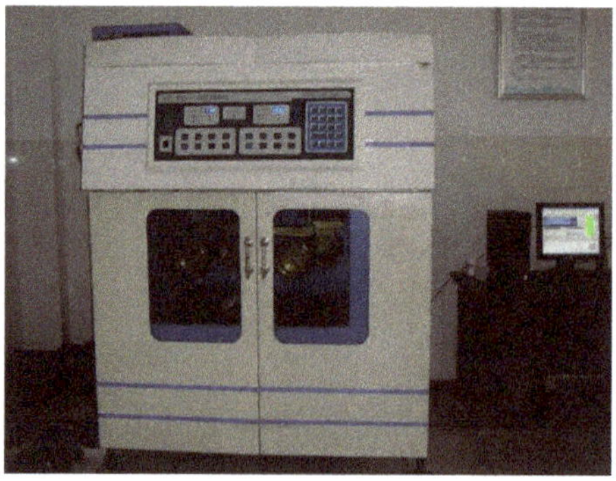

Figure 5. Wheel-tracking test device.

A high DS of the AC mixture is indicative of an excellent performance of its resistance to permanent deformation at high temperature.

3. Results and Discussion

3.1. Results of Vibrating and Marshall Compaction Tests

The proportions of various coarse aggregates in the three AC mixes and corresponding VCA values are shown in Table 5. Vibrating compaction tests were repeated three times for each AC mixture, and the values of VCA are represent as means. Based on the testing density parameters of Marshall compacted samples for the three AC mixtures, the calculated results of VMA, AV and VCA_{mix} as a function of the PP of BS are presented in Figures 6–8. The average values of the four duplicates are obtained.

Table 5. Results of the vibrating compaction test of coarse aggregates.

AC Mixes	Proportions of Different Coarse Aggregates					VCA (%)
	20–30 mm	10–20 mm	10–15 mm	5–10 mm	3–5 mm	
AC-13	–	–	29	51	20	36.01
AC-20	–	65	–	20	15	35.65
AC-25	25	45	–	15	15	34.12

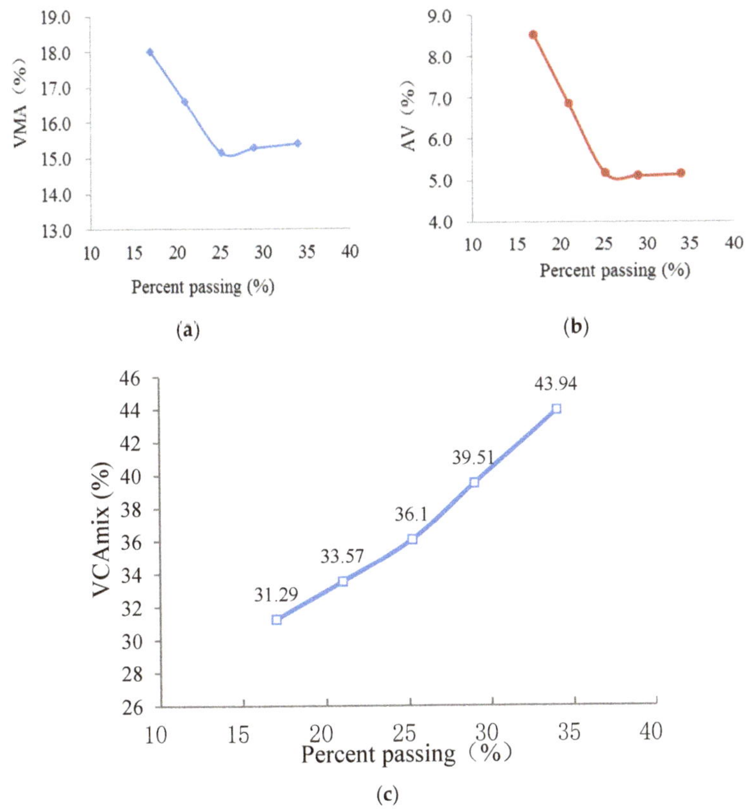

Figure 6. VPs of AC-13 samples versus PP. (**a**) VMA; (**b**) AV; (**c**) VCA_{mix}.

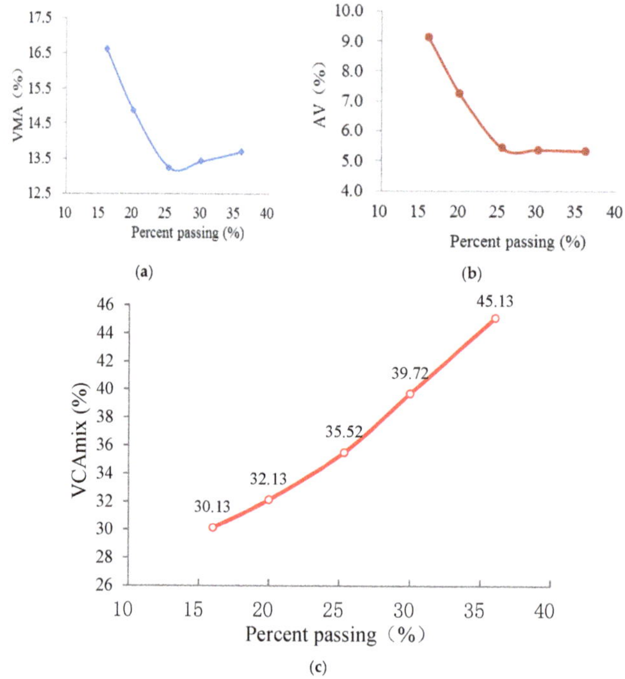

Figure 7. VPs of AC-20 samples versus PP. (**a**) VMA; (**b**) AV; (**c**) VCAmix.

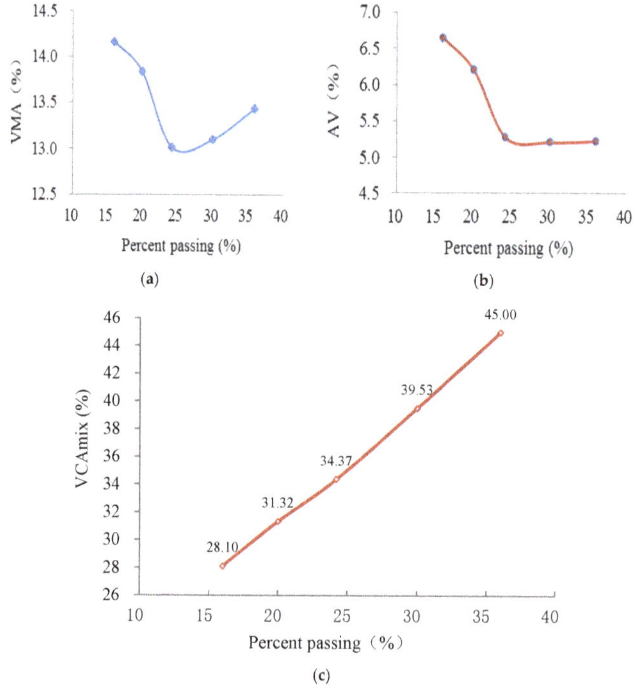

Figure 8. VPs of AC-25 samples versus PP. (**a**) VMA; (**b**) AV; (**c**) VCAmix.

As shown in Figures 6–8, the AG has crucial effects on the VPs (i.e., VMA, AV and VCA_{mix}) of AC mixtures. The values of VMA first decline and then elevate with increasing percentage passing of the BS, and they exhibit an upward-opening parabola, whereas those of AV reduce monotonically and appear to be constant with the increasing percentage passing of the BS. The values of VCA_{mix} show a linear increase with the increasing percentage passing of the BS. In addition, the VPs of different AGs of AC-25, AC-20 and AC-13 have the same rule of change. VMA and AV are close to the smallest value when the PP of BS is around 25. It is speculated that the aggregate structure of the AC mixture changes as a function of the PP of the BS. The inner forms of AC mixtures are skeleton-gap structures when the content of the coarse aggregate is larger (the PP is <25). Meanwhile, the coarse skeleton structures of AC mixtures are disrupted by excessive fine aggregate when the content of fine aggregate is larger (the PP is >25). Hence, the aggregate skeleton structure is in its densest state when the proportion of fine and coarse aggregates is appropriate.

3.2. *Results of Wheel-Tracking Tests*

Wheel-tracking tests were performed in triplicate for each AC mixture, and the values of DS for AC-25, AC-20 and AC-13 are displayed in Figure 9 (the error bars indicate standard deviation). To perform a comparative analysis of the key parameter (VMA) and RP of the three AC mixtures, the values of VMA are also plotted in Figure 9.

As shown in Figure 9, the values of DS first elevate and then decline with the increasing percentage passing of the BS, which is opposite to the trend of VMA. The DS of different AGs of AC-25, AC-20 and AC-13 have the same rule of change. However, the values of DS are not influenced by the nominal maximum aggregate sizes. In addition, statistical tests were conducted by employing the analysis of variance (ANOVA) method. One-way ANOVA is a commonly used technique for determining the effect of gradation types on the RP of AC mixtures according to the percentage passing of the BS. Table 6 demonstrates the ANOVA results of the DS of AC mixtures. Based on the *F*-statistics and *F*-critical, AG is an important factor affecting the values of DS of the three AC mixtures with a 95% significance level ($\alpha = 0.05$). It also can be seen that the peak values of DS nearly correspond to the smallest values of VMA (Figure 9). It is speculated that AC mixtures have the best RP when the coarse skeleton structure is in its densest state, which is desirable for the design of the AC mixture. That condition indicates that the rutting resistance of the AC mixture mainly depends on the grading structure of the aggregate. Therefore, it is important to optimize AG for enhancing the high-temperature RP of AC mixtures.

3.3. *Discussion*

As we all know, one of the most defining characteristics of an SMA is the concept of a stone-on-stone skeleton. This is where a large proportion of coarse aggregate particles are in contact with each other to form a skeleton or framework with relatively large voids. The sand-sized particles, filler and binder are then accommodated within the voids in the coarse aggregate skeleton [27]. Supposing that the skeleton of the AC mixture is composed of aggregates with a particle size larger than 2.36 mm, the skeletal dense structure (i.e., stone-on-stone skeleton with the densest state) can be formed as in the case of SMA. Referring to the method for assessing the stone-on-stone aggregate skeleton of SMA, the parameters of VCA and VCA_{mix} can also be applied to determine the skeletal dense structure of the AC mixture. Based on the above experimental results (see Table 5 and Figures 6–8), it can be seen that the values of VCA_{mix} (34.37, 35.52 and 36.1 for AC-25, AC-20 and AC-13, respectively) are very close to those of VCA (34.12, 35.65 and 36.01 for AC-25, AC-20 and AC-13, respectively) when the percentage passing of the 2.36 mm sieve is ≈25 for each AC mixture. Moreover, at this percentage passing, the VMA and AV almost have the minimum values, which demonstrate that the coarse skeleton structure is in its dense state. Hence, the skeletal dense structure of the AC mixture can be evaluated by comparing the values of VCA_{mix} and VCA. Additionally, the reference values of the percentage passing of BS are presented for guiding the gradation design of the AC mixture.

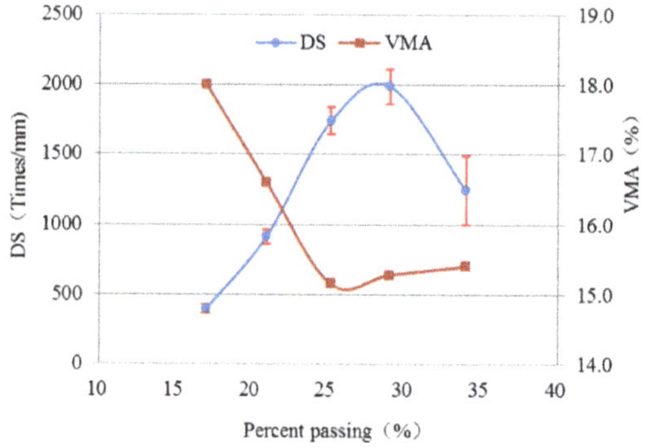

(**a**)

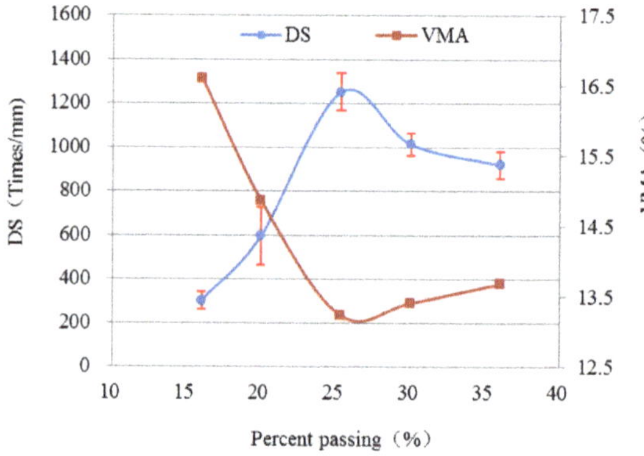

(**b**)

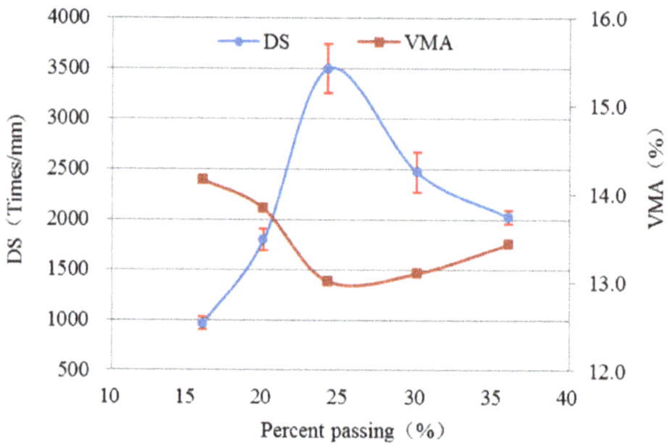

(**c**)

Figure 9. The values of DS and VMA of various AC mixtures versus PP. (**a**) AC-13; (**b**) AC-20; (**c**) AC-25.

Table 6. ANOVA results for the DS of AC mixtures.

Source of Variation	Degree of Freedom	Sum of Squares	Mean Square	*F*-Statistics	*F*-Critical	*p*-Value
AC-13						
Gradation	4	4,885,150	1,221,288	69.83495	3.47805	2.87×10^{-7}
Error	10	174,882	17,488.2			
Total	14	5,060,032				
AC-20						
Gradation	4	1,656,796	414,198.9	64.31926	3.47805	4.26×10^{-7}
Error	10	64,397.33	6439.733			
Total	14	1,721,193				
AC-25						
Gradation	4	10,371,596	2,592,899	107.8359	3.47805	3.52×10^{-8}
Error	10	240,448.7	24,044.87			
Total	14	10,612,044				

4. Summary and Conclusions

According to the results and analyses on the VPs and RP of AC mixtures with various AGs, the main findings and conclusions are summarized as follows:

(1) AG has crucial effects on the VPs of AC mixtures. The values of VMA first decline and then elevate with the increasing percentage passing of the BS; those of AV reduce monotonically and appear to be constant, while those of VCA_{mix} show a linear increase.

(2) Statistical analysis of the wheel-tracking test data indicates that AG is important for enhancing the RP of AC mixtures. The values of DS first elevate and then decline with the increasing percentage passing of the BS, which is opposite to the trend of VMA.

(3) AC mixtures show the best RP when the aggregate skeleton structure is in its densest state, which corresponds to the smallest values of VMA. Therefore, it is necessary to optimize AG to enhance the high-temperature RP of AC mixtures.

(4) Based on the experimental results and analyses, the AC mixture can be designed to be a skeletal dense structure, provided that 2.36 mm is employed as the BS between fine and coarse aggregates and the percentage passing of the BS is appropriate. In addition, the skeletal dense structure of the AC mixture can be evaluated by comparing the values of VCA_{mix} and VCA.

(5) This research is focused on the RP of AC mixtures; the effects of AG on moisture susceptibility and fatigue performance will be researched next. Further studies with other asphalt binders are also needed to verify our findings.

Author Contributions: Conceptualization, W.C. and S.L. (Shutang Liu); Data curation, J.Z.; Formal analysis, W.L.; Funding acquisition, S.L. (Shurong Lou); Investigation, W.C.; Methodology, W.C. and S.L. (Shutang Liu); Project administration, W.L.; Resources, X.R.; Supervision, X.R.; Visualization, S.L. (Shurong Lou) and J.Z.; Writing—original draft, W.C.; Writing—review & editing, S.L. (Shutang Liu). All authors have read and agreed to the published version of the manuscript.

Funding: This work was supported Shandong Hi-Speed Group Transportation Technology Plan (2019QL08).

Institutional Review Board Statement: Not applicable.

Informed Consent Statement: Not applicable.

Data Availability Statement: The data presented in this study are available on request from the corresponding author. The data are not publicly available due to them forming part of an ongoing study.

Conflicts of Interest: The authors declare no conflict of interest.

References

1. Park, K. Optimization of Aggregate Gradation for High-Performing Hot Mix Asphalt. Ph.D. Thesis, University of Rhode Island, Kingston, RI, USA, 2008.
2. Swathi, M.; Andiyappan, T.; Guduru, G.; Reddy, M.A.; Kuna, K.K. Design of asphalt mixes with steel slag aggregates using the Bailey method of gradation selection. *Constr. Build. Mater.* **2021**, *279*, 122426. [CrossRef]
3. Foreman, J. Effect of Voids in the Mineral Aggregate on Laboratory Rutting Behavior of Asphalt Mixtures. Master's Thesis, University of Arkansas, Fayetteville, AR, USA, 2008.
4. Roberts, F.; Kandhal, P.; Brown, E. *Hot Mix Asphalt Materials, Mixture Design and Construction*, 2nd ed.; NAPA Research and Education Foundation: Greenbelt, MD, USA, 1996.
5. The Asphalt Institute. *Mix Design Methods for Asphalt Concrete and Other Hot-Mix Types*; Manual Series No. 2 (MS-2); The Asphalt Institute: Lexington, KY, USA, 1995.
6. The Asphalt Institute. *Superpave Mix Design*; Superpave Series No. 2 (SP-2); The Asphalt Institute: Lexington, KY, USA, 2001.
7. Shen, S.; Yu, H. Characterize packing of aggregate particles for paving materials: Particle size impact. *Constr. Build. Mater.* **2011**, *25*, 1362–1368. [CrossRef]
8. Chen, S.; Liao, M. Evaluation of internal resistance in hot-mix asphalt (HMA) concrete. *Constr. Build. Mater.* **2002**, *16*, 313–319. [CrossRef]
9. Alvarez, A.; Martin, A.; Estakhri, C. A review of mix design and evaluation research for permeable friction course mixtures. *Constr. Build. Mater.* **2011**, *25*, 1159–1166. [CrossRef]
10. Park, D. Effects of Aggregate Gradation and Angularity on VMA and Rutting Resistance. Master's Thesis, Texas A & M University, College Station, TX, USA, 2000.
11. Graziani, A.; Ferrotti, G.; Pasquini, E.; Canestrari, F. An application to the European practice of the Bailey method for HMA aggregate grading design. *Procedia Soc. Behav. Sci.* **2012**, *53*, 991–1000. [CrossRef]
12. Golalipour, A.; Jamshidi, E.; Niazi, Y.; Afsharikia, Z.; Khadem, M. Effect of aggregate gradation on rutting of asphalt pavements. *Procedia Soc. Behav. Sci.* **2012**, *53*, 440–449. [CrossRef]
13. Dong, Z.; Xiao, G.; Gong, X. Analysis on impact of gradation and anti-rutting additive on rutting resistance of asphalt mixture. *J. Highw. Transp. Res. Dev.* **2014**, *31*, 27–46.
14. Fang, M.; Park, D.; Singuranayo, J.; Chen, H.; Li, Y. Aggregate gradation theory, design and its impact on asphalt pavement performance: A review. *Int. J. Pavement Eng.* **2019**, *20*, 1408–1424. [CrossRef]
15. Zhao, W. The Effects of Fundamental Mixture Parameters on Hot-Mix Asphalt Performance Properties. Ph.D Thesis, Clemson University, Clemson, SC, USA, 2011.
16. Sangsefidi, E.; Ziari, H.; Mansourkhaki, A. The effect of aggregate gradation on creep and moisture susceptibility performance of warm mix asphalt. *J. Pavement Eng.* **2014**, *2*, 133–141. [CrossRef]
17. Hafeez, I.; Kamal, M.; Mirza, M. An experimental study to select aggregate gradation for stone mastic asphalt. *J. Chin. Inst. Eng.* **2015**, *1*, 1–8. [CrossRef]
18. Xiao, F.; David, A.; Amirkhanian, S.; He, L. Aggregate gradations on moisture and rutting resistances of open graded friction course mixtures. *Constr. Build. Mater.* **2015**, *85*, 127–135. [CrossRef]
19. Kim, Y.; Park, H.; Aragão, F.; Lutif, J. Effects of aggregate structure on hot-mix asphalt rutting performance in low traffic volume local pavements. *Constr. Build. Mater.* **2009**, *23*, 2177–2282. [CrossRef]
20. Sun, S.; Li, P.; Cheng, L.; Wang, X.; Zhang, W. Analysis of skeleton contact stability of graded aggregates system and its effect on slip creep properties of asphalt mixture. *Constr. Build. Mater.* **2022**, *316*, 125911. [CrossRef]
21. Devulapalli, L.; Sarang, G.; Kothandaraman, S. Characteristics of aggregate gradation, drain down and stabilizing agents in stone matrix asphalt mixtures: A state of art review. *J. Traffic Transp. Eng.* **2022**, *9*, 167–179. [CrossRef]
22. Liu, S.; Cao, W.; Qi, X.; Ren, S. Research and application of statistical law of VCA formed from the packing of basalt coarse aggregates. *Constr. Build. Mater.* **2014**, *71*, 484–491. [CrossRef]
23. Liu, S.; Cao, W.; Li, Y.; Xue, Z. Building and application of VMAa physical model of mineral aggregate without adding asphalt. *China J. Highw. Transp.* **2016**, *8*, 1–10.
24. *JTG F40-2004*; Technical Specifications for Construction of Highway Asphalt Pavements. Ministry of Transport: Beijing, China, 2004.
25. Liu, S.; Shang, Q.; Guo, Z. Estimation of optimum asphalt content range of hot mix asphalt in objective mix design. *J. Highw. Transp. Res. Dev.* **2006**, *2*, 39–42.
26. *JTG E20-2011*; Standard Test Methods of Bitumen and Bituminous Mixtures for Highway Engineering. Ministry of Transport: Beijing, China, 2011.
27. Austroads Ltd. *Development of Aspects of a Design Procedure for Stone Mastic Asphalt*; No. AP-T221-13; Austroads Publication: Sydney, Australia, 2013.

materials

Article

Optimization of Asphalt-Mortar-Aging-Resistance-Modifier Dosage Based on Second-Generation Non-Inferior Sorting Genetic Algorithm

Yang Lv [1], Shaopeng Wu [1,*], Peide Cui [1,2], Serji Amirkhanian [3], Haiqin Xu [1], Yingxue Zou [1] and Xinkui Yang [1]

1 State Key Laboratory of Silicate Materials for Architectures, Wuhan University of Technology, Wuhan 430070, China; lvyang@whut.edu.cn (Y.L.); cuipeide@whut.edu.cn (P.C.); xuhaiqin@whut.edu.cn (H.X.); zouyingxue@whut.edu.cn (Y.Z.); yangxk@whut.edu.cn (X.Y.)
2 School of Transportation, Southeast University, Jiulonghu, Nanjing 211189, China
3 Department of Civil Construction and Environmental Engineering, University of Alabama, Tuscaloosa, AL 35487, USA; samirkhanian@eng.ua.edu
* Correspondence: wusp@whut.edu.cn

Abstract: The use of steel slag powder instead of filler to prepare asphalt mortar was beneficial to realize the effective utilization of steel slag and improve the performance of asphalt concrete. Nevertheless, the anti-aging properties of steel-slag powder–asphalt mortar need to be further enhanced. This study used antioxidants and UV absorbers in steel-slag powder–asphalt mortar to simultaneously improve its thermal-oxidation and UV-aging properties. The dosage of modifier was optimized by second-generation non-inferior sorting genetic algorithm. Fourier-Transform Infrared Spectroscopy, a dynamic shear rheometer and the heavy-metal-ion-leaching test were used to evaluate the characteristic functional groups, rheological properties and heavy-metal-toxicity characteristics of the steel-slag-powder-modified asphalt mortar, respectively. The results showed that there was a significant correlation between the amount of modifier and G*, δ, and the softening point. When the first peak appeared for G*, δ, and the softening point, the corresponding dosages of x_1 were 2.15%, 1.0%, and 1.1%, respectively, while the corresponding dosage of x_2 were 0.25%, 0.76%, and 0.38%, respectively. The optimal value of the modifier dosage x_1 was 1.2% and x_2 was 0.5% after weighing by the NSGA-II algorithm. The asphalt had a certain physical solid-sealing effect on the release of heavy-metal ions in the steel-slag powder. In addition, the asphalt structure was changed under the synergistic effect of oxygen and ultraviolet rays. Therefore, the risk of leaching heavy-metal ions was increased with the inferior asphalt-coating performance on the steel-slag powder.

Keywords: steel-slag powder; asphalt mortar; anti-aging properties; heavy-metal ions

Citation: Lv, Y.; Wu, S.; Cui, P.; Amirkhanian, S.; Xu, H.; Zou, Y.; Yang, X. Optimization of Asphalt-Mortar-Aging-Resistance-Modifier Dosage Based on Second-Generation Non-Inferior Sorting Genetic Algorithm. *Materials* **2022**, *15*, 3635. https://doi.org/10.3390/ma15103635

Academic Editor: Karim Benzarti

Received: 25 April 2022
Accepted: 16 May 2022
Published: 19 May 2022

1. Introduction

Asphalt pavement has been widely used in expressways due to its advantages of flat surface without joints, low noise, fast traffic recovery, and convenient maintenance [1,2]. By the end of 2020, the total highway mileages were 5,198,100 km in China, which is an increase of 185,600 km from the end of the previous year. However, due to complex traffic loads and climatic conditions, physicochemical reactions such as dehydrogenation, polycondensation, and oxidation [3] of asphalt pavement have been produced on account of the coupling of heat and oxygen [4,5]. Its road performance has rapidly attenuated and deteriorated with the occurrence of stresses, such as stripping, potholing, and cracking [6]. Therefore, improving the anti-aging performance of asphalt-pavement materials is crucial for reducing the stress on asphalt pavement, prolonging its service life, and ensuring traffic safety [7–9].

The annual production of steel slag was more than 100 million tons with a low comprehensive-utilization rate of only about 30% [10]. The accumulation of steel slag causes water pollution, heavy-metal release and other problems by occupying a lot of land [11–13]. Therefore, an enormous challenge has been posed to local ecological development and resource reuse. Steel slag has the properties of a rough surface texture, high intensity, wear resistance, and high alkalinity [14–16], which make it an innovative substitute for natural aggregate in asphalt mixtures. Previous research has focused on the use of steel slag as an aggregate in asphalt mixtures while neglecting the replacement of natural mineral powder with steel-slag powder (SSP) [17]. The use of SSP in asphalt concrete is beneficial to the improvement of its effective-utilization rate, and could enhance the performance of bituminous concrete by increasing the stiffness of asphalt mortar [18]. However, the steel-slag powder–asphalt mortar (SSP–AM) has poor anti-aging properties, especially under UV radiation. The porous structure of slag powder leads to multiple reflections of UV radiation, which exhibits easier entry than heat and oxygen and induces more serious aging [19,20].

Nowadays, some researchers have improved the aging resistance by adding nanoparticles or polymer modification to asphalt [21,22]. Hu, ZH et al. [23] investigated the preparation of composite antioxidants from expanded graphite (EG) loaded with $CaCO_3$ nanoparticles and $Mg(OH)_2$ ($EG/CaCO_3/MH$) in order to reduce the thermo-oxidative aging of the binder. The results showed that $EG/CaCO_3/MH$ had a synergistic inhibitory effect on the thermo-oxidative aging of asphalt cement. Yang, J et al. [24] evaluated the UV-aging resistance of titanium dioxide/polystyrene reduced graphene oxide (TiO_2/PS-rGO) on SBS modified asphalt. It was found that the viscosity aging index and ductility retention of the TiO_2/PS-rGO/SBS-modified binder were decreased by 65.36 and increased by 8.79, respectively, which indicates that the UV-aging resistance of the binder was improved by TiO_2/PS-rGO. Rajib, A et al. [25] explored the feasibility of biochar in the retardation of the aging of rubberized and base asphalt binder. The results indicated that biochar could delay the UV aging of asphalt by reducing the rheological and chemical aging-indicators of asphalt. However, only the anti-aging properties of the binder were the concern of most researchers, and the research directions were only thermal-oxygen aging or ultraviolet aging. There have been few reports on the coupling of UV- and thermo-oxidative-aging resistance of asphalt mortar.

For SSP–AM, it was of great significance to explore a new method of simultaneously improving the thermal-oxidative- and UV-aging properties. Based on this, the antioxidants 1098 and UV-5411 were selected for asphalt mortar at the same time to solve this problem. The 1098 antioxidant is a typical high-temperature antioxidant with excellent refractory characteristics. The temperature of the asphalt mixture reaches as high as 160 °C during the mixing process, so the antioxidant needs to have excellent high-temperature resistance. The UV absorbent UV-5411 is an ultraviolet absorber with a high-efficiency light-stabilization effect. The ultraviolet rays are converted into heat by ultraviolet absorption through chemical action, thereby preventing further thermal-oxygen and ultraviolet aging of the asphalt. Nevertheless, in the multi-objective optimization of the modifier dosage, the experimental analysis samples were obtained by the researchers through orthogonal experiments. The combination of sample points obtained by this method did not always fill the total design space of the variables [26]. It was prone to the blind accumulation of sample points, resulting in a waste of time and resources. Therefore, an effective experimental design method was adopted to comprehensively investigate the performance parameter of asphalt modified by the two additives.

The application of numerical simulation, an approximate mathematical model and engineering optimization to solve multi-parameter and multi-objective optimization algorithms has become a research hotspot [27,28]. Kollmann, J et al. [29] used the optimization method of a two-dimensional finite-element model to simulate the generation and propagation of cracks in the asphalt mixture. It was illustrated that damage occurred between each successive element, resulting in relatively unconstrained crack growth. Sivilevicius, H et al. [30] applied numerical simulations of the composition to select the optimum added quantity of new and old materials on the road surface. The results showed that the algorithm can be used in the preliminary stage of the RHMA hybrid structure design. The amount of RAP could not exceed 8.43% when the RHMA mixture AC-16 was applied to the surface layer. The Latin hypercube design (LHD) is a random, multidimensional, stratified sampling method [31]. The n-dimensional space is evenly divided into m intervals, and independent equal probability sampling is performed in each sub-interval [32]. The uniformity of random LHD was improved by the Optimal Latin hypercube design (Opt LHD), making the fit of factors and responses more uniform and random within the design-space region [33].

Steel-slag powder–asphalt mortar (SSP–AM) has poor anti-aging properties. The research on the coupling of UV and thermal-oxidative-aging resistance of SSP–AM is still unclear. In the research on modifier dosage, orthogonal experiments tend to blindly accumulate sample points, resulting in a waste of time and resources, and the multi-objective optimization exploration of modifiers in SSP–AM has not been fully explored. The performance of SSP–AM has not been thoroughly studied. The leaching behavior of heavy metals from SSP–AM is still not known. Based on the above background, the purpose of this research was as follows:

1. The dosage parameters of antioxidants and UV absorbers were selected as the inputs, and the sample points were established based on the Opt LHD design method.
2. The softening point, phase angle, and complex shear modulus of asphalt mortar after thermal-oxygen and ultraviolet aging were selected as the outputs to establish a Kriging model [34].
3. The multi-objective optimization analysis of antioxidants and UV absorbers was performed by using NSGA-II [35] to determine the optimal dosage of each modifier. Hence, a chemically reactive composite-modified asphalt mortar with good comprehensive performance was prepared.
4. The optimized model was verified in accordance with the experimental results. The chemical and phase compositions of SSP were detected by X-ray fluorescence (XRF) and X-ray diffraction (XRD). The characteristic functional groups, rheological properties and heavy-metal-toxicity characteristics of the composite-modified asphalt mortar were investigated by Fourier-Transform Infrared Spectroscopy (FTIR), a dynamic shear rheometer (DSR) and the heavy-metal-ion-leaching test (TCLP), respectively.

2. Materials

2.1. Steel-Slag Powder

An experimental study was conducted on SSP of Baotou Iron and Steel Co., Ltd., Baotou, Inner Mongolia, China. The jaw crusher was used to grind the steel slag to make its particle size less than 0.075 mm. Its basic properties are shown in Table 1, and it can be seen that the indexes met the requirements of the test methods of aggregate for highway engineering. Meanwhile, XRF was used to detect the chemical composition as shown in Table 2. The main components of the SSP were Al, Si, Fe, Ca, Mg and Mn respectively.

Table 1. Basic properties of SSP.

Properties	Tested Value	Specifications
Density (g/cm^3)	3.57	≥2.9
Los Angeles abrasion (%)	8.5	≤28
Specific surface area (m^2/g)	1.95	-
Water absorption ratio (%)	0.68	≤1.0
Crush values	12.7	≤26

Table 2. Chemical composition of SSP.

Compounds	CaO	Fe_2O_3	SiO_2	MgO	Al_2O_3	MnO	TiO_2	P_2O_5	Others
Steel slag	35.5	33.3	14.0	4.9	2.4	4.5	1.1	2.2	2.1

2.2. Asphalt

The AH-70 base asphalt was produced by Guochuang Co., Ltd., Wuhan, Hubei, China. The essential performance indexes of asphalt were tested in accordance with Standard Test Methods of Bitumen and Bituminous Mixtures for Highway Engineering (JTG E20-2011). The test results met the requirements as shown in Table 3.

Table 3. Physical information of asphalt.

Information	Unit	Result	Requirements
Penetration (25 °C, 5 s, 100 g)	0.1 mm	67.0	60.0–80.0
Softening point	°C	48.3	≥45.0
Ductility (5 °C, 5 cm/min)	cm	156.7	≥100.0
Density (15 °C)	g/cm^3	1.035	-
Solubility (Trichloroethylene)	%	99.8	≥99.0

2.3. Antioxidant

The German BASF 1098 antioxidant used in this article was a nitrogen-containing hindered phenolic antioxidant with low volatility, resistance to extraction, non-toxicity, heat-resistant oxidation, and other properties. The 1098 antioxidant in this paper was selected because it is a typical high temperature antioxidant with excellent refractory characteristics. The asphalt mixtures were mixed at 160 °C, so the antioxidant should have excellent high-temperature resistance. The photothermal aging resistance of polymer materials was effectively improved on account of its functions such as decomposing hydroperoxide, trapping free radicals, and trapping singlet oxygen. Its addition to asphalt was expected to trap thermal or photoinduction-free radicals based on the light-stabilization effect, thereby preventing further aging reactions of the asphalt. Its decomposition chemical equation and basic physical properties are presented in Figure 1 and Table 4, respectively.

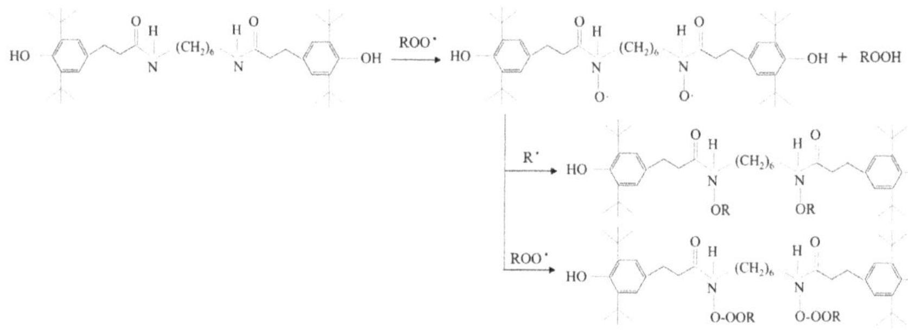

Figure 1. The decomposition chemical equation of antioxidant 1098.

Table 4. Basic physical properties of antioxidant 1098.

Information	Unit	Result
Melting point	°C	156–161
Flashing point	°C	282
Density (20 °C)	g/cm^3	1.045
Solubility (Water)	%	0.01

2.4. Ultraviolet Absorbent

The BASF UV-5411 used in this study was a hydroxyphenyl benzotriazole ultraviolet absorber, which is an ultraviolet absorber with high-efficiency light stabilization. The ultraviolet rays were converted into heat by UV-5411 through chemical action, thereby preventing further thermal-oxygen and ultraviolet aging of the asphalt. Its decomposition chemical equation and basic physical properties are shown in Figure 2 and Table 5, respectively.

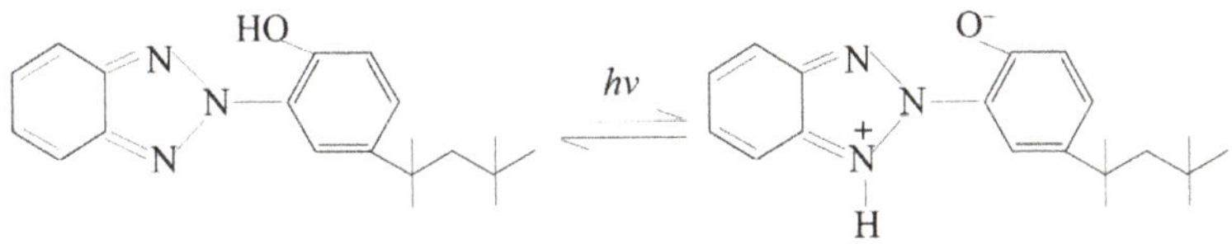

Figure 2. The decomposition chemical equation of UV-5411.

Table 5. Basic physical properties of UV-5411.

Information	Unit	Result
Melting point	°C	103.0–105.3
Flashing point	°C	>150
Density (20 °C)	g/cm^3	1.18
Solubility (Water)	%	<0.01

2.5. Preparation of Steel-Slag-Powder-Modified Asphalt Mortar

In this paper, asphalt mortar was prepared by using steel-slag powder instead of natural mineral powder. The density of mineral powder and steel-slag powder is quite different. Taking limestone mineral powder as an example, its apparent relative density is 2.7, while the apparent relative density of steel-slag powder is 3.2. Therefore, the volume ratio was used to mix the asphalt mortar, and the volume percentage of asphalt: steel-slag powder was 1:0.3. In addition, Wei et al. [36] determined that the amount of asphalt and fillers was determined by the qualified asphalt concrete mix proportion, while the asphalt aggregate ratio was 4.9%. Therefore, the volume ratio of fillers to asphalt was determined as 0.3.

The preparation process was shown in Figure 3 as follows: First, the steel-slag powder and asphalt were placed in an oven to be heated (AH-70 asphalt at 135 °C and steel-slag powder at 150 °C). Secondly, the asphalt was heated in an oil bath at 160 °C, and then the steel-slag powder (the volume ratio was 0.3) and modifiers (UV-5411 was 0–2.4 wt% and antioxidant 1098 was 0–4.8 wt%) were added with shearing at a low speed of 500 rpm for 15 min. Then, the machine was sheared at a high speed of 4000 rpm for 30 min to fully mix the steel-slag powder and modifiers with the asphalt. Finally, the prepared steel-slag-powder-modified asphalt mortar was transferred to a storage tank for use.

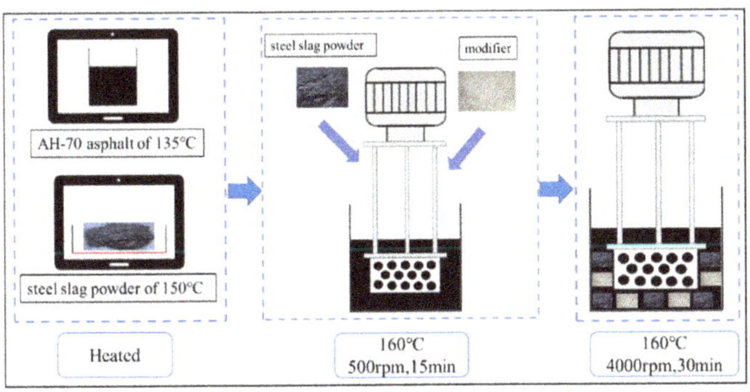

Figure 3. The preparation of steel-slag-powder-modified asphalt mortar.

Since the density of steel-slag powder was much higher than that of 70# base asphalt, the steel-slag powder would gradually segregate to the lower part of the asphalt mortar during the storage process, which resulted in poor uniformity. Therefore, the principle of current production was adopted during the use of steel-slag-powder-modified asphalt mortar in order to reduce the error caused by the experiment.

3. Experimental Methods

3.1. Optimization Test of Dosage Simulation

3.1.1. Experimental Design Based on Opt LHD Sampling

The basic principle of LHD sampling is to evenly divide each dimensional coordinate interval $[x_k^{min}, x_k^{max}]$, i $\in$ [1, n] into m intervals in the n-dimensional space, and each small interval is denoted as $[x_k^{i-1}, x_k^{i}]$, i $\in$ [1, m]. However, the uniformity of sampling points could not be considered in this sampling method, so the uniform space filling of sample points was not filled. Therefore, the Opt LHD sampling method was adopted to improve the uniformity of random LHD. The fit of factors and responses was more uniform and random within the design-space region, resulting in higher computational accuracy for the design of the experimental methods. In this paper, the dosage of antioxidant 1098 $x_1 \in (0, 4.8)$ and the UV absorber 5411 $x_2 \in (0, 2.4)$ were evenly divided into 25 sample spaces. The distribution of LHD and Opt LHD sampling designs are shown in Figure 4a,b, respectively. It can be seen that some regional sampling points were too concentrated in LHD sampling, while others were lost. However, the distribution of test points was more uniform in the Opt LHD.

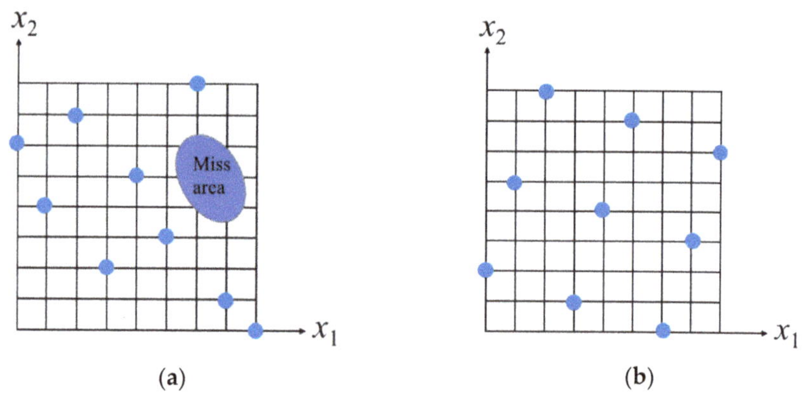

Figure 4. The distribution of LHD and Opt LHD sampling designs. (**a**) Random LHD sampling, (**b**) Opt LHD sampling.

3.1.2. Experimental Design Based on Kriging Model

Kriging, also known as spatial local interpolation, is an unbiased optimal estimation method for regional variables in a finite area based on variogram theory and structural analysis. The x_1 and x_2 were set as the inputs; G*, δ, and the softening point were set as the outputs. A Kriging model was established for 25 sample data with its basic principle as follows:

x_0 is set as the unobserved point that needs to be estimated, $x_1, x_2, \ldots, x_N$, are set as the observed points around it, and the observed values are correspondingly $y(x_1)$, $y(x_2)$, $\ldots$, $y(x_N)$. The estimate of the unmeasured point is denoted as $\tilde{y}(x_0)$, which is obtained by the weighted sum of the known observations of the adjacent observation points:

$$\tilde{y}(x_0) = \sum_{i=1}^{N} \lambda_i y(x_i) \tag{1}$$

Among them, λ_i is the undetermined weighting coefficient. The key to the Kriging interpolation is to calculate the weight coefficient λ_i, which must satisfy two conditions:

1 The unbiased estimates: the true value of the evaluation point is set to $y(x_0)$. The $y(x_i)$, $\widetilde{y}(x_0)$ and $y(x_0)$ can be regarded as a random variable.

$$E[\widetilde{y}(x_0) - y(x_0)] = 0, \sum_{i=1}^{N} \lambda_i = 1 \tag{2}$$

2 The variance of the difference between the true value $y(x_0)$ and the estimated value is minimal.

$$D[\widetilde{y}(x_0) - y(x_0)] = \min \tag{3}$$

$$D[\widetilde{y}(x_0) - y(x_0)] = -\sum_{i=1}^{N} \sum_{j=1}^{N} \lambda_i \lambda_j \gamma(x_i, x_j) + 2\sum_{i=1}^{N} \lambda_i \gamma(x_i, x_0) \tag{4}$$

where, $\gamma(x_i, x_j)$ represents the semivariance value of the parameter when the distance between the two points x_i and x_j is used as the distance h. The $\gamma(x_i, x_0)$ is the semivariance value of the parameter when the distance between the two points x_i and x_0 is used as the distance h.

3.1.3. The Multi-Objective Optimization Based on NSGA-II

The simultaneous optimization of multiple sub-objectives is a multi-objectives optimization problem. In most cases, it is generally impossible to concurrently achieve the optimum for multiple sub-goals. The ultimate goal of solving multi-objectives optimization problems is to harmonize various objectives, hence each sub objective is as optimal as possible. In the non-dominated ranking, NSGA-II was selected to approach the individual of the Pareto front, which enhanced its ability to advance. In the Pareto optimal relationship, the individuals in the group were compared according to their target function vector and divided into multiple frontier layers that were controlled in sequence. NSGA-II is widely used due to its advantages of solving the Pareto solution set with good accuracy and dispersion. NSGA-II was used to optimize the Kriging approximation model in this research. The constraints, optimization objectives, lower and upper bounds were set according to the established mathematical model to obtain a set of solutions that all satisfy the conditions. The main process was in Figure 5 as follows:

1. The initial population P_0 was randomly generated and a new population Q_0 was generated through selection, intersection and variation. The population R_0 WAs obtained by merging P_0 and Q_0.
2. R_t was sorted by non-inferiority to obtain non-inferior front segments $F_1, F_2, \ldots$.
3. F_i was sorted by crowding distance, and the better individuals and the previous segments $F_1, F_2, \ldots, F_{i-1}$ were selected to form N individuals and population P_{t+1}.
4. Population P_{t+1} was replicated, crossed and deformed to form population Q_{t+1}. If the termination conditions were met, then it ended. Otherwise, it proceeded to step 2 to continue execution.

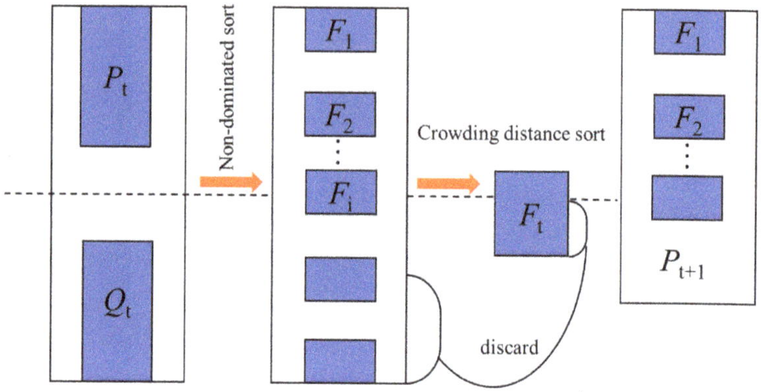

Figure 5. Flowchart for solving the Pareto solution set.

3.1.4. Optimize Experimental Design

The sampling points were established based on Opt LHD, as shown in Table 6. The steel-slag-powder-modified asphalt mortar (SSP-MAM) with different contents of antioxidant and UV absorber was prepared to undergo TFOT short-term aging and seven-day UV aging. The complex shear modulus (G*), softening point and phase angle (δ) of the aged SSP-MAM were recorded by observation.

Table 6. Opt LHD experimental design and results.

Number	Modifier Dosage		Measured Value		
	x_1 (%)	x_2 (%)	G* (64 °C)/kPa	δ (°)	Softening Point (°C)
1	0	1.2	9.099	87.3	58.3
2	0.6	1.5	11.324	85.8	59.9
3	3.2	0.6	10.073	86.8	59.0
4	0.4	0.9	8.983	88.6	58.2
5	1.2	0	10.434	85.5	59.3
6	2.4	2.4	13.368	83.3	61.5
7	0.2	1.3	10.454	84.6	59.3
8	2.2	1.8	11.437	84.0	60.0
9	0.8	1.1	11.675	84.1	60.2
10	3.8	0.3	11.371	85.9	60.0
11	3.6	2.2	13.341	83.0	61.4
12	4.6	1.0	13.314	83.4	61.4
13	1.8	0.7	12.064	83.4	60.5
14	2.6	2.1	13.721	82.7	61.7
15	4.2	0.5	12.475	83.9	60.8
16	4.8	2.3	15.631	82.5	63.1
17	2.8	0.8	13.388	83.2	61.5
18	1.0	0.2	11.946	83.6	60.4
19	1.6	0.4	13.511	82.9	61.6
20	4.4	1.4	15.901	82.2	63.3
21	3.0	1.6	16.760	81.8	64.0
22	3.4	2.0	16.083	81.7	63.5
23	2.0	0.1	12.395	83.3	60.7
24	4.0	1.7	16.020	81.8	63.4
25	1.4	1.9	17.193	81.3	64.3

Based on the experimental basis and the literature of the laboratory, the content of antioxidant 1098 was in the range of 0–4.8%, represented by x_1; the content of UV-5411 was in the range of 0–2.4%, represented by x_2.

3.2. Properties of SSP

Empyrean XRD (Panaco, The Netherlands) and Axios XRF (Almelo, The Netherlands) were used to characterize the physicochemical properties of the SSP.

3.3. Aging-Test Method

3.3.1. Short-Term Thermal-Oxygen-Aging Test

The 85-type asphalt film oven was used for the TFOT aging test to simulate the short-term thermal-oxygen-aging behavior of SSP-MAM in the mixing, paving, and compaction processes. The test temperature was controlled at 163 ± 1 °C, the rotating speed of the turntable was set at 5.5 ± 1 r/min, the tilt angle of the turntable and the horizontal plane was not more than 3°, and the aging time was 5 h.

3.3.2. Accelerated-UV-Aging Test

The LHX-205 type intelligent numerical-control ultraviolet-aging test box was used to simulate accelerated-ultraviolet-aging experiments. The SSP-MAM sample after short-term thermal-oxygen aging was placed on the horizontal turntable of the ultraviolet-aging box. The ultraviolet radiation intensity on the surface of the asphalt sample was 30 W/cm^2 by adjusting the height of the horizontal turntable. The wavelength of ultraviolet light was 365 nm, the experiment temperature was set to 60 °C, and the experiment time was seven days.

3.4. Asphalt Mortar Characteristic Analysis Method

3.4.1. Functional Group Characteristics of Asphalt Mortar

In the FTIR, the molecular structure was observed by measuring the vibrational and rotational spectra of molecules. A Nexus intelligent FTIR was adopted to analysis the functional groups of SSP-MAM. First, the asphalt/carbon disulfide solution with a mass concentration of 5% was configured. Then the asphalt sample was dissolved in the carbon-disulfide solution, and the prepared solution was dropped onto the KBr wafer. The SSP-MAM sample film could be obtained after the complete volatilization of the carbon-disulfide solution. The scanning beam range of the infrared spectrometer was 4000–400 cm^{-1}, and the number of scans was 64.

3.4.2. Rheological Properties of Asphalt Mortar

The dynamic shear rheometer (DSR) is an instrument applied to analyze the rheological properties of polymer materials. It has been widely used due to its good experimental accuracy and repeatability, since SHRP recommended using it to test the rheological properties of asphalt in 1993. The rheological properties of SSP-MAM samples were tested by MCR-102 DSR (Anton Paar, Graz, Austria). The rheological parameters at different temperatures were scanned under the strain-control mode. The scanning frequency was controlled as 10 rad/s, the temperature scanning range was 30–80 °C, the upper and lower plate diameter was 25 mm, the heating rate was 2 °C/min, and the plate spacing (asphalt film thickness) was 1 mm.

3.5. Leaching Behavior of Asphalt Mortar

TCLP developed by the US Environmental Protection Agency (EPA) was adopted for the toxicity-extraction experiment of steel slag in this research. The leaching agent was used to adjust the pH of the solid waste for tumble-extraction experiments in order to implement the Resource Conservation and Regeneration Act (RCRA) for the management of hazardous and solid waste. Simultaneously, the leaching content of inorganic components in solid waste was simulated by judging its contamination capacity.

In this experiment, the leaching behavior of heavy-metal ions from SSP and SSP-MAM was investigated. SSP-MAM was specially prepared into particles smaller than 9.5 mm to meet the requirements of the test samples in TCLP. The specific experimental steps in Figure 6 were as follows: First, 50.0 g of the particle was weighed and placed in a 2 L polyethylene bottle for later use. Secondly, the acetate buffer solution with pH value of 2.88 ± 0.05 was prepared, and the corresponding amount of acetate buffer solution was added according to the liquid- solid ratio of 20:1 (L/kg). Then, the polyethylene bottle was placed on the inversion shaker, the rotation speed of the shaker was set to 30 rpm, the shaking time was 18 ± 2 h, and the temperature was 23 ± 2 °C. Finally, the mixture was allowed to stand after shaking, and the clarified filtrate was collected in a centrifuge tube with filter paper for next step of testing. The leaching concentration of heavy-metal elements was determined by atomic absorption spectrometry.

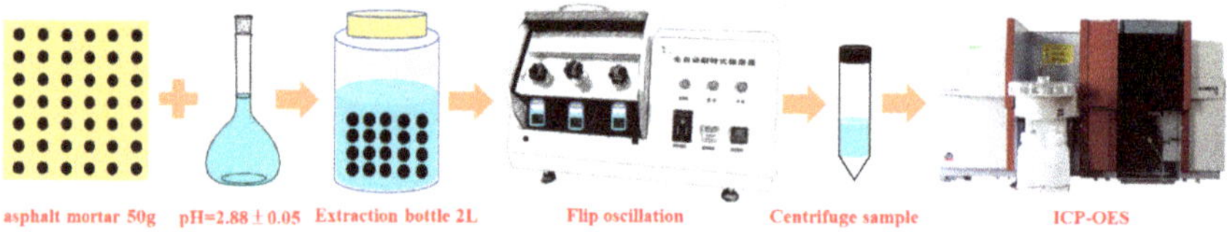

Figure 6. Procedures of TCLP.

3.6. Experiment Plan

The experimental plan was shown in Figure 7.

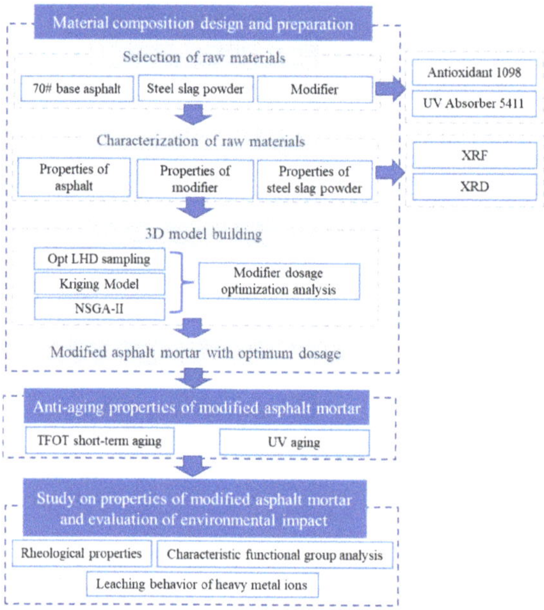

Figure 7. Experiment plan.

4. Results and Discussions

4.1. Optimization Test of Dosage Simulation

The partial-approximation model, contour plot, and R-Squared accuracy of G* and x_1, x_2 are presented in Figure 8a–c, and so on for phase angle and softening point. From Figure 8c,f,i, the values of R^2 can be obtained as 0.96652, 0.95828, and 0.95732, respectively, where the blue line represents the actual average value, indicating that the fitting accuracy and reliability were high. The fit of anisotropy and exponential functions were selected as the Kriging model by comparing the output R^2 values. Figure 8b,e,h are contour maps of the projections of Figure 8a,d,g on the horizontal plane, respectively. The denser the contour lines and the greater the slope of the fitted surface, the more significant the influence of this factor was. The contour lines in the figure were dense and steep, indicating that there was significant correlation between the amount of modifier and G*, δ, and the softening point. When the first peak appeared, for G*, δ, and the softening point, the corresponding dosages of x_1 were 2.15%, 1.0%, and 1.1%, respectively. It showed that the effect of antioxidants on δ was more significant, that is, less antioxidant could change δ. Meanwhile, the corresponding dosages of x_2 were 0.25%, 0.76%, and 0.38%, respectively. It was clarified that the effect of the UV absorber on G* was more significant, that is, less UV absorber could change G*.

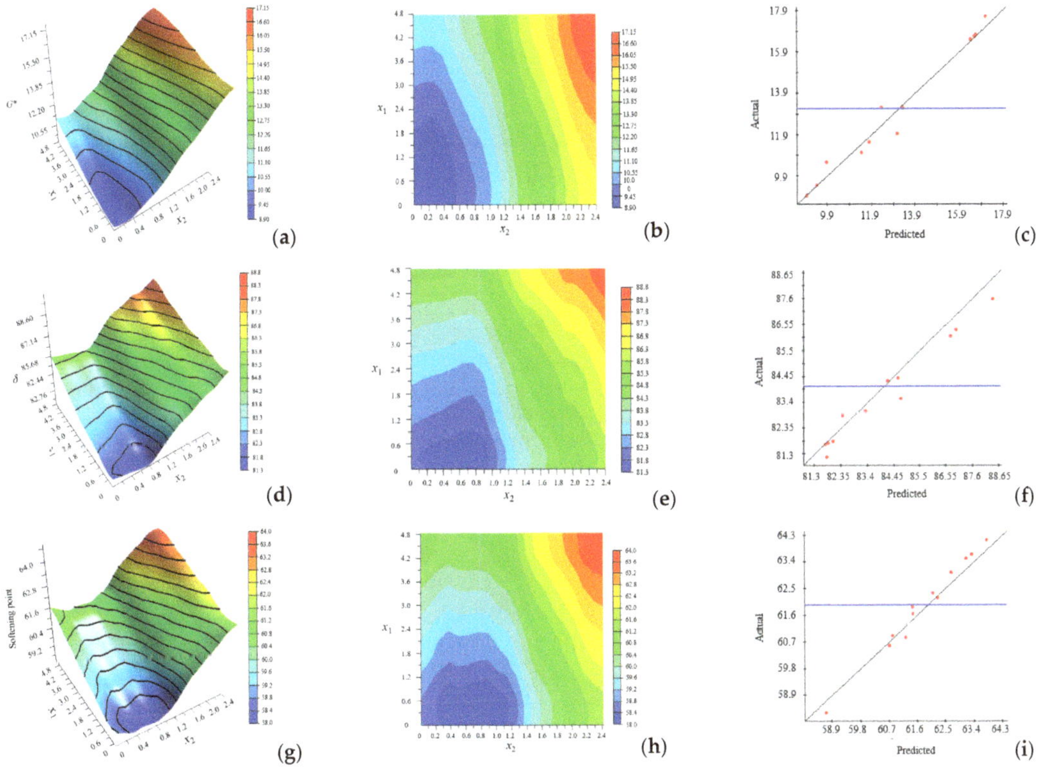

Figure 8. Partial-approximation model, contour plot, and R-squared accuracy of output and input. (**a**) Partial-approximation model of G*, (**b**) Contour plot of G*, (**c**) R-Squared accuracy of G*, (**d**) Partial-approximation model of δ, (**e**) Contour plot of δ, (**f**) R-Squared accuracy of δ, (**g**) Partial-approximation model of softening point, (**h**) Contour plot of softening point, (**i**) R-Squared accuracy of softening point.

In this paper, the softening point and phase angle were selected as research objects as examples. The optimized Pareto front is shown in Figure 9. The parameters of the NSGA-II algorithm were set as: population size was 100, evolutionary generation was 100, hybridization-distribution coefficient was 20, hybridization probability was 0.9, mutation-distribution coefficient was 100, and mutation probability was 0.9. All the points in Figure 9 were qualified non-inferior solutions. It was necessary to select the appropriate solution in order to determine the final optimization scheme. The individuals with larger crowding distances had greater advantages in accordance with the optimization strategy of the NSGA-II algorithm. The optimal value of the modifier dosage x_1 was 1.2%, and x_2 was 0.5% after weighing by the NSGA-II algorithm.

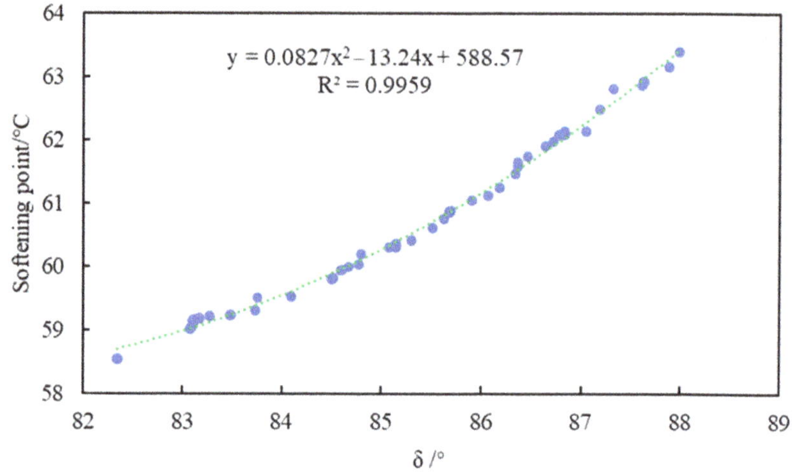

Figure 9. Pareto front.

4.2. Properties of SSP

The mineral phase of SSP was characterized by XRD as shown in Figure 10. It can be seen from the diffraction pattern that there were a variety of diffraction peaks with serious overlap, which indicates that the types of minerals contained in the steel slag were quite complex. The strongest diffraction peaks and three strong peaks appearing at 2θ of 32.961 and 33.067 belonged to silicate minerals, indicating that C_3S and C_2S were the main mineral phases of steel slag.

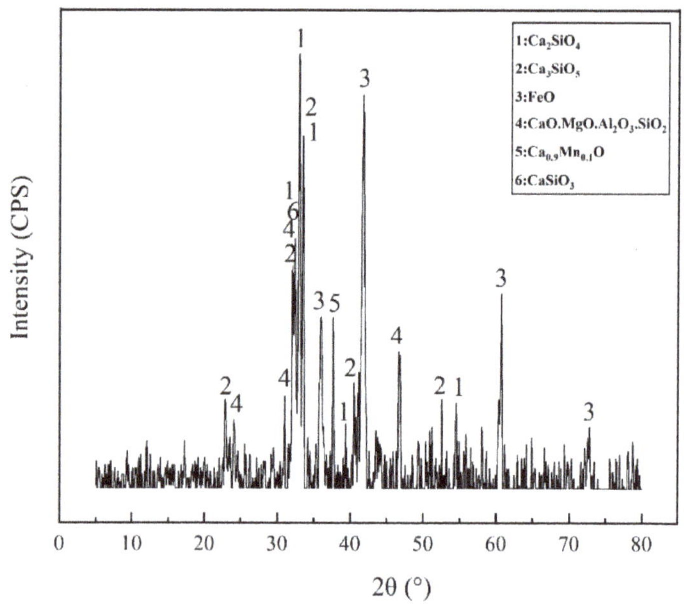

Figure 10. The XRD of SSP.

4.3. Functional-Group Characteristics of Asphalt Mortar

Infrared spectroscopy tests were carried out on modified asphalt mortar (MAM) and unmodified asphalt mortar (UAM) to investigate the change characteristics of functional groups during thermal-oxygen and ultraviolet aging. The characteristic peaks of oxygen-containing functional groups were more distinct due to the absorption of oxygen during aging. The formula to calculate the index of the characteristic functional group carbonyl (C=O) and sulfoxide (S=O) of asphalt were as follows:

$$I_{C=O} = \frac{S_{1700\ cm^{-1}}}{S_{2000\sim600\ cm^{-1}}} \times 100\% \tag{5}$$

$$I_{S=O} = \frac{S_{1030\ cm^{-1}}}{S_{2000\sim600\ cm^{-1}}} \times 100\% \tag{6}$$

where $S_{1700\ cm^{-1}}$ is the area of the carbonyl peak band centered around 1700 cm^{-1}, $S_{1030\ cm^{-1}}$ is the area of the sulfoxide peak band centered around 1030 cm^{-1}, and $S_{2000\sim600\ cm^{-1}}$ is the area of the spectral bands between 2000 and 600 cm^{-1}.

In the UAM, the value of $I_{S=O}$ and $I_{C=O}$ were 1.56% and 1.3%, respectively, while the values of $I_{S=O}$ and $I_{C=O}$ in the MAM were 0.9% and 1.1%, respectively. It revealed that the content of oxygen-containing functional groups and the degree of photo-oxidation of the MAM were declined. Therefore, the thermal-oxygen- and ultraviolet-aging resistance of the MAM was improved.

In the Figure 11, the characteristic absorption peaks of $I_{S=O}$ and $I_{C=O}$ in UAM appeared at 1033 cm^{-1} and 1692 cm^{-1}, respectively, while in MAM they appeared at 1676 cm^{-1} and 998 cm^{-1}, respectively. The reason for the different degrees of bathochromic shift of the two characteristic absorption peaks was that the antioxidants and UV absorbers contained auxiliary color groups such as non-bonded-electron heteroatom-saturated groups. Simultaneously, the absorption peaks being shifted to long wavelengths under the combined action of conjugation and auxochrome groups was attributed to the presence of aromatic heterocycles in antioxidants and UV absorbers.

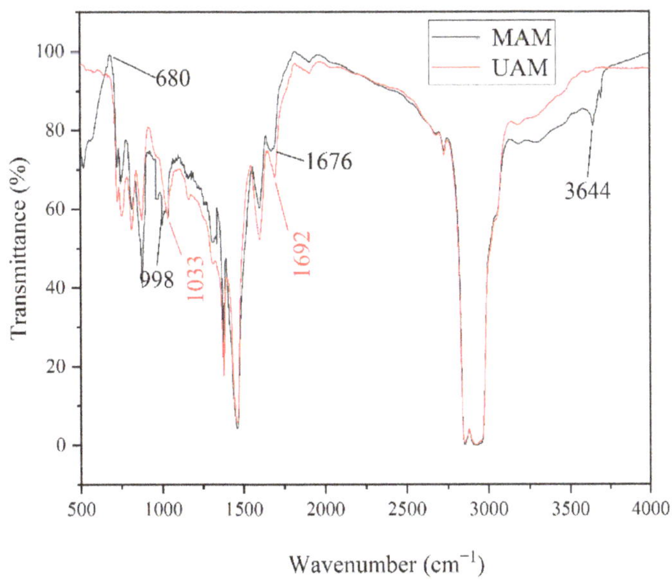

Figure 11. Infrared analysis of asphalt mortar.

In addition, the characteristic peaks of C-H stretching vibration of N-H and aromatic hydrocarbon benzene ring in MAM were present at 3644 cm^{-1} and 680 cm^{-1}, respectively, but they did not appear in UAM. This illustrated that antioxidants and UV absorbers had chemical reactions with asphalt to generate new functional groups during the aging process of asphalt, rather than simple physical blending [37].

4.4. Rheological Properties of Asphalt Mortar

Figure 12 showed that G* decreased with the increase in temperature, indicating that the rheological properties of SSP-MAM had obvious temperature dependence. The transition of asphalt from a highly elastic state at low temperature to a viscous fluid state at high temperature due to the increase in the free volume of asphalt with increasing temperature. Therefore, the decrease in the maximum shear stress and the increase in the maximum shear strain of the asphalt led to the decrease in G*. Meanwhile, the G* of the UAM was larger, which was due to the oxidation reaction of the asphalt film in contact with the oxygen in the air at high temperature. The content of aromatics in asphalt was decreased, while pectin and asphaltene was increased. Therefore, a molecular weight migration from small to large occurred between the asphalt components. The asphalt was hardened by the reduction in plastic flow deformability based on the increase in the overall molecular weight of the asphalt. The G* in MAM was smaller on account of the modifier chemically reacting with oxygen, thereby preventing the contact of the bitumen film with oxygen.

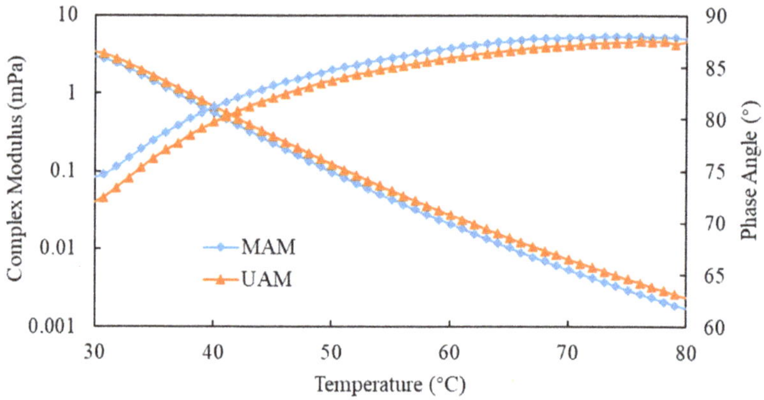

Figure 12. Rheological properties of asphalt mortar.

The anti-aging ability of bitumen was enhanced with the increase in phase-angle ratio and the decrease in viscosity loss during the aging process. It can be observed that the phase angle increased with the increase in temperature, which was larger in MAM than in UAM. It illustrated that the free radicals generated by the asphalt monomer due to thermal oxygen and light were captured by antioxidants, while the ultraviolet rays were converted into heat by ultraviolet absorbent through chemical action, thus preventing further thermal-oxygen and ultraviolet aging of the asphalt.

4.5. Leaching Behavior of Asphalt Mortar

Figure 13 presents the leaching results of heavy-metal ions. It can be clearly discerned that the leaching concentrations of As, Cr^{6+} and Cu were higher, followed by Zn, Pb, Ni, and finally Cd. Meanwhile, the leaching concentration of heavy-metal ions in the SSP was the largest, followed by the aged SSP-MAM, and finally the unaged SSP-MAM.

A possible cause was that the SSP was coated with chemically stable bitumen to prevent it from reacting with the acidic extractant. On the other hand, the high-polarity water could not fully contact the SSP on account of the low polarity of the asphalt. Therefore, asphalt had a certain physical solid-sealing effect on the release of heavy-metal ions in the SSP, and further reduced the leaching heavy-metal-ion content by reducing the chance of contact between the SSP and the extraction solution. In addition, the encapsulation ability of the asphalt to the heavy-metal ions in the SSP was also imparity. The leaching amount of As and Pb decreased the most, which were 47.8% and 46.4%, respectively, followed by Cr^{6+} (22.6%) and Cd (21.2%), Zn (10.6%), and finally, Ni (7.4%) and Cu (5.2%).

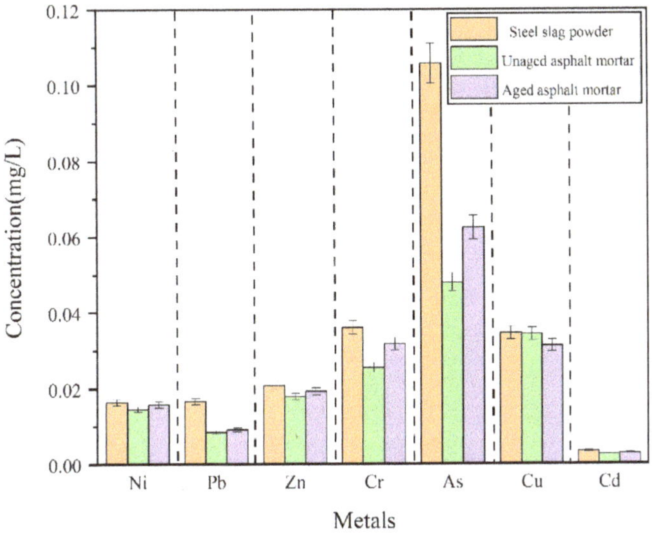

Figure 13. Heavy-metal leaching from asphalt mortar.

Obviously, the aged SSP-MAM was larger than the unaged SSP-MAM in terms of ionic leaching concentration. The reason was that the asphalt structure was changed under the synergistic effect of oxygen and ultraviolet rays. The flexibility and ductility of the asphalt became depraved with the decrease in light components and the increase in asphaltene. At this point, the risk of leaching heavy-metal ions increased with the inferior asphalt coating performance on SSP.

5. Conclusions

SSP was used to replace natural mineral to prepare asphalt mortar in this research, and the thermal-oxidation- and ultraviolet-aging resistance of SSP-MAM was evaluated. The multi-objective optimization analysis of antioxidants and UV absorbers was carried out using the NSGA-II. A chemically reactive composite-modified asphalt mortar with good comprehensive properties was prepared. The optimization model was verified according to the experimental results. Based on the above research content, the main conclusions of this paper were as follows:

1. The denser the contour lines and the greater the slope of the fitted surface, the more significant the influence of this factor was. The contour lines were dense and steep, indicating that there was significant correlation between the dosage of modifier and G*, δ, and the softening point. When the first peak appeared, for G*, δ, and the softening point, the corresponding dosages of x_1 were 2.15%, 1.0%, and 1.1%, respectively, while the corresponding dosages of x_2 were 0.25%, 0.76%, and 0.38%, respectively. All the points in the Pareto front were qualified non-inferior solutions. The individuals with larger crowding distances had greater advantages in accordance with the optimization

strategy of the NSGA-II algorithm. The optimal value of the modifier dosage x_1 was 1.2%, and x_2 was 0.5% after weighing by the NSGA-II algorithm.

2. In the UAM, the value of $I_{S=O}$ and $I_{C=O}$ were 1.56% and 1.3%, respectively, while the values of $I_{S=O}$ and $I_{C=O}$ in the MAM were 0.9% and 1.1%, respectively. This revealed that the content of oxygen-containing functional groups and the degree of photo-oxidation of the MAM were declined. The characteristic absorption peaks of sulfoxide and carbonyl group of the MAM had different degrees of bathochromic shift.
3. The complex shear modulus was smaller and the phase angle was larger in MAM. The contact of the bituminous membrane with oxygen was prevented as the modifier reacted chemically with oxygen. This illustrated that the free radicals generated by the asphalt monomer due to thermal oxygen and light were captured by antioxidants, while the ultraviolet rays were converted into heat by ultraviolet absorbent through chemical action, thus preventing further thermal-oxygen and ultraviolet aging of the asphalt.
4. The leaching concentration of heavy-metal ions in the SSP was the largest, followed by the aged SSP-MAM, and finally the unaged SSP-MAM. The asphalt had a certain physical solid-sealing effect on the release of heavy-metal ions in SSP. In addition, the encapsulation ability of asphalt for heavy-metal ions in SSP was also insufficient. The leaching amount of As and Pb decreased the most, which were 47.8% and 46.4%, respectively. The asphalt structure was changed under the synergistic effect of oxygen and ultraviolet rays. Therefore, the risk of leaching heavy-metal ions was increased with the inferior asphalt coating performance on SSP.

The UV-5411 and antioxidant 1098 increased the thermal oxygen and UV-aging resistance of SSP-MAM. Therefore, the mixing performance of the asphalt mixture was improved and its service life was prolonged. The solid-sealing effect of asphalt on SSP reduced the risk of leaching of heavy-metal ions, thus improving the application potential of SSP instead of mineral powder. However, due to the high economic cost of UV absorbers and antioxidants, their large-scale applications are limited. Therefore, future research can begin with reducing the dosage of modifiers, finding cheap replacement materials with excellent performance, or introducing new materials to reduce the dosage of UV absorbers and antioxidants.

Author Contributions: Conceptualization, Y.L., S.A., H.X., Y.Z. and X.Y.; data curation: S.W., S.A., H.X. and Y.Z.; methodology, Y.L., P.C., S.A. and H.X.; formal analysis, Y.L., P.C., Y.Z. and X.Y.; investigation, Y.L., S.A. and X.Y.; writing—original draft, Y.L.; writing—review and editing, P.C.; supervision, S.W.; project administration, S.W.; funding acquisition, S.W. All authors have read and agreed to the published version of the manuscript.

Funding: The work presented in this paper is financially supported by the National Key R&D Program of China (No. 2018YFB1600200), National Natural Science Foundation of China (No. 51978547), Technological Innovation Major Project of Hubei Province (2019AEE023) and Key R&D Program of Hubei Province (2020BCB064).

Institutional Review Board Statement: Not applicable.

Informed Consent Statement: Not applicable.

Data Availability Statement: The data presented in this study were available on request from the corresponding author.

Acknowledgments: The authors acknowledge Wuhan University of Technology for materials and experimental instrument support sincerely.

Conflicts of Interest: The authors declare no conflict of interest.

Abbreviations

Acronyms	Full Name
SSP	Steel-slag powder
SSP–AM	Steel-slag powder–asphalt mortar
LHD	Latin hypercube design
Opt LHD	Optimal Latin hypercube design
NSGA-II	Second-generation non-inferior sorting genetic algorithm
SSP-MAM	Steel-slag-powder-modified asphalt mortar
MAM	Modified asphalt mortar
UAM	Unmodified asphalt mortar

References

1. Wu, H.; Li, P.; Nian, T.; Zhang, G.; He, T.; Wei, X. Evaluation of asphalt and asphalt mixtures' water stability method under multiple freeze-thaw cycles. *Constr. Build. Mater.* **2019**, *228*, 117089. [CrossRef]
2. Wan, P.; Liu, Q.; Wu, S.; Zhao, Z.; Chen, S.; Zou, Y.; Rao, W.; Yu, X. A novel microwave induced oil release pattern of calcium alginate/nano-Fe3O4 composite capsules for asphalt self-healing. *J. Clean. Prod.* **2021**, *297*, 126721. [CrossRef]
3. Zhu, K.; Qin, X.; Wang, Y.; Lin, C.; Wang, Q.; Wu, K. Effect of the oxygen concentration on the combustion of asphalt binder. *J. Anal. Appl. Pyrolysis* **2021**, *160*, 105370. [CrossRef]
4. Mainieri, J.J.G.; Singhvi, P.; Ozer, H.; Sharma, B.K.; Al-Qadi, I.L. Fatigue Tolerance of Aged Asphalt Binders Modified with Softeners. *Transp. Res. Rec.* **2021**, *2675*, 1229–1244. [CrossRef]
5. Zhao, Y.; Chen, M.; Wu, S.; Jiang, Q.; Xu, H.; Zhao, Z.; Lv, Y. Effects of waterborne polyurethane on storage stability, rheological properties, and VOCs emission of crumb rubber modified asphalt. *J. Clean. Prod.* **2022**, *340*, 130682. [CrossRef]
6. Yan, C.; Yuan, L.; Yu, X.; Ji, S.; Zhou, Z. Characterizing the fatigue resistance of multiple modified asphalts using time sweep test, LAS test and elastic recovery test. *Constr. Build. Mater.* **2022**, *322*, 125806. [CrossRef]
7. Cheraghian, G.; Wistuba, M.P.; Kiani, S.; Behnood, A.; Afrand, M.; Barron, A.R. Engineered nanocomposites in asphalt binders. *Nanotechnol. Rev.* **2022**, *11*, 1047–1067. [CrossRef]
8. Wang, F.; Xie, J.; Wu, S.; Li, J.; Barbieri, D.M.; Zhang, L. Life cycle energy consumption by roads and associated interpretative analysis of sustainable policies. *Renew. Sustain. Energy Rev.* **2021**, *141*, 110823. [CrossRef]
9. Li, N.; Jiang, Q.; Wang, F.; Cui, P.; Xie, J.; Li, J.; Wu, S.; Barbieri, D.M. Comparative Assessment of Asphalt Volatile Organic Compounds Emission from field to laboratory. *J. Clean. Prod.* **2021**, *278*, 123479. [CrossRef]
10. Chinnu, S.N.; Minnu, S.N.; Bahurudeen, A.; Senthilkumar, R. Recycling of industrial and agricultural wastes as alternative coarse aggregates: A step towards cleaner production of concrete. *Constr. Build. Mater.* **2021**, *287*, 123056. [CrossRef]
11. Cui, P.; Wu, S.; Xiao, Y.; Liu, Q.; Wang, F. Hazardous characteristics and variation in internal structure by hydrodynamic damage of BOF slag-based thin asphalt overlay. *J. Hazard. Mater.* **2021**, *412*, 125344. [CrossRef] [PubMed]
12. Cui, P.; Wu, S.; Xiao, Y.; Hu, R.; Yang, T. Environmental performance and functional analysis of chip seals with recycled basic oxygen furnace slag as aggregate. *J. Hazard. Mater.* **2021**, *405*, 124441. [CrossRef] [PubMed]
13. Lv, Y.; Wu, S.; Cui, P.; Liu, Q.; Li, Y.; Xu, H.; Zhao, Y. Environmental and feasible analysis of recycling steel slag as aggregate treated by silicone resin. *Constr. Build. Mater.* **2021**, *299*, 123914. [CrossRef]
14. Liu, Y.; Tang, M.; Hu, Q.; Zhang, Y.; Zhang, L. Densification behavior, microstructural evolution, and mechanical properties of TiC/AISI420 stainless steel composites fabricated by selective laser melting. *Mater. Des.* **2020**, *187*, 108381. [CrossRef]
15. Yang, C.; Wu, S.; Cui, P.; Amirkhanian, S.; Zhao, Z.; Wang, F.; Zhang, L.; Wei, M.; Zhou, X.; Xie, J. Performance characterization and enhancement mechanism of recycled asphalt mixtures involving high RAP content and steel slag. *J. Clean. Prod.* **2022**, *336*, 130484. [CrossRef]
16. Li, J.; Yu, J.; Wu, S.; Xie, J. The Mechanical Resistance of Asphalt Mixture with Steel Slag to Deformation and Skid Degradation Based on Laboratory Accelerated Heavy Loading Test. *Materials* **2022**, *15*, 911. [CrossRef]
17. Ho, J.C.M.; Liang, Y.; Wang, Y.H.; Lai, M.H.; Huang, Z.C.; Yang, D.; Zhang, Q.L. Residual properties of steel slag coarse aggregate concrete after exposure to elevated temperatures. *Constr. Build. Mater.* **2022**, *316*, 125751. [CrossRef]
18. Ziaee, S.A.; Behnia, K. Evaluating the effect of electric arc furnace steel slag on dynamic and static mechanical behavior of warm mix asphalt mixtures. *J. Clean. Prod.* **2020**, *274*, 123092. [CrossRef]
19. Xu, H.; Wu, S.; Li, H.; Zhao, Y.; Lv, Y. Study on Recycling of Steel Slags Used as Coarse and Fine Aggregates in Induction Healing Asphalt Concretes. *Materials* **2020**, *13*, 889. [CrossRef]
20. Li, Y.; Feng, J.; Yang, F.; Wu, S.; Liu, Q.; Bai, T.; Liu, Z.; Li, C.; Gu, D.; Chen, A.; et al. Gradient aging behaviors of asphalt aged by ultraviolet lights with various intensities. *Constr. Build. Mater.* **2021**, *295*, 123618. [CrossRef]
21. Broering, W.B.; de Melo, V.S.; Manfro, A.L. Incorporation of nanoalumina into a polymeric asphalt matrix: Reinforcement of the nanostructure, improvement of phase stability, and amplification of rheological parameters. *Constr. Build. Mater.* **2022**, *320*, 126261. [CrossRef]
22. Huang, H.; Chen, Z.; Li, J.; Wang, J.; Ding, S.; Zhang, H.; Wu, C. Aging Resistances Evaluation of Multi-dimensional Nanomaterials Modified Asphalt by Characterizing Binders Recovered from Aged Asphalt Mixtures. *J. Test. Eval.* **2022**, *50*, 105–116. [CrossRef]

23. Hu, Z.; Xu, T.; Liu, P.; Oeser, M.; Wang, H. Improvements of Developed Graphite Based Composite Anti-Aging Agent on Thermal Aging Properties of Asphalt. *Materials* **2020**, *13*, 4005. [CrossRef] [PubMed]
24. Yang, J.; Muhammad, Y.; Yang, C.; Liu, Y.; Su, Z.; Wei, Y.; Li, J. Preparation of TiO2/PS-rGO incorporated SBS modified asphalt with enhanced resistance against ultraviolet aging. *Constr. Build. Mater.* **2021**, *276*, 121461. [CrossRef]
25. Rajib, A.; Saadeh, S.; Katawal, P.; Mobasher, B.; Fini, E.H. Enhancing Biomass Value Chain by Utilizing Biochar as A Free Radical Scavenger to Delay Ultraviolet Aging of Bituminous Composites Used in Outdoor Construction. *Resour. Conserv. Recycl.* **2021**, *168*, 105302. [CrossRef]
26. Wang, S.; Yang, J.; Chen, H.; Chu, K.; Yu, X.; Wei, Y.; Zhang, H.; Rui, M.; Feng, C. A Strategy for the Effective Optimization of Pharmaceutical Formulations Based on Parameter-Optimized Support Vector Machine Model. *Aaps Pharmscitech* **2022**, *23*, 66. [CrossRef]
27. Deng, H.; Sheng, G.; Zhao, H.; Meng, F.; Zhang, H.; Ma, J.; Gong, J.; Ruan, J. Integrated optimization of fracture parameters for subdivision cutting fractured horizontal wells in shale oil reservoirs. *J. Pet. Sci. Eng.* **2022**, *212*, 110205. [CrossRef]
28. Zong, C.; Li, Q.; Li, K.; Song, X.; Chen, D.; Li, X.; Wang, X. Computational fluid dynamics analysis and extended adaptive hybrid functions model-based design optimization of an explosion-proof safety valve. *Eng. Appl. Comput. Fluid Mech.* **2022**, *16*, 296–315. [CrossRef]
29. Kollmann, J.; Lu, G.; Liu, P.; Xing, Q.; Wang, D.; Oeser, M.; Leischner, S. Parameter optimisation of a 2D finite element model to investigate the microstructural fracture behaviour of asphalt mixtures. *Theor. Appl. Fract. Mech.* **2019**, *103*, 102319. [CrossRef]
30. Sivilevicius, H.; Vislavicius, K. Simulation of composition of recycled hot-mix asphalt mixture produced in asphalt mixing plant. *Constr. Build. Mater.* **2019**, *214*, 17–27. [CrossRef]
31. Samadhiya, A.; Namrata, K. Probabilistic screening and behavior of solar cells under Gaussian parametric uncertainty using polynomial chaos representation model. *Complex Intell. Syst.* **2021**, *8*, 989–1004. [CrossRef]
32. Zhu, H.; Liu, L.; Long, T.; Peng, L. A novel algorithm of maximin Latin hypercube design using successive local enumeration. *Eng. Optim.* **2012**, *44*, 551–564. [CrossRef]
33. Kim, H.; Lee, T.H. A robust elastic net via bootstrap method under sampling uncertainty for significance analysis of high-dimensional design problems. *Knowl. Based Syst.* **2021**, *225*, 107117. [CrossRef]
34. Cao, R.; Sun, Z.; Wang, J.; Guo, F. A single-loop reliability analysis strategy for time-dependent problems with small failure probability. *Reliab. Eng. Syst. Saf.* **2022**, *219*, 108230. [CrossRef]
35. Mohamed, K.T.; Abdel-razak, M.H.; Haraz, E.H.; Ata, A.A. Fine tuning of a PID controller with inlet derivative filter using Pareto solution for gantry crane systems. *Alex. Eng. J.* **2022**, *61*, 6659–6673. [CrossRef]
36. Wei, M.; Wu, S.; Xu, H.; Li, H.; Yang, C. Characterization of Steel Slag Filler and Its Effect on Aging Resistance of Asphalt Mastic with Various Aging Methods. *Materials* **2021**, *14*, 869. [CrossRef]
37. Feng, Z.-G.; Yu, J.-Y.; Zhang, H.-L.; Kuang, D.-L.; Xue, L.-H. Effect of ultraviolet aging on rheology, chemistry and morphology of ultraviolet absorber modified bitumen. *Mater. Struct.* **2013**, *46*, 1123–1132. [CrossRef]

materials MDPI

Article

Investigation of the Rheological Properties and Chemical Structure of Asphalt under Multiple Aging Conditions of Heat, UV and Aqueous Solution

Yingxue Zou [1], Ling Pang [1], Shi Xu [2,3,*], Shaopeng Wu [1], Miao Yuan [4], Serji Amirkhanian [5], Haiqin Xu [1], Yang Lv [1] and Xiang Gao [1]

1 State Key Laboratory of Silicate Materials for Architectures, Wuhan University of Technology, Luoshi Road 122, Wuhan 430070, China
2 School of Civil Engineering and Architecture, Wuhan University of Technology, Luoshi Road 122, Wuhan 430070, China
3 Faculty of Civil Engineering and Geosciences, Delft University of Technology, Stevinweg 1, 2628 CN Delft, The Netherlands
4 Foshan Transportation Science and Technology Co., Ltd., Foshan 528315, China
5 Department of Civil Construction and Environmental Engineering, University of Alabama, Tuscaloosa, AL 35487, USA
* Correspondence: xushi@whut.edu.cn

Abstract: During the service period, asphalt materials are affected by various natural factors, including heat, ultraviolet light, oxygen and moisture, etc., resulting in the reduction of pavement performance, the increase of pavement distress and shortening of service life. This study aims to investigate the aging performance of asphalt under multiple aging conditions of heat, UV and aqueous solution. Thermal-oxygen aging, UV aging and hydrostatic erosion tests were carried out sequentially on asphalt. The rheological properties, chemical structure and element composition of asphalt were characterized before and after aging, and the effect mechanism of multiple conditions was discussed. The results show that the multiple conditions of heat and UV can increase the rutting resistance and weaken the cracking resistance of asphalt. However, the effect degree of UV decreases gradually with the deepening of aging degree. Additionally, the effect of water on the physicochemical properties is less than that of UV; however, water can increase the sensitivity of physicochemical properties to UV. In summary, this study explored the short-term cycling effect of heat, light and water on asphalt and provided an idea for simulation test of asphalt under multiple aging condition.

Keywords: asphalt; thermal-oxygen aging; UV aging; moisture damage; rheological performance; chemical structure

Citation: Zou, Y.; Pang, L.; Xu, S.; Wu, S.; Yuan, M.; Amirkhanian, S.; Xu, H.; Lv, Y.; Gao, X. Investigation of the Rheological Properties and Chemical Structure of Asphalt under Multiple Aging Conditions of Heat, UV and Aqueous Solution. *Materials* **2022**, *15*, 5711. https://doi.org/10.3390/ma15165711

Academic Editor: Simon Hesp

Received: 20 July 2022
Accepted: 15 August 2022
Published: 19 August 2022

1. Introduction

Asphalt concrete, due to its advantages of easy construction, driving comfort and low noise, is widely used as a paving material throughout the world [1,2]. The appearance of asphalt concrete distress is mainly due to the occurrence of aging under the conditions of heat, oxygen, ultraviolet and water [3]. Thermal-oxygen aging occurs mainly during the construction period, including mixing, paving, etc. Its mechanism mainly includes the volatilization and oxidation of light components, the condensation of saturates and the degradation of macromolecules in asphalt [4]. UV radiation has the most significant effect on asphalt concrete aging during service. The molecular chains in asphalt absorb enough energy from the UV wavelengths to cause the bonds to break [5]. Additionally, water and the oxygen in the water can cause the alternating occurrence of asphalt oxidation, dissolution and migration [6]. In different regions with different climatic conditions, asphalt concretes are subjected to the diverse service environment [7]. Asphalt concrete in different areas are exposed to various corrosive media, such as salt, acid and alkali [8–11]. Asphalt aging is the main cause of asphalt concrete aging [12]. Therefore, there are a large number of studies are concerned with the aging of asphalt, but preliminary research mainly focused on single factor of asphalt aging due to the complexity of asphalt composition [13–15].

With the deepening of research, multiple factors are gradually being considered in the simulation of asphalt aging. Abouelsaad and White pointed out that the performance of hot mix asphalt mixture gradually weakened with the increase of coupling aging time [16]. Tan and Li investigated the coupled effect of thermal-oxygen–UV on asphalt and found the effect of coupled aging was more obvious than thermal-oxygen aging, which showed a rapid decay of asphalt properties [17]. Li et al. supposed that the coupled ageing mechanism of UV and different aqueous solution mainly includes three channels, including the excitation and break of asphalt molecules and the dissolution and separation of the organic component in asphalt [18]. It was Ilaria and Eyad who found that the combined effect of UV–heat–oxygen–moisture caused a portion of the asphalt to become soluble and dissolved, while the rest of the asphalt showed cracking [19]. Zhang et al. concluded that coupled aging effect of heat–UV–water was considered to have a significant effect on the viscoelasticity and high-temperature performance of warm mix asphalt [20]. These atmospheric factors had a very important role in the degradation and microstructural evolution of asphalt. However, the above studies simulate the simultaneous action of multiple factors on asphalt, and the interaction and contribution of each factor is not clear. Because the aging mechanism of the three kinds of aging methods is various, so is the effect behavior on properties of asphalt. The study of asphalt aging mechanism is important not only for predicting the longevity of asphalt, but information about asphalt aging can help correctly restore asphalt properties with recycled asphalt pavement (RAP). The industries all over the world are looking for solutions to use 100% RAP, which can significantly improve economic and ecological outcomes [21]. Therefore, it is of great interest to investigate the aging performance of asphalt under multiple conditions of heat, UV and aqueous solution.

Changes in asphalt properties are mainly caused by changes in the internal chemical composition and structure. The combination of EA and FTIR can complement each other and facilitate a more accurate analysis of the microstructure of asphalt after aging [22]. Therefore, the aging performance of asphalt under multiple conditions of heat, UV and aqueous solution was investigated in this study using the characterization of the rheological properties, element composition and chemical structure of asphalt, and the performance evolution law was discussed. The research program is illustrated in Figure 1. This study adopted sequentially Thin Film Oven Test (TFOT), UV aging test and hydrostatic immersion test to conduct different aging test on 70 A. The multiple aging conditions are divided into three levels, including heat–UV coupling, UV–solution coupling and UV–water cycle. After aging, Dynamic Shear Rheometer (DSR) test and Bending Beam Rheometer (BBR) test were employed to observe the change of rheological properties, including high-temperature rutting resistance and low-temperature cracking resistance. Additionally, the major element composition and the characteristic functional groups of asphalt were detected by EA test and FTIR test, respectively. The results of EA test and FTIR test were combined to analyze the changes in chemical composition before and after aging and to investigate the aging mechanism of the properties changes.

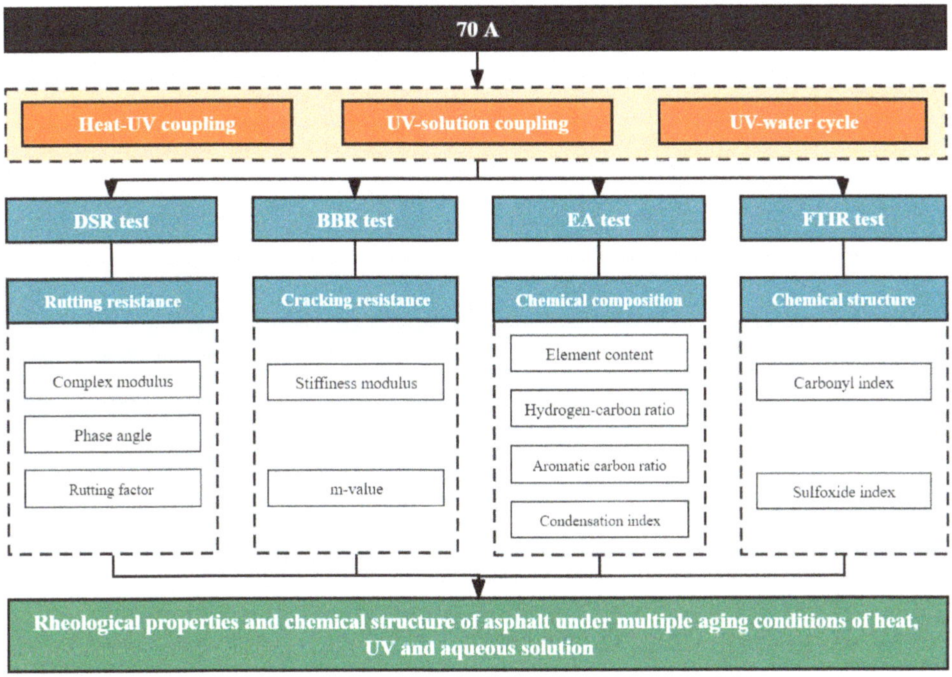

Figure 1. Research program.

2. Materials and Experiments

2.1. Materials

2.1.1. Asphalt

Base asphalt with 60/80 penetration grade (simply referred as 70 A) employed in this study was obtained from Hubei Guochuang Road Material Technology Co., Ltd. (Wuhan, China). The basic physical properties were illustrated in Table 1.

Table 1. Physical properties of 70 A.

Physical Properties	Units	70 A	Standards
Penetration (25 °C, 100 g, 5 s)	0.1 mm	72.4	ASTM D-5 [23]
Softening point	°C	49.6	ASTM D-36 [24]
Ductility (10 °C/5 °C)	cm	>100	ASTM D-113 [25]
Solubility (trichloroethylene)	%	99.5	ASTM D-2042 [26]

2.1.2. Preparation of Aqueous Solution

Four kinds of aqueous solutions with different media were prepared, including distilled water, 10 wt% NaCl saline solution, pH 3 acid solution and pH 11 alkali solution, to simulate asphalt immersed in various aqueous solution. The 10 wt% NaCl saline solution was obtained by dissolving solid sodium chloride with distilled water. The distilled water was used to dilute the mixed solution of sulfuric acid and nitric acid with a molar ratio of 9:1, and the pH 11 alkali solution was prepared by dissolving a certain amount of solid sodium hydroxide. The pH value of which was monitored by a precision pH meter.

2.2. *Aging Simulation Test of Asphalt*

Figure 2 illustrates the sample preparation procedure under the multiple conditions of heat, UV and aqueous solution following six steps:

i. The 50 g of 70 A was poured on a dry aging tray with a diameter of 140 mm, then the aging tray was put into a thermal film oven at 163 °C for 5 h to obtain TFOT aged samples;
ii. The TFOT aged samples were placed in UV aging oven at 50 W/m^2 of irradiation intensity and 60 °C for 5 days, 10 days and 15 days to obtain UV aged samples;
iii. The UV-aged samples were treated by hydrostatic immersion experiments in a water bath at 60 °C for 5 days with distilled water, 10 wt% NaCl salt solution, pH 3 acid solution and pH11 alkali solution to obtain the samples (UV 5d + immersion 5d);
iv. The sample (UV 5d + water 5d) were subjected to UV aging for another 5 days, according to step iii, to obtain the samples (UV 5d + water 5d + UV 5d);
v. Trichloroethylene was selected as a solvent to dissolve the upper part of the samples at each stage for 90 s, and the trichloroethylene-asphalt solution was poured into a clean container then placed in fume cupboard for 72 h to allow the trichloroethylene to evaporate completely. Finally, the residues were collected as the aged asphalt samples.

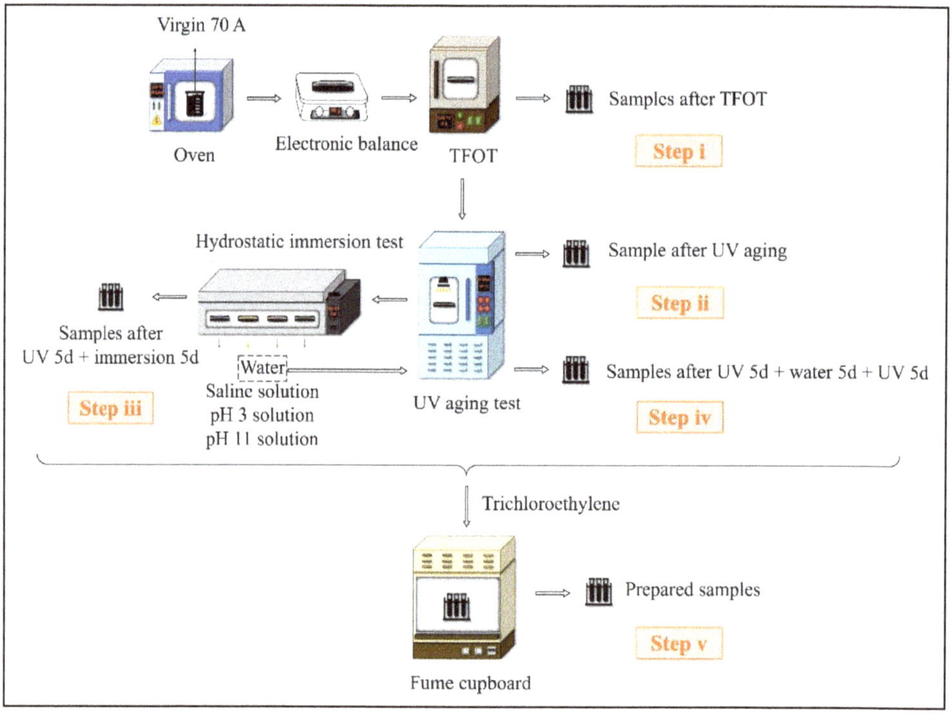

Figure 2. Sample preparation procedure.

2.3. *Characterization of Asphalt*

The rheological properties, chemical structure and element composition of asphalt were characterized by DSR, BBR, EA and FTIR, and the related information of the instruments as shown in the Table 2. These tests were performed in accordance with relevant specifications.

Table 2. Relevant information of instruments.

Test	Instruments	Origin	Test Parameters
DSR [27]		Smartpave 102 Anton Paar Co., Ltd. Ostfildern, Germany	Stain: 0.5% Frequency: 10 rad/s Temperature: 30–80 °C Heating rate: 2 °C/min Plate diameter: 25 mm Plate gap: 1 mm
BBR [28]		SYD-0627 Shanghai Changji Geological Instrument Co., Ltd. Shanghai, China	Load: 980 ± 50 mN Temperature: −6, −12, −18 °C Span length: 102 mm
EA [29]		Vario EL Cube Elementar Analysensysteme GmbH Langenselbold, Germany	Mode: C/H/N/S
FTIR [30]		Nicolet 6700 Thermo Fisher Scientifific Waltham, MA, USA	Chip: KBr Scanning range: 4000–400 cm^{-1} Scan time: 64 times

2.3.1. DSR Test

As a kind of viscoelastic material, asphalt is sensitive to changes in temperature and load. The rheological property of asphalt has a vital effect on its processability and pavement performance of asphalt mixture [31]. The high-temperature rheological property of asphalt samples was characterized by a temperature sweep using DSR with strain control mode. About 0.8 g of asphalt sample was used to prepare the cylinder with a diameter of 25 mm and a height of 1 mm. The complex modulus (G*) and phase angle (δ) of asphalt are the main parameters. The high-temperature rheological property is usually characterized by the rutting factor (G*/sinδ) to evaluate the rutting resistance [32]. The ratio of G*/sinδ before and after aging is defined as the Rutting factor Aging Index (RAI), which can quantify the influence of aging conditions on the high-temperature performance of asphalt, as shown in Equation (1) [33].

$$\mathrm{RAI} = \frac{G^*_{\mathrm{aging}}/\sin\delta_{\mathrm{aging}}}{G^*_{\mathrm{virgin}}/\sin\delta_{\mathrm{virgin}}} \quad (1)$$

where G^*_{virgin} and G^*_{aging} represent the G* of asphalt before and after aging, respectively, Pa; and δ_{virgin} and δ_{aging} represent the δ of asphalt before and after aging, respectively. The greater the RAI, the more distinct the aging effect.

2.3.2. BBR Test

As the evaluation parameters of cracking resistance, the creep Stiffness modulus (S) and creep rate (m-value) of asphalt were obtained by BBR test at low temperature, as shown in Equations (2) and (3) [34]. The sample size was 127 mm × 12.7 mm × 6.35 mm. The higher the S, the worse the low temperature ductility. The m-value represents the change rate of S, and the greater the value, the higher the relaxation rate and the more excellent the low temperature performance.

$$S(t) = \frac{PL^3}{4bh^3\Delta(t)} \quad (2)$$

$$m(t) = B + 2C[\lg(t)]^2 \quad (3)$$

where $S(t)$ represents the creep stiffness at 60 s, MPa; P represents the test load, mN; L, b and h represent the span length, width and depth of samples, mm; $\Delta(t)$ represents the deflection of samples at 60 s; and B and C represent the regression coefficients of $\lg[S(t)]$ and $\lg[m(t)]$.

2.3.3. EA Test

The variation of the element composition of asphalt was detected by elemental analyzer under multiple condition. The analysis principle is dynamic adsorption–desorption TCD measurement program [35]. The C, H, N, S and O elements are the main elements of asphalt, and the test mode of C, H, N and S elements was selected. The asphalt with a weight of 5 mg is burned under pure oxygen conditions, and then the gas produced by the combustion is measured. After homogenization, the gas is separated in chromatography within the separation zone and finally measured in the detection zone. Since the content of hetero atoms in asphalt is too low to be ignored, the content of O is obtained by subtracting the content of C, H, N and S elements from 100%.

The density method was employed to analyze the structural composition of asphalt. Three important indexes, the molar ratio of hydrogen–carbon (n(H)/n(C)), aromatic carbon ratio (f_A) and condensation index (C_I), were selected. The larger the n(H)/n(C), the more saturated hydrocarbons in the asphalt and the less the aging degree and vice versa, as shown in Equation (4) [36].

$$\mathrm{n(H)/n(C)} = 11.92 \times [\omega(\mathrm{H})/\omega(\mathrm{C})] \tag{4}$$

where $\omega(\mathrm{H})$ and $\omega(\mathrm{C})$ represent the mass fraction of hydrogen and carbon atoms in all elements of asphalt, respectively, %.

The f_A reflects the ratio of aromatic carbon atoms to total carbon atoms. The larger the f_A, the more ring structure, especially the aromatic rings, which means that there are more macromolecules and a greater degree of aging, as shown in Equations (5)–(8).

$$\rho = 1.4673 - 0.0431\omega(\mathrm{H}) \tag{5}$$

$$M_{c/\rho} = 1201/[\rho \times \omega(\mathrm{C})] \tag{6}$$

$$\left(M_{c/\rho}\right)_c = M_{c/\rho} - 6[100 - \omega(\mathrm{C}) - \omega(\mathrm{H})]/\omega(\mathrm{C}) \tag{7}$$

$$f_A = 0.09\left(M_{c/\rho}\right)_c - 1.15\mathrm{n(H)/n(C)} + 0.77 \tag{8}$$

where ρ represents the density of asphalt, g/cm^3, and $M_{c/[yellow]\rho}$ and $\left(M_{c/\rho}\right)_c$ represent the molar volume of individual carbon atom before and after correction, respectively, L/mol.

The C_I represents the molecule condensation degree, and the larger the C_I, the greater the molecule condensation degree and the more complex the ring structure, as shown in Equation (9).

$$C_I = 2 - \mathrm{n(H)/n(C)} - f_A \tag{9}$$

2.3.4. FTIR Test

Under the multiple aging condition, the chemical structure of asphalt was detected by FTIR with OMNIC 6.2 software (Thermo Fisher). Firstly, asphalt with a weight of 0.1 g was dissolved by CS_2 to prepare 5 wt% asphalt–CS_2 solution. Then, two drops were placed on the KBr chip with a glue dropper, and the CS_2 was completely vaporized by infrared light. Finally, the prepared samples were undertaken for the FTIR test. The carbonyl group and sulfoxide group are the main products of asphalt oxidation, and their indexes ($I_{C=O}$ and $I_{S=O}$) are usually used to quantify the aging degree of asphalt, which can be calculated using Equations (10) and (11) [37].

$$I_{C=O} = A_{1700\mathrm{cm}^{-1}}/\Sigma A_{2000-600\mathrm{cm}^{-1}} \tag{10}$$

$$I_{S=O} = A_{1030\mathrm{cm}^{-1}}/\Sigma A_{2000-600\mathrm{cm}^{-1}} \tag{11}$$

where $A_{1700cm^{-1}}$ and $A_{1030cm^{-1}}$ represent the area of carbonyl group and sulfoxide group at the 1700 cm^{-1} and 1030 cm^{-1}, respectively, and $\Sigma A_{2000-600cm^{-1}}$ represents the total area from 2000 cm^{-1} to 600 cm^{-1}. The higher the indexes, the greater the degree of aging.

3. Results and Discussion

3.1. High-Temperature Rutting Resistance

The G* and δ of asphalt are the main parameters of the high-temperature rheological property of asphalt. The greater the G*, the greater the shear deformation resistance and vice versa. The δ can evaluate the ratio of elastic components and viscous components. The greater the δ, the greater the viscous components. Conversely, the smaller the δ, the greater the elastic components. Figure 3 shows that the G* and δ of asphalt after multiple aging through heat, UV and aqueous solution. As can be seen from Figure 3a, the G* and δ of asphalt aged with TFOT increased and decreased, respectively, indicating the increase in shear deformation resistance and elastic components. The trend was further deepened after UV aging, manifesting the effect of UV was accumulated on the sample after thermal-oxygen aging. As can be seen from Figure 3b, the G* of asphalt samples declined after immersion in water, saline solution and acid solution, while the G* of asphalt samples increased after immersion in alkali solution. The δ of asphalt decreased after immersion in the four kinds of solutions. This could be explained by that there were some oxidation products on the surface of samples (UV 5d), and the existence of moisture resulted in dissolution and migration of these, leading to the reduction of the shear deformation resistance during immersion in three kinds of solution except alkali solution [38]. However, the reduction of other components might lead to an increase in the relative content of crystalline wax, resulting in an increment in the elastic components [39]. Chloride ions in saline solution could promote the emulsification of asphalt, and the esterification reaction between acid and olefin in asphalt generated long chain isomerized alkanes. The chemical reactions increased the content of saturates, resulting in the further reduction of the shear deformation resistance. Among these, the effect of salt solution on the high-temperature rheological property of asphalt was greater than that of water and acid solution, which might be due to the crystallization of salt in asphalt [40]. Moreover, the saponification of alkali solution with asphalt accelerated the asphalt oxidation to generate more asphaltenes, resulting in an increasing of the shear deformation resistance and elastic components [41]. After aging cycle of UV and water, the G* and δ of asphalt are illustrated in Figure 3c. The order of G* from small to large was the following order, (UV 5d + water 5d) < (UV 5d) < (UV 10d) < (UV 15d) < (UV 5d + water 5d + UV 5d). It suggested that the effect of water and UV on shear deformation resistance was opposite due to asphalt dissolution and migration, but water can increase the sensitivity of the shear deformation resistance to UV. The order of δ from small to large is the following: (UV 5d + water 5d + UV 5d) < (UV 15d) < (UV 10d) < (UV 5d + water 5d) < (UV 5d); manifesting water can also increase sensitivity of the viscoelasticity to UV.

Figure 4 shows the RAI of asphalt after the multiple aging. From Figure 4a, it can be seen that the RAIs of TFOT and UV aging were greater than 1 and increased with the increase of UV time. It indicates that the thermo-oxidative aging and UV aging have a positive effect on the rutting resistance of asphalt, and the positive effect could be accumulated. The RAI increased over temperature, suggesting that effect degree of thermal-oxygen and UV on the rutting resistance increased with the increment of temperature. Figure 4b,c illustrates the data graphs with the G*/sinδ of sample (UV 5d) as the $G^*_{virgin}/\sin\delta_{virgin}$. From Figure 4b, it can be seen that the RAIs of asphalt samples exposed to water, saline solution and acid solution were less than 1, while that of sample immersed in alkali solution was greater than 1. This demonstrates that the first three solutions had a negative impact on the rutting resistance of asphalt, while alkali solution improved the rutting resistance of asphalt. The RAIs of samples suffered the aqueous solutions except the alkali solution and were not affected by temperature, indicating they had similar effects on the rutting resistance at different temperatures. Additionally, the RAI of the asphalt treated by alkali solution increased with temperature, manifesting it had greater effects on the rutting resistance at high temperatures. From Figure 4c, it can be seen that the RAI of samples (UV 5d + water 5d) was less than that of samples (UV 10d), while the RAI of samples (UV 5d + water 5d + UV 5d) was greater than that of samples (UV 15d). It indicates that the effect of water on asphalt rutting resistance was less than UV, but it could increase the sensitivity of rutting resistance to UV. The RAI of samples (UV 5d + water 5d), (UV 10d) and (UV 15d) were not affected by temperature, manifesting water and UV had similar effects on the rutting resistance at different temperatures after UV aging for 5d. The RAI of samples (UV 5d + water 5d + UV 5d) increased with temperature, demonstrating that water made the rutting resistance more sensitive to temperature.

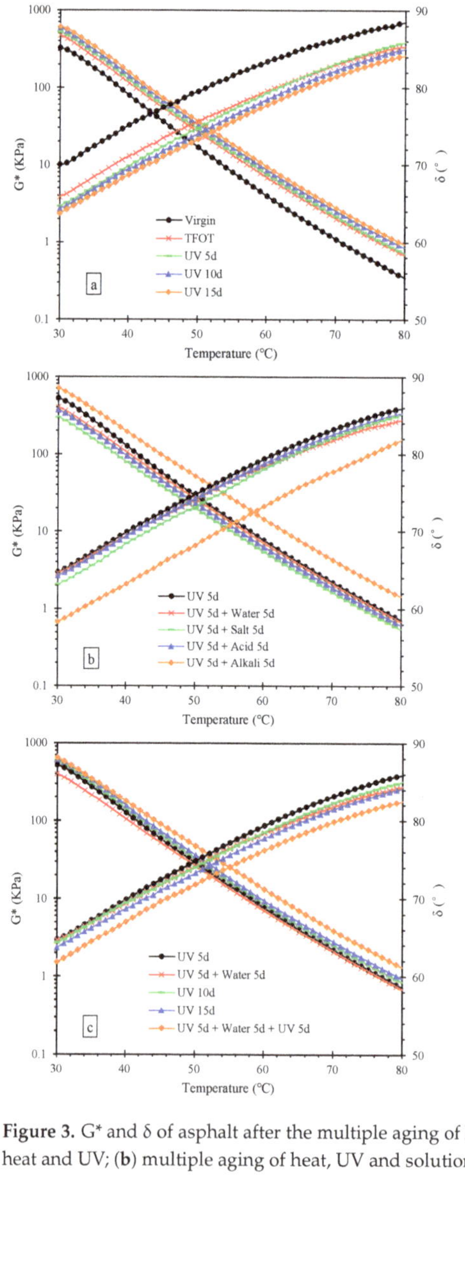

Figure 3. G* and δ of asphalt after the multiple aging of heat, UV and solution. (**a**) multiple aging of heat and UV; (**b**) multiple aging of heat, UV and solution; (**c**) aging cycle of UV and water.

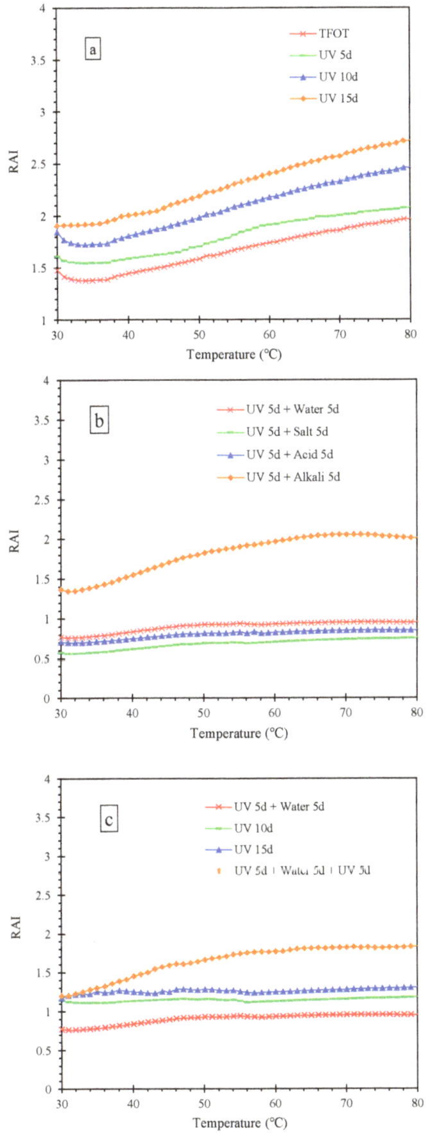

Figure 4. RAI of asphalt after the multiple aging of heat, UV and solution. (**a**) multiple aging of heat and UV; (**b**) multiple aging of heat, UV and solution; (**c**) aging cycle of UV and water.

3.2. Low-Temperature Cracking Resistance

Figure 5 shows that the S- and m-value of asphalt after the multiple aging of heat, UV and water. It can be seen that the S of asphalt increased and the m-value gradually decreased with the decrease of test temperature. It indicated that the low-temperature ductility and relaxation rate of the asphalt gradually decreased with the decreasing of temperature and the asphalt started to become hard and brittle. After thermal-oxygen aging, the S of samples gradually increase, while the m-value declined, manifesting the low-temperature, cracking resistance was weakened. The addition of UV aging deepened the change trend, meaning that thermal-oxygen and UV could accumulatively weaken the low-temperature cracking resistance. After the multiple aging of UV and aqueous solution, the S of asphalt exposed to water, saline solution and acid solution decreased and the m-value increased, while the opposite pattern appeared in samples suffered from alkali solution. This might be because the dissolution and migration of polar components in asphalt led to the softening of asphalt and enhancement of flexibility, leading to better cracking resistance at low temperatures. However, the polar components increased during immersion in alkali solution, hardening the asphalt samples and weakening the crack resistance at low temperature [42]. The rangeability order of S from small to large was the following: (UV 5d + water 5d) < (UV 5d) < (UV 10d) < (UV 15d) < (UV 5d + water 5d + UV 5d); and that of m-value was the opposite. It suggested that the effect of water was positive on cracking resistance, but water could increase the sensitivity of the cracking resistance to UV.

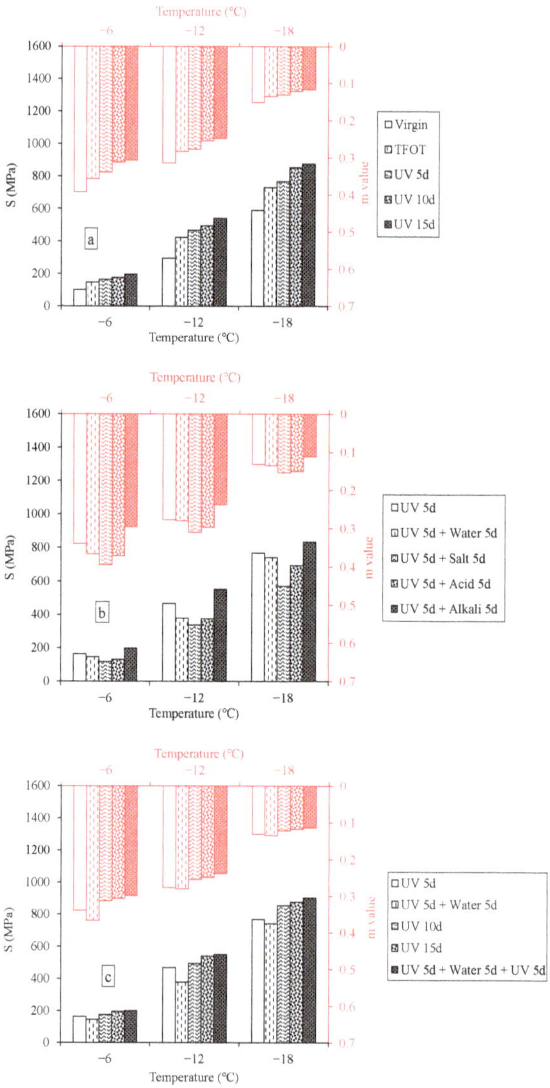

Figure 5. S and m-value of asphalt after the multiple aging of heat, UV and solution. (**a**) multiple aging of heat and UV; (**b**) multiple aging of heat, UV and solution; (**c**) aging cycle of UV and water.

3.3. Element Composition

Asphalt is a complex compound with a variety of polymeric hydrocarbons and their non-metallic derivatives, whose main constituent elements are carbon, hydrogen, oxygen, sulfur, nitrogen, etc. The effect of heat and UV on the element composition of asphalt samples were investigated to discuss the change mechanism in properties, as illustrated in Table 3. For 70 A, the elemental content of C and H was higher than that of N, S and O. After TFOT, the content of O increased, while that of others reduced, manifesting that oxidation reaction occurred to introduce oxygen atoms in air into the asphalt, therefore the relative content of other elements decreased. Moreover, the change trend of element composition increased gradually with the extension of UV aging time, indicating that the influence of UV accumulates gradually over time. The n(H)/n(C), f_A and C_I of asphalt after the multiple aging are displayed in Figure 6. The n(H)/n(C) declined, while the f_A and C_I increased after TFOT. The results showed that heat could decrease saturated hydrocarbon and increase the aromatic ring substance and its condensation degree, causing the existence of the more complex ring structure. The polymer molecule with the kind of aromatic ring as the main chain could not be rotated internally, resulting in the increasing of rigidity and the weakening of flexibility. As a result, the high-temperature rutting resistance was improved and the low temperature cracking resistance was weakened, as reflected in the DSR and BBR results. The trend was further driven by UV aging. Table 4 illustrated the comparison of rangeability in C_I after different aging methods. It could be seen that the rangeability caused by TFOT was greater than that caused by UV 5d, and the rangeability gradually decreased with the extension of UV aging time, indicating that the sensitivity of elemental composition to UV decreased with the deepening of aging.

Table 3. Element content of asphalt after the multiple aging of heat and UV (%).

Samples	C	H	N	S	O
Virgin	82.98	10.27	0.73	4.43	1.59
TFOT	82.94	10.24	0.72	4.38	1.72
UV 5d	82.86	10.21	0.72	4.37	1.84
UV 10d	82.86	10.20	0.72	4.36	1.86
UV 15d	82.67	10.17	0.72	4.34	2.10

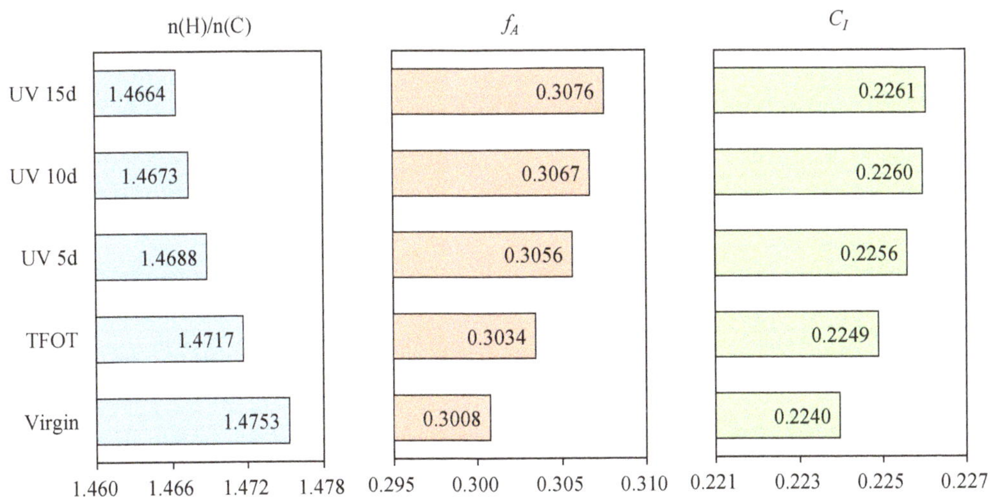

Figure 6. Important indexes (n(H)/n(C), f_A and C_I) of asphalt after the multiple aging of heat and UV.

Table 4. Rangeability of C_I after the multiple aging.

Samples	ΔC_I
Virgin/TFOT	0.0009
TFOT/UV 5d	0.0007
UV 5d/UV 10d	0.0004
UV 10d/UV 15d	0.0001
UV 5d/UV 5d + water 5d	0.0001
UV 5d + water 5d/UV 5d + water 5d + UV 5d	0.0005

Table 5 illustrated the element composition of asphalt samples after the multiple aging of UV and aqueous solution. Compared with UV 5d, the C and H remain essentially unchanged and the N and S declined to a certain extent, while the content of O increased after immersion in different aqueous solution. The n(H)/n(C) declined, while the f_A and C_I increased after immersion in different aqueous solution, as shown in Figure 7. It indicated that the oxidation during immersion caused the existence of molecules with high condensation degree, in which the water-soluble heterocyclic compounds containing N and S (including anhydrides, lactones and cyclic lactams) were easier to dissolve and migrate [42]. This resulted in the situation where N and S decreased but C and H remain essentially unchanged. The presence of solute could accelerate the trend, and the order of the effect degree is as follows: alkali > acid > salt > water.

Table 5. Element content of asphalt after the multiple aging of UV and aqueous solution (%).

Samples	C	H	N	S	O
UV 5d	82.86	10.21	0.72	4.37	1.84
UV 5d + water 5d	82.86	10.21	0.71	4.34	1.88
UV 5d + salt 5d	82.86	10.20	0.71	4.34	1.89
UV 5d + acid 5d	82.85	10.19	0.70	4.33	1.93
UV 5d + alkali 5d	82.84	10.18	0.69	4.32	1.97

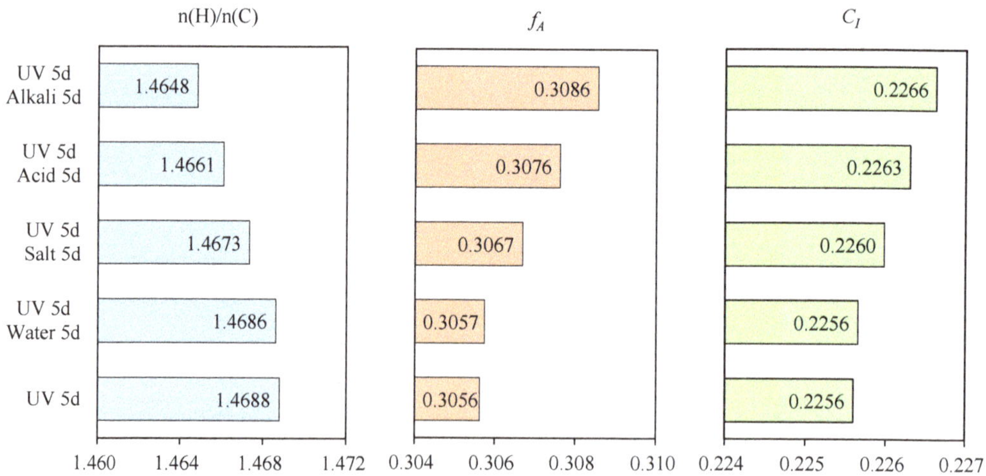

Figure 7. Important indexes (n(H)/n(C), f_A and C_I) of asphalt after the multiple aging of UV and aqueous solution.

The element composition of asphalt was observed after the aging cycle of UV and water, as shown in Table 6. Compared to the sample (UV 10d), the sample (UV 5d + water 5d) had the same content of C, higher content of H and O and the lower content of N and

S. The sample (UV 5d + water 5d + UV 5d) had the lower content of C and H, the same content of N and the higher content of S and O than the sample (UV 15d). Figure 8 describes the n(H)/n(C), f_A and C_I of asphalt after the aging cycle. The sample (UV 5d +water 5d) had the smaller n(H)/n(C), the greater f_A and C_I than the sample (UV 10d), meaning that the effect of water on element composition was less than UV. Compared to the sample (UV 15d), the n(H)/n(C) of sample (UV 5d +water 5d + UV 5d) was smaller but the f_A and C_I were greater. In addition, it could be seen from Table 4 that the rangeability of C_I between the sample (UV 5d + water 5d and UV 5d) and the sample (UV 5d + water 5d) was 1.25 times that between the sample (UV 10d) and the sample (UV 5d), demonstrating that water could increase the sensitivity of element composition to UV.

Table 6. Element content of asphalt after the aging cycle of UV and water (%).

Samples	C	H	N	S	O
UV 5d	82.86	10.21	0.72	4.37	1.84
UV 5d + water 5d	82.86	10.21	0.71	4.34	1.88
UV 10d	82.86	10.20	0.72	4.36	1.86
UV 15d	82.67	10.17	0.72	4.34	2.10
UV 5d + water 5d + UV 5d	82.61	10.16	0.72	4.35	2.16

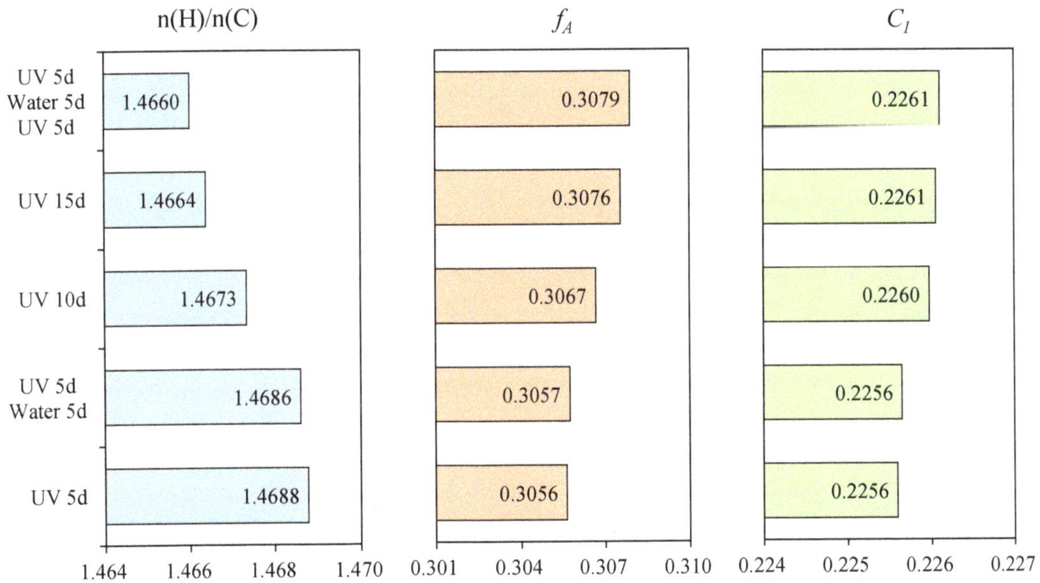

Figure 8. Important indexes (n(H)/n(C), f_A and C_I) of asphalt after the aging cycle of UV and water.

3.4. Chemical Structure

The change of element composition and chemical structure are the fundamental reason for the change of rheological properties. The effect of multiple aging on chemical structure were characterized by FTIR test. Figure 9 illustrates that the $I_{C=O}$ and $I_{S=O}$ of asphalt after the multiple aging of heat and UV. The $I_{C=O}$ of virgin asphalt was close to 0, while the $I_{S=O}$ was far greater than the $I_{C=O}$. It shows that the virgin asphalt was not oxidized and some parts of S element belong to S=O of asphalt itself [43]. Compared with the virgin, the $I_{C=O}$ of samples increased by 0.06, 0.11, 0.15 and 0.17, while the $I_{S=O}$ samples increased by 0.21, 0.35, 0.45 and 0.53 after TFOT aging, UV 5d, UV 10d and UV 15d. The coupling of thermal-oxygen and UV could accumulate to promote the asphalt oxidation. It indicated that asphalt aging increased polar functional groups, such as C=O and S=O, which had permanent dipoles and generate electrostatic force, resulting in the increase in the intermolecular friction resistance of asphalt [44]. The high-temperature rutting resistance was improved, and the low-temperature cracking resistance was weakened, which was shown in the DSR and BBR results. The rangeability of $I_{C=O}$ and $I_{S=O}$ under different aging methods was compared in Table 7. It could be seen that the rangeability caused by TFOT was greater than that caused by UV 5d, and the rangeability gradually decreased with UV aging time, indicating the sensitivity of the asphalt oxidation to UV decreased with the deepening of the aging degree. Because in the short-term aging, asphalt aging to generate carbonyl and sulfoxide group. During UV aging, the carbonyl and sulfoxide groups decompose to form long chains or rings, and the aromatics and colloid converted to asphaltenes, promoting the further aging of asphalt [45]. Therefore, the preventing formation of C=O and S=O and the synthesis of long chain can delay the asphalt aging to contribute to the sustainable development of the asphalt pavement [46].

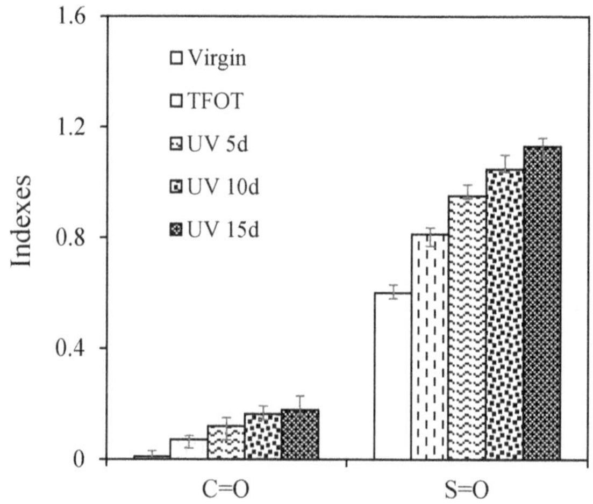

Figure 9. $I_{C=O}$ and $I_{S=O}$ of asphalt after the multiple aging of heat and UV.

Table 7. Rangeability of $I_{C=O}$ and $I_{S=O}$ after the multiple aging.

Samples	$\Delta I_{C=O}$	$\Delta I_{S=O}$
Virgin/TFOT	0.06	0.21
TFOT/UV 5d	0.05	0.14
UV 5d/UV 10d	0.04	0.10
UV 10d/UV 15d	0.02	0.08
UV 5d/UV 5d + water 5d	0.03	0.07
UV 5d + water 5d/UV 5d + water 5d + UV 5d	0.10	0.45

After the multiple aging of UV and aqueous solution, the $I_{C=O}$ and $I_{S=O}$ of asphalt are illustrated in Figure 10. After immersion in water, saline solution, acid solution and alkali solution, the $I_{C=O}$ increased by 25.31%, 50.38%, 67.08% and 136.42%, while the $I_{S=O}$ increased by 7.22%, 21.02%, 29.60% and 90.27%. This increment was greater than the increment when UV and aqueous solution are treated asphalt simultaneously, indicating that sequential treatment has a greater impact on asphalt than simultaneous treatment [18]. Aqueous solution could accelerate the asphalt oxidation after UV aging, and the order of the effect degree is as follows: alkali > acid > salt > water. Figure 11 shows the $I_{C=O}$ and $I_{S=O}$ of asphalt after aging cycle of UV and water. The sample (UV 5d + water 5d) had the lower $I_{C=O}$ and $I_{S=O}$ than the sample (UV 10d), denoting water had the less effect on chemical structure than UV. However, the $I_{C=O}$ and $I_{S=O}$ of sample (UV 5d + water 5d + UV 5d) were greater than that of sample (UV 15d). Moreover, from Table 7, it can be seen that the rangeability $I_{C=O}$ and rangeability $I_{S=O}$ between sample (UV 5d +water 5d + UV 5d) and sample (UV 5d + water 5d) was 2.5 and 4.5 times as much as that between the sample (UV 10d) and the sample (UV 5d), respectively. It shows that water could increase the sensitivity of asphalt chemical structure to UV. That agreed with the result of existing studies [47].

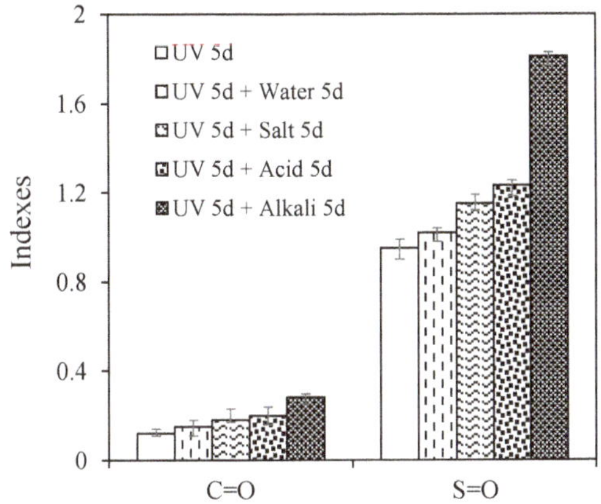

Figure 10. $I_{C=O}$ and $I_{S=O}$ of asphalt after the multiple aging of UV and aqueous solution.

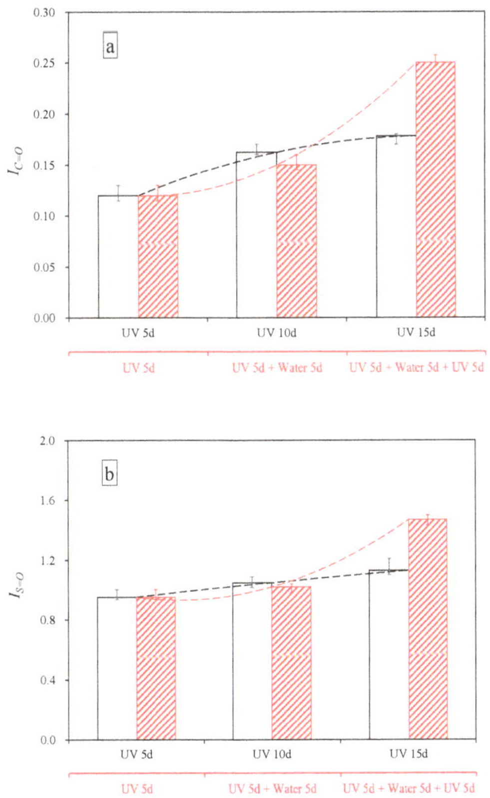

Figure 11. $I_{C=O}$ and $I_{S=O}$ of asphalt after the aging cycle of UV and water. (**a**) $I_{C=O}$, (**b**) $I_{S=O}$.

4. Conclusions

The high-temperature rutting resistance and low-temperature cracking resistance of asphalt under multiple conditions of heat, UV and aqueous solution were investigated by DSR test. The related mechanisms of changes in aging performance were discussed by the characterization of the chemical structure and element composition.

- Heat can increase the rutting factor, RAI and S, and decline the m-value to improve the high-temperature rutting resistance and weaken the low-temperature cracking resistance. According to the results of the EA test and FTIR test, the f_A, C_I, $I_{C=O}$ and $I_{S=O}$ increased after thermo-oxidative aging. It can be explained by the fact that thermo-oxidative aging causes the saturated hydrocarbon and the aromatic ring substance with the higher condensation degree increase to form more polar components. The addition of UV further deepens asphalt aging, and the sensitivity of physicochemical properties to UV decreased with the deepening of aging.
- Aqueous solution can further affect the rheological properties of samples aged by UV. Due to the dissolution and migration of polar components, the rutting resistance of asphalt samples was weakened but the cracking resistance was improved slightly after immersion in water, saline solution and acid solution, whereas the opposite pattern appears in samples suffered from alkali solution due to the saponification reaction. The order of influence of the degree of aqueous solution on UV-aged asphalt is as follows: alkali > acid > salt > water.
- Water has the smaller effect on element composition and chemical structure, but water can increase the sensitivity of physicochemical properties to UV.

This study explored the short-term cycling effect of heat, light and water on asphalt to provide an idea for simulation testing and anti-aging technology of asphalt under multiple aging conditions. The findings of this study also help to restore the lost properties of the binder from RAP, showing the significant economic and environmental outcomes. In future research, we will use the SARA test to investigate the influence of multiple aging conditions on the four components, and further explore the evolution of asphalt performance under multiple aging conditions.

Author Contributions: Y.Z.: Conceptualization, Experiment, Writing—Original Draft. L.P.: Editing, Formal Analysis. S.X.: Methodology, Formal Analysis. S.W.: Writing—Review and Editing. M.Y.: Validation, Data curation. S.A.: Editing, Supervision. H.X.: Data Curation, Software. Y.L.: Visualization, Software. X.G.: Experiment, Validation. All authors have read and agreed to the published version of the manuscript.

Funding: This research was funded by the National Natural Science Foundation of China [52108416]; the Independent Innovation Foundation of Wuhan University of Technology [223131001]; and Hubei Province Supports Technological Innovation and Development Projects for High-Tech Enterprises (2021BAB074).

Institutional Review Board Statement: Not applicable.

Informed Consent Statement: Not applicable.

Acknowledgments: This work was supported by the National Natural Science Foundation of China [52108416]; the Independent Innovation Foundation of Wuhan University of Technology [223131001]; and Hubei Province Supports Technological Innovation and Development Projects for High-Tech Enterprises (2021BAB074).

Conflicts of Interest: The authors declare no conflict of interest.

References

1. Li, N.; Jiang, Q.; Wang, F.; Cui, P.; Xie, J.; Li, J.; Wu, S.; Barbieri, D. Comparative assessment of asphalt volatile organic compounds emission from field to laboratory. *J. Clean. Prod.* **2021**, *278*, 123479. [CrossRef]
2. Wan, P.; Liu, Q.; Wu, S.; Zhao, Z.; Yu, X. A novel microwave induced oil release pattern of calcium alginate/nano-Fe_3O_4 composite capsules for asphalt self-healing. *J. Clean. Prod.* **2021**, *297*, 126721. [CrossRef]
3. Li, Q.; Zeng, X.; Wang, J.; Luo, S.; Meng, Y.; Gao, L.; Wang, X. Aging performance of high viscosity modified asphalt under complex heat-light-water coupled conditions. *Constr. Build. Mater.* **2022**, *325*, 126314. [CrossRef]
4. Wang, F.; Xie, J.; Wu, S.; Li, J.; Barbieri, D.M.; Zhang, L. Life cycle energy consumption by roads and associated interpretative analysis of sustainable policies. *Renew. Sust. Energ. Rev.* **2021**, *141*, 110823. [CrossRef]
5. Li, J.; Yu, J.; Wu, S.; Xie, J. The mechanical resistance of asphalt mixture with steel slag to deformation and skid degradation based on laboratory accelerated heavy loading test. *Materials* **2022**, *15*, 911. [CrossRef]
6. Xu, H.; Wu, S.; Chen, A.; Zou, Y. Influence of erosion factors (time, depths and environment) on induction heating asphalt concrete and its mechanism. *J. Clean. Prod.* **2022**, *363*, 132521. [CrossRef]
7. Cui, P.; Wu, S.; Xiao, Y.; Hu, R.; Yang, T. Environmental performance and functional analysis of chip seals with recycled basic oxygen furnace slag as aggregate. *J. Hazard. Mater.* **2021**, *405*, 124441. [CrossRef] [PubMed]
8. Sun, X.; Wang, Y.; Li, H.; Yang, X.; Sun, L.; Wang, X.; Wang, T.; Wang, W. Organic acids in cloud water and rainwater at a mountain site in acid rain areas of South China. *Environ. Sci. Pollut. Res.* **2016**, *23*, 9529–9539. [CrossRef] [PubMed]
9. Liu, F.; Zhang, T.; Luo, T.; Zhou, M.; Ma, W. Study on the deterioration of concrete under dry-wet cycle and sulfate attack. *Materials* **2020**, *13*, 4095. [CrossRef]
10. Shao, T.; Zhao, J.J.; Liu, A.; Long, X.; Rengel, Z. Effects of soil physicochemical properties on microbial communities in different ecological niches in coastal area. *Appl. Soil Ecol.* **2020**, *150*, 103486. [CrossRef]
11. Magdy, M.; Gar Alalm, M.; El-Etriby, H.K. Comparative life cycle assessment of five chemical methods for removal of phenol and its transformation products. *J. Clean. Prod.* **2021**, *291*, 125923. [CrossRef]
12. Yang, C.; Wu, S.; Cui, P.; Amirkhanian, S.; Zhao, Z.; Wang, F.; Zhang, L.; Wei, M.; Zhou, X.; Xie, J. Performance characterization and enhancement mechanism of recycled asphalt mixtures involving high RAP content and steel slag. *J. Clean. Prod.* **2022**, *336*, 130484. [CrossRef]
13. Zeng, W.; Wu, S.; Wen, J.; Chen, Z. The temperature effects in aging index of asphalt during UV aging process. *Constr. Build. Mater.* **2015**, *93*, 1125–1131. [CrossRef]
14. Li, H.; Yu, J.; Wu, S.; Liu, Q.; Li, Y.; Wu, Y.; Xu, H. Investigation of the effect of induction heating on asphalt binder aging in steel fibers modified asphalt concrete. *Materials* **2019**, *12*, 1067. [CrossRef] [PubMed]
15. Hamzah, M.O.; Valentin], J. A review on moisture damages of hot and warm mix asphalt and related investigations. *J. Clean. Prod.* **2015**, *99*, 39–58.
16. Abouelsaad, A.; White, G. The combined effect of ultraviolet irradiation and temperature on hot mix asphalt mixture aging. *Sustainability* **2022**, *14*, 5942. [CrossRef]
17. Tan, Z.; Li, Q. Study on the effect of thermo-oxygen-light coupling aging on asphalt performance. *Highw. Automot. Appl.* **2016**, *176*, 71–72+93.
18. Li, Y.; Li, H.; Nie, S.; Wu, S.; Liu, Q.; Li, C.; Shu, B.; Li, C.; Song, W.; Zou, Y.; et al. Negative impacts of environmental factors (UV radiation, water and different solutions) on bitumen and its mechanism. *Constr. Build. Mater.* **2020**, *265*, 120288. [CrossRef]
19. Ilaria, M.; Eyad, M. The influence of moisture on the evolution of the microstructure of asphalt binders with aging. *Road Mater. Pavement Des.* **2018**, *21*, 331–346.
20. Zhang, X.; Huang, G.; Liu, Z.; Feng, M.; Li, C. Influence of coupled aging condition of heat, light and water on performance of warm mix asphalt. *J. Highw. Trans. Res. Dev.* **2018**, *36*, 10–19.
21. Rathore, M.; Haritonovs, V.; Meri, R.M.; Zaumanis, M. Rheological and chemical evaluation of aging in 100% reclaimed asphalt mixtures containing rejuvenators. *Constr. Build. Mater.* **2022**, *318*, 126026. [CrossRef]
22. Yang, X.; Mills-Beale, J.; You, Z. Chemical characterization and oxidative aging of bio-asphalt and its compatibility with petroleum asphalt. *J. Clean. Prod.* **2017**, *142*, 1837–1847. [CrossRef]
23. *ASTM D-5*; Standard Test Method for Penetration of Bituminous Materials. ASTM International: West Conshohocken, PA, USA, 2013.
24. *ASTM D36*; Standard Test Method for Softening Point of Bitumen (Ring-and-Ball Apparatus). ASTM International: West Conshohocken, PA, USA, 2009.
25. *ASTM D-113*; Standard Test Method for Ductility of Bituminous Materials. ASTM: West Conshohocken, PA, USA, 2007.
26. *ASTM D-2042*; Standard Test Method for Solubility of Asphalt Materials in Trichloroethylene. ASTM International: West Conshohocken, PA, USA, 2015.
27. Joshi, C. Determining the rheological properties of asphalt binder using a dynamic shear rheometer (DSR). *Int. J. Res. Eng.* **2013**, *2*, 192–196.
28. *ASTM D6648-08(2016)*; Standard Test Method for Determining the Flexural Creep Stiffness of Asphalt Binder Using the Bending Beam Rheometer (BBR). ASTM: West Conshohocken, PA, USA, 2016.
29. Li, Y.; Wu, S.; Liu, Q.; Xie, J.; Li, H.; Dai, Y.; Li, C.; Nie, S.; Wei, S. Aging effects of ultraviolet lights with same dominant wavelength and different wavelength ranges on a hydrocarbon-based polymer (asphalt). *Polym. Test.* **2019**, *75*, 64–75. [CrossRef]

30. Zhao, Z.; Wu, S.; Liu, Q.; Yang, C.; Zou, Y.; Wan, P. Feasibility assessment of CeO2 nanoparticles as aging-resistant agent of asphalt. *Constr. Build. Mater.* **2022**, *330*, 127245. [CrossRef]
31. Qian, G.; Yu, H.; Gong, X.; Zhao, L. Impact of Nano-TiO_2 on the NO_2 degradation and rheological performance of asphalt pavement. *Constr. Build. Mater.* **2019**, *218*, 53–63. [CrossRef]
32. Jiang, Q.; Li, N.; Yang, F.; Ren, Y.; Wu, S.; Wang, F.; Xie, J. Rheology and volatile organic compounds characteristics of warm-mix flame retardant asphalt. *Constr. Build. Mater.* **2021**, *298*, 123691. [CrossRef]
33. Zhao, Z.; Wu, S.; Liu, Q.; Xie, J.; Yang, C.; Wang, F.; Wan, P. Recycling waste disposable medical masks in improving the performance of asphalt and asphalt mixtures. *Constr. Build. Mater.* **2022**, *337*, 127621. [CrossRef]
34. Ren, H.; Qian, Z.; Huang, W.; Li, H.; Liu, Y. Low-temperature thermal cracking performance of waterborne epoxy asphalt emulsion mastic based on bending beam rheometer (BBR). *Constr. Build. Mater.* **2022**, *334*, 127461. [CrossRef]
35. Alhaddad, A. Investigation of rheological properties of Iraqi asphalt cement. *IOP Conf. Ser. Mater. Sci. Eng.* **2020**, *737*, 012134. [CrossRef]
36. Li, J.; Jiang, Y.; Chen, Q.; Liu, Y.; Wu, C. Microscopic properties of 5U north American rock asphalt and its modified asphalts based on E-M-D method. *Guangxi Sci.* **2015**, *22*, 83–86.
37. Zhang, N.; Fan, G.; Lv, S.; He, F.; Fan, X.; Peng, X.; Liu, T.J.; Liu, H. The influence of SBS modification on the rheological property of asphalt before and after regeneration. *Constr. Build. Mater.* **2021**, *310*, 125239. [CrossRef]
38. Yang, H.; Pang, L.; Zou, Y.; Liu, Q.; Xie, J. The effect of water solution erosion on rheological, cohesion and adhesion properties of asphalt. *Constr. Build. Mater.* **2020**, *246*, 118465. [CrossRef]
39. Speidel, M.; Malte, K.; Andreas, H.; Klaus, R.; Angelika, E.; Rolf, D.; Oliver, B. Structural and tribometric characterization of biomimetically inspired synthetic "insect adhesives". *Beilstein J. Nanotechnol.* **2017**, *8*, 45–63. [CrossRef] [PubMed]
40. Wang, Y.; Ye, J.; Liu, Y.; Qiang, X.; Feng, L. Influence of freeze–thaw cycles on properties of asphalt-modified epoxy repair materials. *Constr. Build. Mater.* **2013**, *41*, 580–585. [CrossRef]
41. Wanfen, P.; Shen, C.; Tang, X.; Wei, B.; Zhao, N.; Mei, Z. Research advances about oil-water interfacial dilational viscoelasticity in chemical flooding. *Oilfield Chem.* **2018**, *3*, 562–570.
42. Zou, Y.; Amirkhanian, S.; Xu, S.; Li, Y.; Wang, Y.; Zhang, J. Effect of different aqueous solutions on physicochemical properties of asphalt binder. *Constr. Build. Mater.* **2021**, *286*, 122810. [CrossRef]
43. Kai, Y.; Rui, L.A. Characterization of bonding property in asphalt pavement interlayer: A review. *J. Traffic Transp. Eng.* **2021**, *8*, 374–387.
44. Liu, L.; Lu, Y.; Liu, A.; Li, Y.; Guo, R. Analysis of asphalt aging behavior evaluation method based on infrared spectrum. *IOP Conf. Ser. Earth Environ. Sci.* **2021**, *787*, 012044. [CrossRef]
45. Li, Y.; Feng, J.; Yang, F.; Wu, S.; Liu, Q.; Bai, T.; Liu, Z.; Li, C.; Gu, D.; Chen, A.; et al. Gradient aging behaviors of asphalt aged by ultraviolet lights with various intensities. *Constr. Build. Mater.* **2021**, *295*, 123618. [CrossRef]
46. Yan, K.; Liu, W.; You, L.; Ou, J.; Zhang, M. Evaluation of waste cooling oil and European Rock Asphalt modified asphalt with laboratory tests and economic cost comparison-ScienceDirect. *J. Clean. Prod.* **2021**, *310*, 127364. [CrossRef]
47. Chen, H.; Tan, Z.Y.; Qiang, L.I.; Maintenance, D.O. The analysis of impacts on asphalt aging under complex condition of heat-water-light. *J. Transp. Sci. Eng.* **2016**, *32*, 45–48.

materials

Article

Effect of Phosphogypsum Based Filler on the Performance of Asphalt Mortar and Mixture

Jiuming Wan [1,2], Tao Han [1], Kaifei Li [1], Suxun Shu [1,*], Xiaodi Hu [1], Wenxia Gan [1] and Zongwu Chen [3]

1 School of Civil Engineering and Architecture, Wuhan Institute of Technology, Wuhan 430205, China
2 Key Laboratory of Road Structure and Material of Ministry of Transport (Changsha), Changsha University of Science & Technology, Changsha 410114, China
3 Local Joint Engineering Laboratory of Traffic Civil Engineering Materials, Chongqing Jiaotong University, Chongqing 400074, China
* Correspondence: sxshu@wit.edu.cn

Abstract: This study introduced phosphogypsum coupled with steel slag powder to prepare the phosphogypsum based filler (PF) for asphalt mixture. Penetration, penetration index, softening point, ductility, equivalent softening point, moisture stability of asphalt mortars with different steel slag powder content, filler-asphalt ratio, and PF content were studied. Mechanical properties of PF based asphalt mortar (P-AM) were then analyzed to determine the optimum steel slag content in PF. Overall desirability method was used to determine the optimum replacement ratio of PF content in limestone filler. Rheological properties of P-AM were also analyzed through dynamic shear rheometer. Volumetric performance, high-temperature performance, low-temperature performance, and moisture stability tests were carried out on PF based AC-20 asphalt mixture. Results showed that P-AM presented the optimum performance when the content of steel slag powder was 23% by mass of phosphogypsum. Fatigue and rutting factor of asphalt mortar were enhanced by PF. The optimum PF content in replacing limestone filler was 75% through overall desirability evaluation. PF developed the high-temperature performance and moisture stability of asphalt mixture. Additionally, volumetric and low-temperature performance were not significantly affected by PF. It is suggested that using PF which is based on phosphogypsum as a filler of asphalt mixture to partially replace traditional limestone filler was adequate.

Keywords: phosphogypsum; steel slag powder; asphalt mortar; asphalt mixture; overall desirability

Citation: Wan, J.; Han, T.; Li, K.; Shu, S.; Hu, X.; Gan, W.; Chen, Z. Effect of Phosphogypsum Based Filler on the Performance of Asphalt Mortar and Mixture. *Materials* **2023**, *16*, 2486. https://doi.org/10.3390/ma16062486

Academic Editor: Giovanni Polacco

Received: 22 February 2023
Revised: 14 March 2023
Accepted: 17 March 2023
Published: 21 March 2023

1. Introduction

Construction of asphalt pavement usually consumes a large amount of asphalt, aggregate and filler [1]. Among the natural resources, aggregate constitutes the dominant part of the asphalt mixture [2]. However, great consumption of natural minerals for this purpose resulted in environmental and resource pressure. Therefore, solid wastes have been increasingly used to replace the aggregate and filler in order to reduce consumption of natural mineral resources. Phosphogypsum is a solid waste which is produced from phosphoric chemical industrial processes [3]. Its global annual production is about 280 million tons, which also leads to environmental pollution [4–8]. Therefore, consuming and recycling phosphogypsum as much as possible can benefit environment protection.

It has typically been used as a filler of pavement subgrade and to prepared modified asphalt [9–12]. Shen et al. [13] prepared phosphogypsum-steel slag powder-flyash as a solidified material of pavement subgrade. Results showed that its early strength and long-term strength were higher than those of cement stabilized granular materials. Rakesh et al. [14] also found the solidified material, which contained fly ash and 8% lime and 2% phosphogypsum, showed adequate unconfined compressive strength, split strength, slake durability criteria and California bearing ratio.

On the other hand, Amrani et al. [15] used 5 wt% phosphogypsum as modifier to prepare modified asphalt. It was found that the modified asphalt showed higher improvements in stiffness and deformation resistance than the values found with fly ash and phosphate sludge wastes. Cuadri et al. [16] used phosphogypsum coupled with sulfuric acid to prepare modified foamed asphalt. Results showed that foamed asphalt based on phosphogypsum has higher rutting resistance and foaming properties compared with the natural gypsum.

However, using phosphogypsum as filler of asphalt mixture has not been systematically studied. Additionally, how to alleviate its negative effect on moisture stability of asphalt material also requires further investigation, since this has hindered its application in road engineering [17]. Alkaline materials should neutralize acidic substances in phosphogypsum, which could improve its water stability and asphalt adhesion. Steel slag is a by-production of steel manufacture, and it has alkalinity and good mechanical properties [18–20]. Steel slag based mixtures showed adequate permanent deformation and durable performance [21]. In addition, the asphalt mixture with steel slag powder fillers showed better resistance to moisture damage, and better low-temperature crack resistance, than asphalt mixtures with limestone filler [22]. The steel slag powder is presumed to neutralize acidic substances in the phosphogypsum, thereby enhancing the moisture stability of asphalt mixtures in which phosphogypsum is employed as a filler.

The concept of overall desirability has been used to integrate optimization based on multiple objectives into a single objective [23]. This method facilitates the integration of indicators with different data ranges into a desirability measure, and can be used to calculated the composition for achieving the optimum performance [24]. Different performance values have been combined into overall desirability to determine the optimum content, with desirability values varying from 0 to 1. Hence, this method has been used to investigate the optimum PF content in this study.

The study aimed to develop a phosphogypsum based filler (PF) which contained steel slag powder, in order to put this waste product to good use. Firstly, the optimum dosage of steel slag powder as modifier was determined through analysis and characterization of PF based asphalt mortar (P-AM). Secondly, the optimized PF was used to partly replace limestone filler in asphalt mortar. Asphalt mortars containing mixed fillers of PF and limestone filler (PL-AM) at different filler-asphalt ratios were prepared and tested. Thirdly, penetration, softening point, ductility and penetration index were used to determine the PF content in replacing traditional limestone filler based on overall desirability. Finally, asphalt mixtures that employed the mixed filler containing PF were fabricated and tested to verify the feasibility of using phosphogypsum as filler after PF composition and content in the mixed filler were determined. The results of this study provide an original approach for consuming the phosphogypsum in asphalt mixture. This approach is positive for reducing phosphogypsum and steel slag, which is beneficial for environmental protection.

2. Materials and Methods

2.1. Raw Material

2.1.1. Phosphogypsum and Steel Slag Powder

Figure 1 show the appearance of phosphogypsum and steel slag powder. Table 1 shows the main chemical components of phosphogypsum and steel slag powder; the elements are given in form of their oxides. According to XRF analysis, the main component of phosphogypsum is $CaSO_4 \cdot 2H_2O$ and its density is 2.371 g/cm^3. The physically adsorbed moisture of phosphogypsum will be removed at 100 °C heating. Studies reported that phosphogypsum will then turn into hemihydrate phosphogypsum ($CaSO_4 \cdot 0.5H_2O$) when the temperature reaches about 130 °C. Hemihydrate phosphogypsum will turn into anhydrous hard phosphogypsum ($CaSO_4$) when the temperature exceeds 1200 °C.

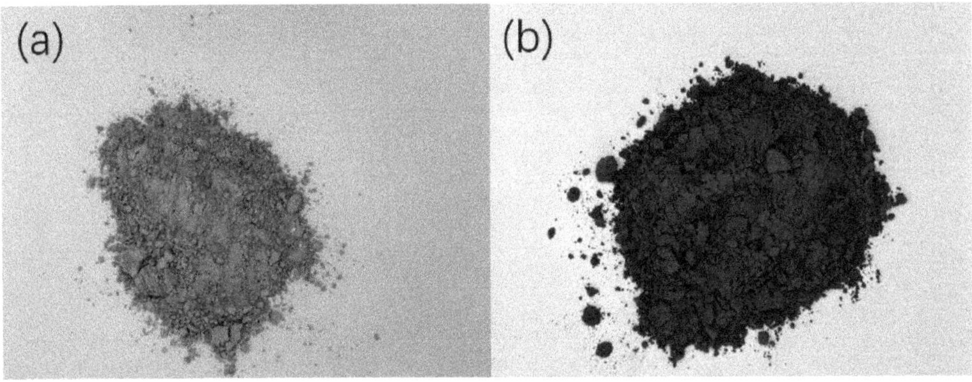

Figure 1. (**a**) Phosphogypsum powder, (**b**) steel slag powder.

Table 1. The chemical composition of phosphogypsum and steel slag powder.

Composition	SO_3	Al_2O_3	SiO_2	CaO	P_2O_5	Fe_2O_3	MgO
Phosphogypsum	44.5%	0.9%	9.5%	31.1%	2.5%	0.7%	/
Steel slag powder	/	22.2%	43.9%	17.8%	/	2.9%	5.7%

Through XRF analysis, the main component of steel slag is Al_2O_3, and its density is 3.498 g/cm^3. The steel slag powder had been exposed in the air for over 1 year. It was found that the free CaO (f-CaO) of steel slag powder was 2.1%, which was below the upper limit of f-CaO. This indicated that steel slag powder will not lead to volume instability.

2.1.2. Asphalt

Table 2 shows the neat asphalt with penetration range of 60–80 used in this study. Mechanical properties of the asphalt are characterized according to standard testing specification for asphalt and mixture testing (JTG E20-2011, Beijing, in Chinese) and technical specification for construction of highway asphalt pavements (JTG F40-2004, Beijing, in Chinese).

Table 2. Mechanical properties of asphalt.

Technical Index	Test Results	Requirements
Penetration (25 °C, 0.1 mm)	70.7	60–80
Softening point (°C)	49.0	≥46
Ductility (15 °C, cm)	>100	≥100
Viscosity (135 °C, Pa·s)	0.46	/

2.1.3. Aggregate and Filler

Table 3 shows the properties of the aggregate and filler, which were tested in accordance with the standards for aggregate testing of highway engineering (JTG E42-2005, Beijing, in Chinese). The results indicated that limestone filler can meet the specification requirements.

Table 3. The basic properties of the aggregate and filler.

Technical Index			Test Results	Requirements
Aggregate	Apparent specific density		2.851	≥2.5
	Crush value (%)		20.7	≤28
	Water absorption (%)		0.8	≤3.0
	Adhesion level		5	5
Filler	Granularity range (%)	<0.6 mm	100	100
		<0.15 mm	91.5	90–100
		<0.075 mm	79	75–100
	Apparent specific density		2.786	/
	Appearance		Agglomerate free of caking	Agglomerate free of caking

2.2. Experimental Methods

Figure 2 illustrates the outline of this study. Firstly, this study introduces the preparation of PF. The content of steel slag powder as modifier of PF was set as 0%, 20%, 40%, 60%, 80% and 100%, respectively. The optimum composition of steel slag powder and phosphogypsum was then investigated through characterization of PF based asphalt mortar (P-AM). Secondly, the optimized PF was used to partly replace limestone powder, which formed mixed filler containing PF and limestone powder. PF-L-AM (PL-AM) with filler-asphalt mass ratios of 0.8, 1.0 and 1.2 were prepared. Optimum PF content of the mixed filler was calculated by overall desirability based on the mechanical properties of PL-AM. How filler-asphalt ratio affect PL-AM's rheological properties was also considered. Finally, asphalt mixtures that employed the mixed filler containing PF were fabricated and tested. The feasibility of using phosphogypsum as filler was then verified after the PF composition and content in mixed filler were determined.

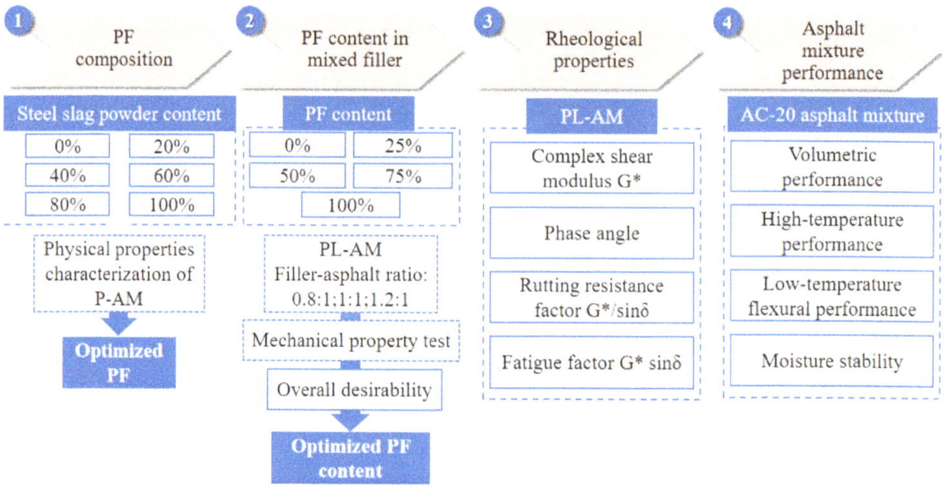

Figure 2. Outline of study.

2.2.1. Preparation of PF

This study employed steel slag powder as the modifier in PF since phosphogypsum was acidic and highly hydrophilic, which could lead to poor moisture stability when using phosphogypsum as filler of asphalt mixture alone. Steel slag powder is generally alkaline which should neutralize the acidity of phosphogypsum in a certain degree, benefiting the adhesion between asphalt and aggregate. Phosphogypsum and steel slag powder were first crushed and screened. Both had a particle size less than 0.075 mm.

Both phosphogypsum and steel slag powder were first heated at 135 °C for 5 h. The phosphogypsum ($CaSO_4{\cdot}2H_2O$) was thus turned into hemihydrate phosphogypsum ($CaSO_4{\cdot}0.5H_2O$) as pretreatment before preparing PF. The dosage of steel slag powder in PF was designed as 0%, 20%, 40%, 60%, 80% and 100%, respectively, based on preliminary tests. Therefore, 6 kinds of PF were prepared. The optimum content of steel slag powder in PF should be determined according to the mechanical characterization of P-AM.

2.2.2. Preparation of Asphalt Mortar

A high-speed mixing device was used to prepare asphalt mortar, and the mixing temperature was 150 °C. The mixing time was 45 min to ensure the uniformity of materials, and during this time, the rotational speed was varied. The initial rotational speed and time were set as 800 r/min for 15 min after PF or limestone filler were put in with the asphalt. Then, the rotational speed was increased to 1800 r/min for another 15 min. The rotational speed was increased to 3600 r/min for the final 15 min.

PFs of different composition were used to prepare P-AM; their volume was adjusted with that of the limestone filler to avoid the effect of volume difference. The filler-asphalt mass ratios of P-AM were 1:1. Additionally, limestone filler was used to prepare a control group. Thus, by replacing limestone filler at different proportions in equal volume, six kinds of P-AM were prepared to investigate the optimum dosage of steel slag powder based on their mechanical properties.

The optimum PF content in replacing limestone filler was then investigated. The PF content in the mixed filler was 0%, 25%, 50%, 75% and 100%. PL-AM was thus introduced. PL-AM's filler-asphalt mass ratios were 0.8, 1.0 and 1.2, respectively. The total volume of the mixed filler containing PF and limestone filler was unchanged when preparing PL-AM. Through characterization and analysis of properties, the optimum replacing proportion of PF for limestone filler can be ascertained. Limestone filler asphalt mortar (L-AM) was also prepared as control group.

2.2.3. Experiments Using Asphalt Mortar

Determination of P-AM Composition

The influence of steel slag powder content on the penetration, softening point and ductility of P-AM was firstly analyzed according to (JTG E20-2011). The results were then analyzed to figure out the optimum content of steel slag powder.

In addition, it is necessary to characterize the effect of PF on adhesion between asphalt and limestone, since phosphogypsum is acidic and shows poor moisture stability. How steel slag powder affects the moisture stability of asphalt mixture can hence be investigated. A boiling test investigating the adhesion between P-AM and aggregate was introduced. Limestone aggregate with the size of 26.5 to 31.5 mm was selected. Aggregates were heated at 150 °C and then put in P-AM at 135 °C for 10 s so that they could be fully covered with P-AM. They were then put in boiling water for 30 min after cooling for 24 h at 25 °C. Afterwards, the boiled aggregate was put in a drying box at 80 °C for 5 h to remove moisture. Finally, the mass loss ratio of P-AM on aggregate can be described as Equation (1) [25].

$$MLR = \frac{m_1 - m_2}{m_1 - m_0} \times 100\% \quad (1)$$

where: MLR = mass loss ratio of P-AM on aggregate, (%);
m_0 = mass of original aggregate, (g);
m_1 = mass of aggregate covered by P-AM;
m_2 = mass of the boiled aggregate.

Determination of PL-AM

The penetration, ductility and softening point of different PL-AM mixes were characterized. The penetration index (PI) and equivalent softening point T_{800} were also included to indicate their mechanical properties. Equation (2) is a unary linear equation which was used to fit the functional relationship between penetration and temperature. Equations (3) and (4) were used to calculate PI and T_{800} [26]. The equations used are as follows and all indexes are dimensionless:

$$l_gP = K + A_{l_gPen} \times T \quad (2)$$

$$PI = \frac{20 - 500A_{l_gPen}}{1 + 50A_{l_gPen}} \quad (3)$$

$$T_{800} = \frac{l_g800 - K}{A_{l_gPen}} = \frac{2.9031 - K}{A_{l_gPen}} \quad (4)$$

where: l_gP = Penetration logarithm at different temperatures;
K = Constant of the linear equation;
A_{l_gPen} = Slope of the linear equation;
T = Testing temperature of penetration;
PI = Penetration index;
T_{800} = Equivalent softening point.

This study adopted the overall desirability method which can combine comprehensive indicators and express the final effect by overall desirability. Introduction of comprehensive indicators in the overall desirability method facilitates the integration of indicators with different data ranges into desirability data; thus, the optimum value of asphalt mortar considering each factor control range can be calculated. Using the overall desirability method, dimensional indicators such as penetration, softening point, and ductility are standardized and converted into corresponding dimensionless desirability values between 0 and 1 through linear transformation. The geometric mean of desirability for each indicator can be calculated. Consequently, the desirability of overall evaluation can be obtained. The closer the desirability of the general evaluation to "1", the better the comprehensive performance of the asphalt mortar will be.

Penetration, penetration index (PI), ductility, and softening point were taken as the calculation indexes of the overall desirability method. Firstly, for each performance indicator (γ), the maximum and minimum value obtained in this study were noted. Then, the desirability (γ_n^*) was calculated by linear transformation of performance indicators, using Equations (5) and (6). The equations specify the calculation of desirability based on γ whose values positively or negatively determine the performance of asphalt mortar. Finally, the desirabilities of each performance indicator were used to calculate their geometric mean value by Equation (7), which was the overall desirability. The calculation method of overall desirability is shown below [27,28]:

$$\gamma_{max}^* = \frac{\gamma - \gamma_{min}}{\gamma_{max} - \gamma_{min}} \quad (5)$$

$$\gamma_{min}^* = \frac{\gamma_{max} - \gamma}{\gamma_{max} - \gamma_{min}} \quad (6)$$

$$OD = (\gamma_1^* \gamma_2^* \gamma_3^* \ldots \gamma_n^*)^{\frac{1}{n}} \quad (7)$$

where: OD = overall desirability, dimensionless;
γ_n^* = desirability index after linear transformation, dimensionless;
γ = performance indicator value, dimensionless;
n = number of performance indicators used in overall desirability, dimensionless;
γ_{max} = maximum value of corresponding performance indicator in this study, dimensio

γ_{min} = minimum value of corresponding performance indicator in this study, dimensionless;

γ^*_{max} = desirability based on the γ whose value positively determined performance of asphalt mortar, dimensionless;

γ^*_{min} = desirability based on the γ whose value negatively determined performance of asphalt mortar, dimensionless.

Rheological Performance

Rheological performance of PL-AM and limestone filler were also characterized by using a dynamic shear rheometer (DSR) after PF's content and composition were determined. A limestone filler-based asphalt mortar with filler-asphalt mass ratio of 1:1 was prepared as control group. Table 4 shows the setting table of DSR high-temperature scanning parameters.

Table 4. DSR high-temperature scanning parameter setting table.

Temperature	Angular Frequency	Rotor Size	Gap Size	Heating Up Speed
30–80 °C	0.1 Hz	25 mm	1 mm	2 °C/120 s

2.2.4. Pavement Performance of PF Based Asphalt Mixture

Mixed filler containing PF was also used to prepared asphalt mixture to indicate its effect on pavement performance. AC-20 was used as the gradation of asphalt mixture which was specified in JTG E40-2005, and its gradation curve is shown in Figure 3. Limestone filler-based AC-20 asphalt mixture was firstly fabricated. Afterwards, PF based asphalt mixture using the mixed filler containing PF can be prepared. The mass ratio of asphalt to aggregate was 4.25% for both the limestone filler and PF based asphalt mixture. Volumetric performance, high-temperature performance, low temperature performance and moisture stability were tested. Volumetric property refers to void volume (VV) and void in mineral aggregate (VMA).

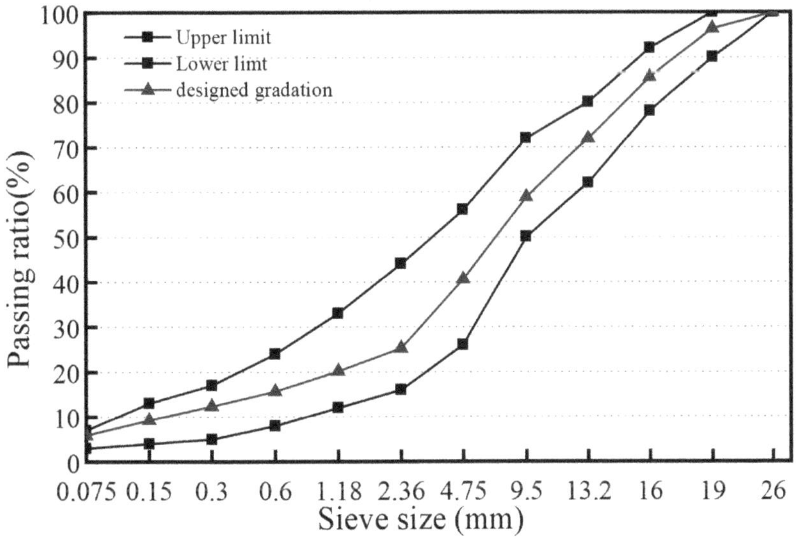

Figure 3. Aggregate gradation curve of AC-20.

High-Temperature Performance

High-temperature performance was evaluated by testing the asphalt mixture's Marshall stability and dynamic stability according to JTG F20-2011. The dynamic stability test should be applied at 60 °C. Track plate specimens and a dynamic stability instrument were used. Stability can be calculated by Equation (8) [29].

$$DS = \frac{(t_2 - t_1) \times 42}{d_2 - d_1} \times c_1 \times c_2 \quad (8)$$

where: DS = dynamic stability of the asphalt mixture, (cycle/mm);

t_1, t_2 = test time, usually 45 min and 60 min;

d_1, d_2 = deformation of specimen surface corresponding to the test specimens t_1 and t_2, (mm).

c_1, c_2 = correction factor of testing machine or specimen, dimensionless.

Low-Temperature Flexural Performance

Low-temperature flexural performance of the asphalt mixtures was evaluated by a three-point bending test. Track plate specimens were cut into beam specimens with a size of 250 mm × 30 mm × 35 mm, according to JTG E20-2011. A Universal Testing Machine (UTM-100) was used for the three-point bending test at −10 °C. The distance between supporting fulcrums was 200 mm. The loading rate of the principal axis was 50 mm/s. Corresponding indictors can be calculated using the following equations [30]:

$$R_B = \frac{3 \times L \times P_B}{2 \times b \times h^2} \quad (9)$$

$$\varepsilon_B = \frac{6 \times h \times d}{L^2} \quad (10)$$

$$S_B = \frac{R_B}{\varepsilon_B} \quad (11)$$

where: R_B = flexural tensile strength (MPa);

ε_B = tensile strain (με);

S_B = tensile stiffness modulus (MPa);

P_B = loading peak (kN);

L = span length of beam (mm);

h = height of midspan section (mm);

b = width of midspan section (mm);

d = midspan deflection at failure (mm).

Moisture Stability

It is important to characterize the moisture stability of the asphalt mixtures as phosphogypsum's high hydrophilicity may lead to poor adhesion between asphalt and aggregate. The immersion Marshall stability ratio (IMS) and freeze-thaw tensile strength ratio (TSR) were used to comprehensively evaluate the moisture stability of asphalt mixture according to (JTG E20-2011). IMS can be calculated by Equation (12) [29].

$$IMS = \frac{MSR_1}{MSR} \times 100\% \quad (12)$$

where: MSR: the average stability of specimen in moisture at 60 °C for 30 min (kN);

MSR_1: the average stability of specimen in moisture at 60 °C for 48 h (kN);

IMS: the average residual stability of specimen in moisture.

On the other hand, TSR can be calculated by Equation (13) [29].

$$TSR = \frac{R_{T2}}{R_{T1}} \times 100 \quad (13)$$

where: TSR: the average strength ratio of the freeze-thaw splitting test;

R_{T1}: splitting tensile strength of the specimens without freeze-thaw cycle (the unconditional);

R_{T2}: splitting tensile strength of specimens after freeze-thaw cycle (the conditional).

3. Results and Discussions

3.1. Determination of PF Composition

3.1.1. Physical Properties of P-AM

Tables 5 and 6 show the results of penetration, ductility and softening point tests for P-AM. The penetration of P-AM showed the lowest value when 20% steel slag powder was added, at which the highest consistence as well as high-temperature performance of P-AM was achieved. The softening point of P-AM was higher than that of L-AM when 0–60% steel slag powder was mixed. P-AM with 20% steel slag powder showed the highest softening point, suggesting the maximum high-temperature performance. These findings indicate that an appropriate mixture of phosphogypsum and steel slag powder results in an improved softening point. However, the ductility values of P-AM were significantly lower than those of limestone asphalt mortar. The content of steel slag powder showed no statistically significant effect on the ductility of P-AM. It was evident that an excessive content of steel slag powder will result in poor physical properties. Furthermore, it was speculated that there could be a coupling effect of phosphogypsum and steel slag powder which could determine the properties of asphalt mortar. Consequently, the optimum content of steel slag powder should be determined after comprehensive consideration of P-AM's physical properties. A functional curve fitting the data of penetration and softening point of P-AM was used to assess the effects of steel slag powder content intuitively, as shown in Figure 4.

Table 5. Physical results of P-AM.

Content of Steel Slag Powder	Penetration (25 °C, 0.1 mm)	Ductility (15 °C, cm)	Softening Point (°C)
0%	54.4	3.6	55.0
20%	45.2	4.8	57.0
40%	49.3	4.3	56.5
60%	53.7	3.9	55.0
80%	57.6	4.5	54.0
100%	62.5	4.7	53.5

Table 6. Results of L-AM.

Penetration (25 °C, 0.1 mm)	Ductility (15 °C, cm)	Softening Point (°C)
51.6	10.7	54.5

Functional curves of the data points were then fitted to show how the content of steel slag powder affected the properties of P-AM. The equation of the fitting curve for the softening point was found to be:

$$y = 22.569x^3 - 40.327x^2 + 16.27x + 55.048$$

The determination coefficient (R^2) was 0.9904, which indicated that the fitting of the softening point was reliable enough. On the other hand, the equation of the fitting curve for penetration was found to be:

$$y = -73.958x^3 + 142.05x^2 - 59.917x + 53.843$$

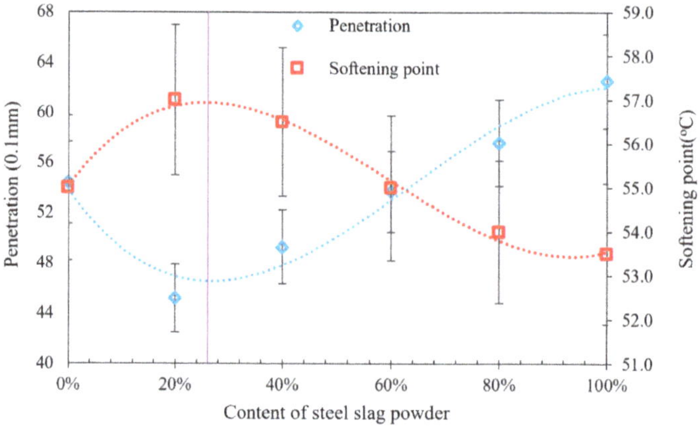

Figure 4. Results of penetration and softening point of P-AM.

Its R^2 was 0.9572, which also suggested that the fitting of penetration was adequate. It was found that the highest softening point and lowest penetration occurred when the content of steel slag powder was 23%. The highest high-temperature performance, stiffness and plasticity can be achieved at this content. Since the content of steel slag powder showed no significant effect on the ductility of P-AM, the optimum content of steel slag powder was determined as 23% of PF volume considering its contribution to the softening point and penetration.

3.1.2. Adhesion Characterization

Images of aggregates covered with P-AM film after the boiling test are shown in Figure 5. It appears that the P-AM film on aggregate with 0% steel slag powder seriously peeled off. However, P-AM films contained steel slag powder showed no obvious spalling after the boiling test. This result indicates that adding steel slag powder to PF can enhance adhesion between the aggregate and the asphalt binder. P-AM without steel slag powder showed poor adhesion with aggregate due to acidity of phosphogypsum. In contrast, alkaline steel slag powder can neutralize acidity of phosphogypsum to a certain degree, so that adhesion between the aggregate and the asphalt binder can be developed. This result further indicates that steel slag was positive for the moisture resistance of the asphalt mixture which is correlated to adhesion between asphalt and aggregate. However, how the content of steel slag powder affected adhesion was hard to conclude from appearances since the spalling was not easily quantifiable (Figure 5).

Figure 5. Aggregates with P-AM after boiling test.

The mass loss percentage of P-AM is shown in Figure 6, and enables quantitive characterization of the adhesion of P-AM. It was found that P-AM without steel slag powder was nearly fully removed by boiling water, showing that using pure phosphogypsum as filler for asphalt mixture was vulnerable to moisture damage. Mass loss percentages of P-AM were significantly reduced for steel slag powder contents over 20% of PF, proving that steel slag powder can effectively enhance adhesion between asphalt and aggregate. The mass loss percentage of P-AM did not show monotonic reduction as more steel slag powder was introduced, however. It is believed that the enhancement effect of steel slag powder on adhesion is limitative, so that continuously increasing the content of steel slag powder cannot further develop adhesion. The content of steel slag powder within PF was hence suggested to be more than 20% to achieve adequate adhesion enhancement. Consequently, considering the optimum content of steel slag powder as illustrated in Figure 4 and adhesion characterization as illustrated in Figure 6, the optimum volume percentage of steel slag in PF was determined as 23%.

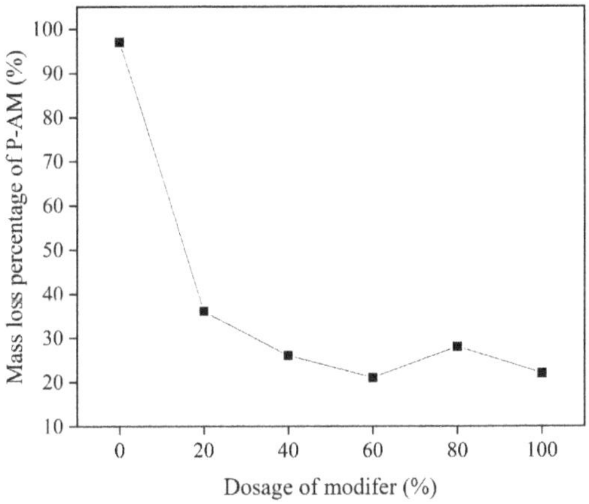

Figure 6. Mass loss percentage of P-AM.

3.2. *Effect of PF Content on PL-AM's Mechanical Properties*

3.2.1. Penetration

Figures 7–9 show the penetration results of PL-AM with filler-asphalt ratios of 0.8, 1.0 and 1.2. Different PF content and testing temperatures were also included. Testing temperature was positively correlated to penetration value due to the viscoelastic characteristics of asphalt. Penetration value showed first a decreasing and then an increasing tendency along with the increase in PF content, independent of temperature. PL-AM presented the lowest penetration results when 75% limestone filler was replaced by an identical volume of PF, regardless of the filler-asphalt ratio and temperature. PL-AM with a higher filler-asphalt ratio shows a lower penetration value since PF is a rigid material. This result suggested that there was a proper composition of filler and PF which achieved the highest stiffness of PL-AM. It was also speculated that there could be a coupling effect of PF and limestone filler so that the effect of PF content on PL-AM's physical property was not monotonic.

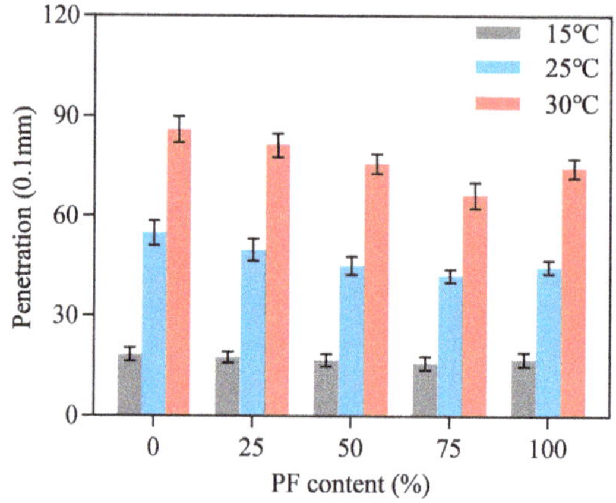

Figure 7. Filler-asphalt ratio of 0.8.

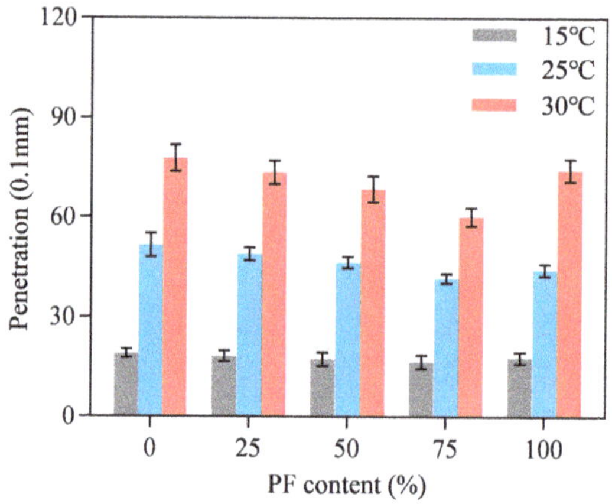

Figure 8. Filler-asphalt ratio of 1.0.

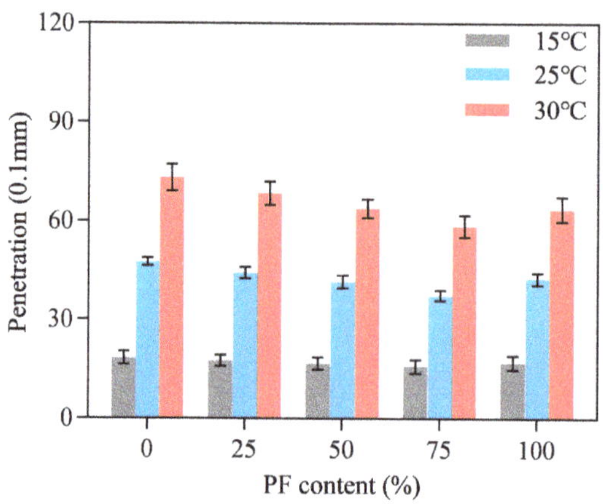

Figure 9. Filler-asphalt ratio of 1.2.

3.2.2. Softening Point

Figure 10 shows PL-AM's softening point at different PF content with filler-asphalt ratios of 0.8, 1.0 and 1.2. It is evident that increased filler-asphalt ratio leads to a higher softening point owing to the enhancement to the filler. The softening point showed a slightly increasing tendency as PF content was raised from 0% to 75%, and decreased when limestone filler was totally replaced by PF, independent of the filler-asphalt ratio. This shows that optimum PF content in the mixed filler was 75% for achieving the highest softening point, which positively determined the high-temperature performance of the corresponding asphalt mixture. On the other hand, it should be noted that the improvement in the softening point caused by replacing limestone filler with PF was not that significant, considering the fact that softening point difference among different PF content was not over 4 °C.

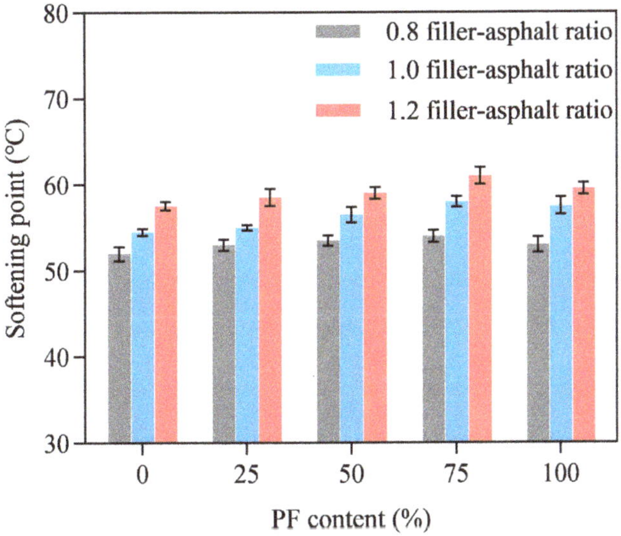

Figure 10. Softening point.

3.2.3. Ductility

Figure 11 presents the ductility of PL-AM at different filler-asphalt ratios and PF content. It was found that higher filler-asphalt ratio led to lower ductility regardless of PF content. Ductility showed a decreasing tendency as the PF content increased from 0% to 50%, but then developed when PF content was over 75%. The lowest ductility value was found when PF content was 50% regardless of the filler-asphalt ratio, and ductility of PL-AM was the second highest when 75% PF was added. The coupling effect of PF and phosphogypsum probably affected the ductility of PL-AM. It illustrated that replacing limestone filler with PF will negatively affect ductility of mortar, which resulted in lower plasticity. However, this negative effect can be reduced by replacing limestone powder with 75% PF. The results suggested that using PF as filler might negatively affect the low-temperature performance of the corresponding asphalt mixture [31].

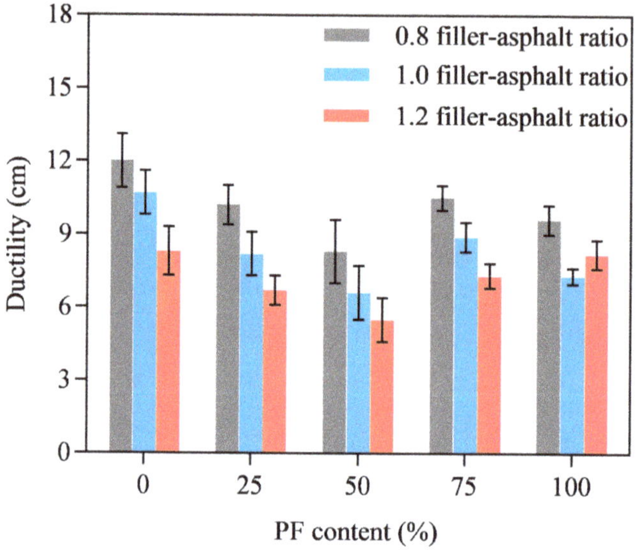

Figure 11. Ductility.

3.2.4. PI and T_{800}

Penetration at 15 °C, 25 °C and 30 °C was used to calculate the penetration index (PI) and equivalent softening point (T_{800}) of asphalt mortar. The logarithmic values of penetration (log P) and temperature (T) were used to calculate PI according to the fitting curves of Equation (2). The linear regression correlation coefficient R^2 in Equation (2) must not be less than 0.997. Equation (3) shows the calculation of PI, which was negatively correlated to asphalt mortar's temperature sensitivity. T_{800} can be calculated based on Equation (4) and illustrated the high-temperature stability of asphalt mortar. Table 7 shows the results for penetration index PI, T_{800}, regression equation and R^2 according to their dependence on PF content and filler-asphalt ratio; PF content 0% is the limestone asphalt mortar that was used as control group. Higher values of PI and T_{800} suggested a stronger ability of asphalt mortar in resisting high-temperature deformation.

Table 7. Results of PI and T800.

Filler-Asphalt Ratio	PF Content	PI	T_{800}	Regression Equation	R^2
0.8	25%	−0.196	54.0	y = 0.0412x + 0.6705	0.9999
0.8	50%	−0.050	55.5	y = 0.0403x + 0.6605	0.9985
0.8	75%	0.256	58.0	y = 0.0385x + 0.6652	0.9999
0.8	100%	0.067	56.5	y = 0.0396x + 0.6727	0.9985
1.0	25%	−0.099	55.5	y = 0.0406x + 0.6586	0.9980
1.0	50%	0.017	56.5	y = 0.0389x + 0.6406	0.9998
1.0	75%	0.363	59.5	y = 0.0378x + 0.6309	0.9999
1.0	100%	−0.180	55.5	y = 0.0411x + 0.6299	0.9989
1.2	25%	0.067	57.0	y = 0.0396x + 0.6510	0.9998
1.2	50%	0.187	58.0	y = 0.0389x + 0.6406	0.9998
1.2	75%	0.381	60.0	y = 0.0378x + 0.6309	0.9999
1.2	100%	0.239	58.5	y = 0.0386x + 0.6523	0.9991
0.8	0%	−0.338	53.0	y = 0.0421x + 0.6767	0.9993
1.0	0%	−0.164	54.5	y = 0.0410x + 0.6711	0.9980
1.2	0%	−0.050	55.5	y = 0.0403x + 0.6610	0.9993

Both PI and T_{800} values of PL-AM were larger than that of L-AM regardless of filler-asphalt ratio. Thus, PF reduced the temperature sensitivity of PL-AM. On the other hand, either PI and T_{800} first increased and then decreased with increase of PF content. Additionally, PI and T_{800} also showed an increasing trend as the filler-asphalt ratio was increased. PL-AM with 75% PF achieved the highest PI and T_{800}, optimally improving the high-temperature performance. It is speculated that these properties were also related to a coupling effect of PF and phosphogypsum because they did not show a monotonic tendency as PF content was increased. This result was consistent with the penetration and softening point of PL-AM, showing that 75% could be the optimum content of PF in replacing limestone filler.

3.3. Determination of PF Content

Table 8 shows the results of the overall desirability calculation. It can be concluded that the total evaluation desirability value of PL-AM had achieved its maximum when PF content was 75%. This proved that its high-temperature performance was optimally enhanced as 75% PF was introduced. PL-AM's penetration and ductility decreased, while its softening point and PI were the highest at 75% PF, independent of the filler-asphalt ratio. Thus, 75% PF is suggested especially concerning high-temperature performance, consistent with the results for penetration, softening point, PI and T_{800}. Therefore, it is suggested that the overall desirability method is a feasible approach to finding the optimum PF content.

Table 8. Overall desirability.

Filler-Asphalt Ratio	PF Content	Penetration (25 °C, 0.1mm)	PI	Ductility (cm)	Softening Point (°C)	*OD* Value
0.8	0%	54.8	−0.338	12.0	52.0	0
	25%	49.8	−0.196	10.2	53.0	0.268
	50%	45.1	−0.050	8.3	53.5	0.364
	75%	42.1	0.256	10.5	54.0	0.576
	100%	44.7	0.067	9.6	53.0	0.402
1.0	0%	51.6	−0.164	10.7	54.5	0.320
	25%	48.9	−0.099	8.2	55.0	0.358
	50%	46.4	0.017	6.6	56.5	0.380
	75%	41.7	0.363	8.9	58.0	0.716
	100%	44.2	−0.180	7.3	57.5	0.391
1.2	0%	47.6	−0.050	8.3	57.5	0.460
	25%	44.2	0.067	6.7	58.5	0.465
	50%	41.5	0.187	5.5	59.0	0
	75%	37.3	0.381	7.3	61.0	0.730
	100%	42.4	0.239	8.2	59.5	0.671

3.4. Rheological Properties

Figures 12–14 shows the results of the DSR high-temperature scanning test for PL-AM with 75% PF at filler-asphalt ratios of 0.8, 1.0 and 1.2, respectively. The complex shear modulus G* of asphalt mortars showed a decreasing trend, while the phase angle showed a rising tendency with increase of temperature. G* of PL-AM was higher than that of L-AM. PL-AM showed higher δ than that of L-AM from 30 to 60 °C, while the contrary result was found from 60 to 80 °C. A higher filler-asphalt ratio resulted in higher G* of PL-AM, while δ was negatively affected by the filler-asphalt ratio from 65 to 80 °C.

PL-AM had a greater rutting factor (G*/sinδ) compared with L-AM when their filler-asphalt ratio was 1.0. This proved that PF was able to increase the hardness of asphalt mortar, so that the ability to resist deformation in high temperatures, namely high-temperature performance, was improved. On the other hand, the fatigue factor (G*sinδ) of PL-AM was also higher than that of L-AM at the same filler-asphalt ratio. Both rutting and fatigue factors were improved by a higher filler-asphalt ratio. Thus, use of a mixed filler containing

PF could help to enhance the rutting and fatigue resistance of PL-AM. The corresponding PF based asphalt mixture's high-temperature and fatigue performance should be higher than a limestone filler based asphalt mixture.

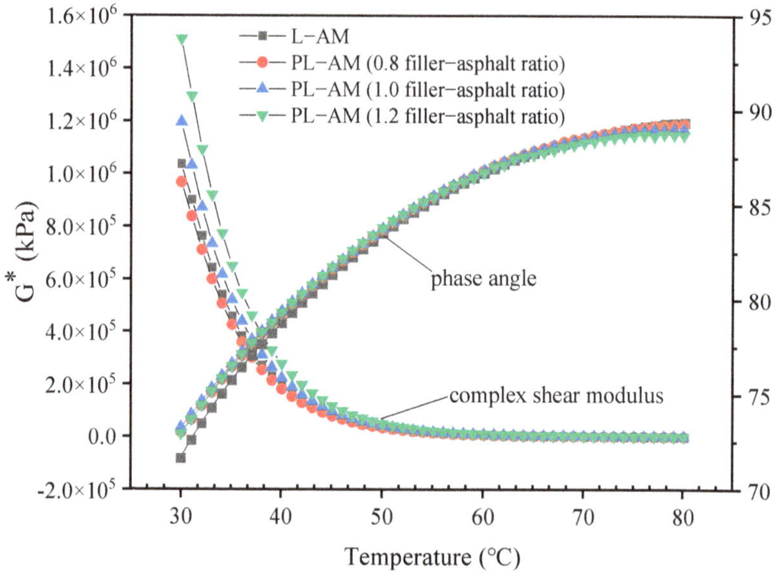

Figure 12. G* and δ of PL-AM and L-AM.

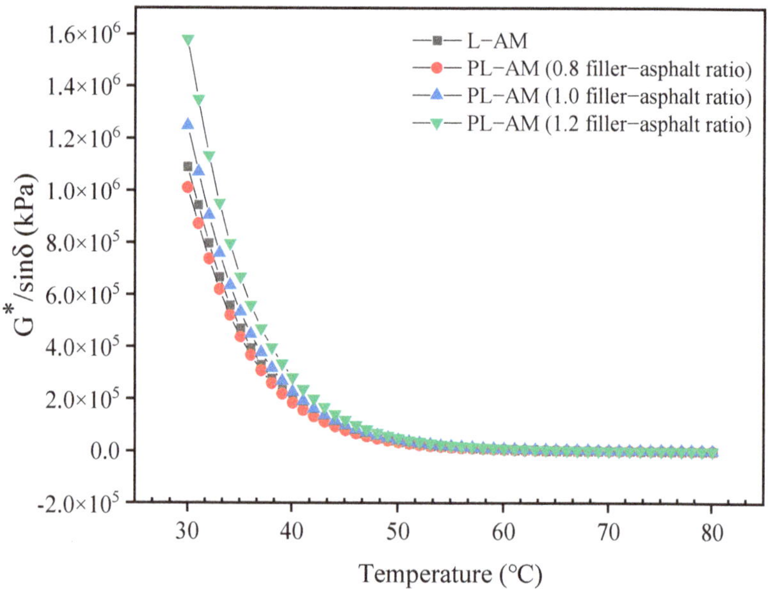

Figure 13. Rutting factor G*/sinδ of PL-AM and L-AM.

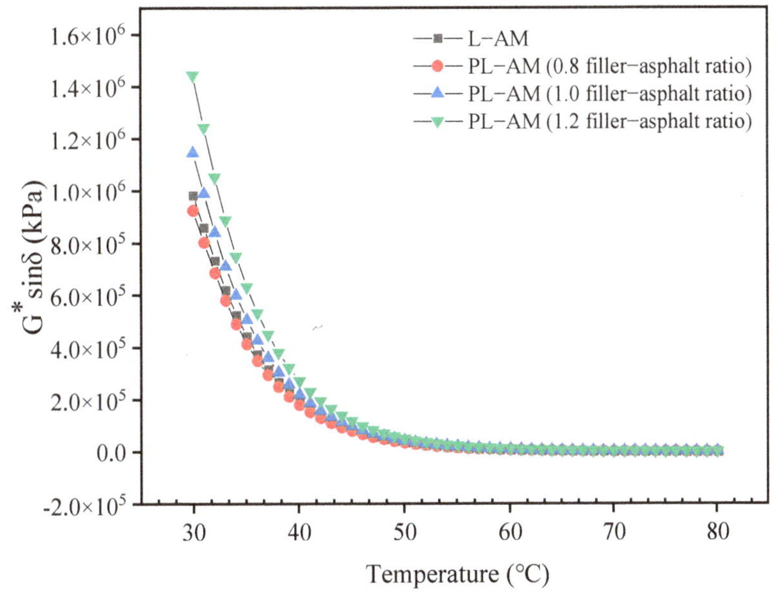

Figure 14. Fatigue factor G*sinδ of PL-AM.

3.5. Pavement Performance

3.5.1. Volumetric Performance

P-AM results suggested that the optimum composition of PF was 23% steel slag powder and 77% phosphogypsum as discussed above. PL-AM results showed that the optimum PF content was 75%. Hence, AC-20 asphalt mixtures with the mixed filler containing PF were prepared. A corresponding asphalt mixture without PF was also fabricated to indicate how PF affects pavement performance. Table 9 shows the optimum asphalt-aggregate mass ratio and volumetric performance of the two kinds of asphalt mixtures. It indicates that using PF to partly replace limestone powder as filler of asphalt mixture showed no clear impact on the optimum asphalt-aggregate mass ratio and volumetric performance.

Table 9. Optimum asphalt-aggregate mass ratio and volumetric performance.

Volume Parameter	PF Based	Limestone Filler Based
Optimum asphalt-aggregate ratio	4.2%	4.2%
VV	4.12%	4.0%
VMA	13.2%	13.0%
VFA	69.3%	69.0%

3.5.2. High-Temperature Performance

Figure 15 shows Marshall stability and dynamic stability of the asphalt mixtures. The dynamic stability values of the two kinds of asphalt mixtures both meet the requirement of (JTG F40-2004), which is not less than 800 cycles/mm. The dynamic stability of limestone filler based asphalt mixture was 922 cycles/mm and that of the PF based asphalt mixture was 1265 times/mm, which was improved by 37.2%. In addition, Marshall stabilities of both asphalt mixtures were higher than the requirement of 8 kN. The PF based asphalt mixture showed a higher Marshall stability than that of the limestone filler based asphalt mixture. This showed clearly that PF could improve the high-temperature performance of the asphalt mixture by partly replacing limestone filler. This result was consistent with the softening point test and DSR high-temperature scanning test results, which also suggested that PF can enhance high-temperature performance.

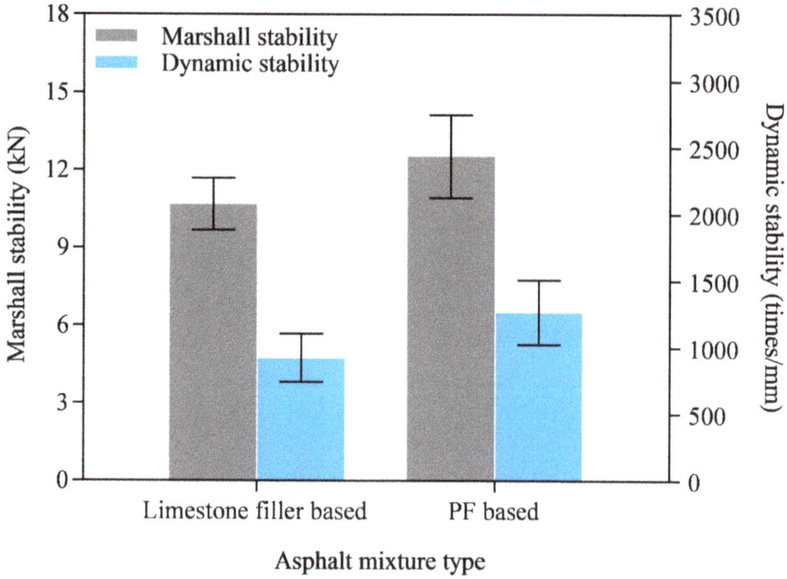

Figure 15. Marshall stability and dynamic stability of asphalt mixture.

3.5.3. Low-Temperature Flexural Performance

Table 10 shows the test results for low-temperature flexural performance. The flexural-tensile strain of the asphalt mixture with limestone filler was 2249, while the strain for the PF based asphalt mixture was 2162. They can meet the requirement in the JTG F40-2004 specification that flexural-tensile strain should not be less than 2000, indicating that their low-temperature flexural performance was adequate. The flexural tensile stiffness modulus and flexural tensile strength stiffness modulus of the PF based asphalt mixture were higher than those of the limestone filler based asphalt mixture. Thus, using PF as filler would not significantly damage the low-temperature performance of the asphalt mixture.

Table 10. Low-temperature flexural performance results.

Types of Asphalt Mixture	Flexural Tensile Strain (με)	Flexural Tensile Strength (MPa)	Flexural Tensile Strength Stiffness Modulus (MPa)
PF based	2162	9.7	4486.59
Limestone filler based	2249	9.3	4135.17

3.5.4. Moisture Stability

Moisture stability was a key performance criterion since phosphogypsum is acidic and has poor moisture stability. Tables 11 and 12 present results of IMS and TSR, respectively. IMS results showed that both the PF based asphalt mixture and the limestone filler based asphalt mixture could meet the JTG F40-2004 specification, requiring not less than 80%. The TSR values of both the PF based asphalt mixture and the limestone filler based asphalt mixture were higher than 75%, which meets the specification requirement. On the other hand, the IMS and TSR of the PF based asphalt mixture were higher than the values for the asphalt mixture without PF, proving that using PF was positive for improving the asphalt mixture's moisture stability. Therefore, it was analyzed that the negative effect of phosphogypsum on moisture stability was offset by the addition of steel slag powder. To conclude, the PF based asphalt mixture showed better high-temperature and moisture stability and adequate low-temperature performance. The results showed that using phosphogypsum based filler containing steel slag powder to partly replace limestone filler was able to develop asphalt mixture's pavement performance.

Table 11. IMS results.

Types of Asphalt Mixture	MSR_1 (60 °C for 48 h, kN)	MSR (60 °C for 30–40 min, kN)	IMS (%) ≥ 80
PF based	11.11	12.53	88.7
Limestone filler based	8.95	10.68	83.8

Table 12. TSR results.

Types of Asphalt Mixture	R_{T1} (MPa)	R_{T2} (MPa)	TSR (%) ≥ 75
PF based	0.990	0.797	80.5
Limestone filler based	0.895	0.698	78.0

4. Conclusions

This study focused on recycling phosphogypsum as an ingredient of an asphalt mixture filler. Phosphogypsum and steel slag powder were mixed to fabricate the phosphogypsum based filler (PF). PF based asphalt mortar (P-AM) was prepared. Penetration, softening point, ductility and boiling tests were firstly conducted to determine the optimum content of steel slag powder in PF. PF-limestone based asphalt mortars were prepared and characterized. The overall desirability method was applied to determine the optimum content of PF in replacing limestone filler in the asphalt mixture. Finally, the PF based asphalt mixture was fabricated and tested to verify the feasibility of using phosphogypsum as an asphalt mixture filler. According to the laboratory test results, the following conclusions can be drawn.

(1) PF enhanced asphalt mortar's softening point and ductility, while penetration was reduced. Steel slag powder clearly improved the adhesion between P-AM and aggregate when its content was over 20%. The highest softening point and lowest penetration occurred when content of steel slag powder was 23%, according to the fitting curve.

(2) PL-AM presented the lowest penetration when 75% limestone filler was replaced by PF. Similarly, the highest softening point, penetration index and equivalent softening point were also found when PF content was 75%. The filler-asphalt ratio was positively correlated to softening point, penetration index and equivalent softening point, while it negatively affected penetration and ductility. The overall desirability achieved the maximum value when PF content was 75% based on PL-AM mechanical performance. Consequently, the optimum PF content was determined as 75% in replacing limestone filler.

(3) The complex shear modulus of PL-AM was improved by PF. PL-AM showed a higher phase angle from 30 to 60 °C, while the contrary result was found from 60 to 80 °C. PF showed no significant effect on volumetric and low-temperature performance. High-temperature and moisture stability performance were improved by PF. Thus, using phosphogypsum based filler containing steel slag powder to partly replace limestone filler improved the asphalt mixture's pavement performance.

Author Contributions: Conceptualization, J.W. and K.L.; methodology, J.W.; software, T.H.; validation, S.S., X.H. and W.G.; formal analysis, T.H.; investigation, K.L., J.W. and T.H.; resources, Z.C.; data curation, T.H.; writing—original draft preparation, J.W.; writing—review and editing, T.H.; visualization, J.W.; supervision, X.H. and S.S.; project administration, J.W. and Z.C.; funding acquisition, J.W. and Z.C. All authors have read and agreed to the published version of the manuscript.

Funding: This research was funded by Open Fund of Key Laboratory of Road Structure and Material of Ministry of Transport (Changsha University of Science & Technology, No. kfj220303), National Natural Science Foundation of China (No. 52108414), Scientific Research Starting Foundation of Wuhan Institute of Technology (No. K202021) and Open Fund of National & Local Joint Engineering Laboratory of Traffic Civil Engineering Materials (No. LHSYS-2020-004).

Institutional Review Board Statement: Not applicable.

Informed Consent Statement: Not applicable.

Data Availability Statement: Not applicable.

Conflicts of Interest: The authors declare no conflict of interest.

References

1. Li, P.; Ding, Z.; Rao, W. Evaluation of deformation properties of asphalt mixture using aggregate slip test. *Int. J. Pavement Eng.* **2016**, *17*, 542–549. [CrossRef]
2. Zhang, X.; Zhang, B.; Chen, H.; Kuang, D. Feasibility evaluation of preparing asphalt mixture with low-grade aggregate, rubber asphalt and desulphurization gypsum residues. *Materials* **2018**, *11*, 1481. [CrossRef]
3. Dai, Q.; Ma, L.; Xie, L.; Yan, B.; Zheng, D.; Yang, J. Analyses on the application of components in the phosphogypsum decomposition residue: Based on the study of migration and conversion rules of elements in the purification procedure of CaO. *Environ. Prog. Sustain. Energy* **2018**, *37*, 1020–1030. [CrossRef]
4. Smaoui-Jardak, M.; Kriaa, W.; Maalej, M.; Zouari, M.; Kamoun, L.; Trabelsi, W.; Ben Abdallah, F.; Elloumi, N. Effect of the phosphogypsum amendment of saline and agricultural soils on growth, productivity and antioxidant enzyme activities of tomato (Solanum lycopersicum L.). *Ecotoxicology* **2017**, *26*, 1089–1104. [CrossRef]
5. Moalla, R.; Gargouri, M.; Khmiri, F.; Kamoun, L.; Zairi, M. Phosphogypsum purification for plaster production: A process optimization using full factorial design. *Environ. Eng. Res.* **2018**, *23*, 36–45. [CrossRef]
6. Islam, G.S.; Chowdhury, F.H.; Raihan, M.T.; Amit, S.S.; Islam, M.R. Effect of phosphogypsum on the properties of Portland cement. *Procedia Eng.* **2017**, *171*, 744–751. [CrossRef]
7. Rashad, A.M. Phosphogypsum as a construction material. *J. Clean. Prod.* **2017**, *166*, 732–743. [CrossRef]
8. Wang, C.-Q.; Xiong, D.-M.; Chen, Y.; Wu, K.; Tu, M.-J.; Wang, P.-X.; Zhang, Z.-J.; Zhou, L. Characteristic pollutant purification analysis of modified phosphogypsum comprehensive utilization. *Environ. Sci. Pollut. Res.* **2022**, *29*, 67456–67465. [CrossRef]
9. Ou, L.; Zhu, H.; Xu, Y.; Chen, R.; Yang, X. Gray correlation entropy analysis of zero shear viscosity and high-temperature rheological parameters of phosphogypsum-modified asphalt. *Case Stud. Constr. Mater.* **2022**, *17*, e01448. [CrossRef]
10. Gregory, C.; Saylak, D.; Ledbetter, W.B. The use of by-product phosphogypsum for road bases and subbases. *Transp. Res. Rec.* **1984**, *998*, 47–52.
11. Li, J.; Zhang, Y.; Mao, L.; Zi, J.; Hu, T. Laboratory research on the application of phosphogypsum as an anti-cracking agent in a cement-treated base. In *CICTP 2014: Safe, Smart, and Sustainable Multimodal Transportation Systems*; American Society of Civil Engineers: Reston, VA, USA, 2014; pp. 990–1002.
12. Cuadri, A.; Navarro, F.; García-Morales, M.; Bolívar, J. Valorization of phosphogypsum waste as asphaltic bitumen modifier. *J. Hazard. Mater.* **2014**, *279*, 11–16. [CrossRef]
13. Shen, W.; Zhou, M.; Ma, W.; Hu, J.; Cai, Z. Investigation on the application of steel slag–fly ash–phosphogypsum solidified material as road base material. *J. Hazard. Mater.* **2009**, *164*, 99–104. [CrossRef]
14. Dutta, R.K.; Kumar, V. Suitability of flyash-lime-phosphogypsum composite in road pavements. *Period. Polytech. Civ. Eng.* **2016**, *60*, 455–469. [CrossRef]
15. Amrani, M.; El Haloui, Y.; Hajikarimi, P.; Sehaqui, H.; Hakkou, R.; Barbachi, M.; Taha, Y. Feasibility of using phosphate wastes for enhancing high-temperature rheological characteristics of asphalt binder. *J. Mater. Cycles Waste Manag.* **2020**, *22*, 1407–1417. [CrossRef]
16. Cuadri, A.; Pérez-Moreno, S.; Altamar, C.; Navarro, F.; Bolívar, J. Phosphogypsum as additive for foamed bitumen manufacturing used in asphalt paving. *J. Clean. Prod.* **2021**, *283*, 124661. [CrossRef]
17. Meskini, S.; Samdi, A.; Ejjaouani, H.; Remmal, T. Valorization of phosphogypsum as a road material: Stabilizing effect of fly ash and lime additives on strength and durability. *J. Clean. Prod.* **2021**, *323*, 129161. [CrossRef]
18. Zhao, M.; Wu, S.; Chen, Z.; Li, C. Production and application of steel slag coarse aggregate in asphalt mixture. *Emerg. Mater. Res.* **2017**, *6*, 219–222. [CrossRef]
19. Chen, J.-S.; Wei, S.-H. Engineering properties and performance of asphalt mixtures incorporating steel slag. *Constr. Build. Mater.* **2016**, *128*, 148–153. [CrossRef]
20. Guo, Y.; Wu, H.; Shen, A.; Yang, X.; Cui, T. Study of the long-term water stability of asphalt mixtures containing steel slag aggregate. *J. Adhes. Sci. Technol.* **2020**, *34*, 877–902. [CrossRef]
21. Wan, J.; Wu, S.; Xiao, Y.; Fang, M.; Song, W.; Pan, P.; Zhang, D. Enhanced ice and snow melting efficiency of steel slag based ultra-thin friction courses with steel fiber. *J. Clean. Prod.* **2019**, *236*, 117613. [CrossRef]

22. Chen, Z.; Leng, Z.; Jiao, Y.; Xu, F.; Lin, J.; Wang, H.; Cai, J.; Zhu, L.; Zhang, Y.; Feng, N. Innovative use of industrially produced steel slag powders in asphalt mixture to replace mineral fillers. *J. Clean. Prod.* **2022**, *344*, 131124. [CrossRef]
23. Ghosh, A.; Mal, P.; Majumdar, A.; Banerjee, D. Optimization of knitted fabric comfort and UV protection using desirability function. *J. Eng. Fibers Fabr.* **2016**, *11*, 155892501601100404. [CrossRef]
24. Alam, M.S.; Ghosh, A.; Majumdar, A. Optimization of bending and shear rigidities of woven fabrics using desirability function. *J. Text. Inst.* **2020**, *111*, 1318–1323. [CrossRef]
25. Wang, F.; Li, Y.; Yu, L.; Pang, W. Study on Influencing Factors of Asphalt-Aggregate Stripping Mechanism. *Adv. Mater. Sci. Eng.* **2021**, *2021*, 1–7. [CrossRef]
26. Zeng, M.; Pan, H.; Zhao, Y.; Tian, W. Evaluation of asphalt binder containing castor oil-based bioasphalt using conventional tests. *Constr. Build. Mater.* **2016**, *126*, 537–543. [CrossRef]
27. Anahas, A.M.P.; Muralitharan, G. Central composite design (CCD) optimization of phytohormones supplementation for enhanced cyanobacterial biodiesel production. *Renew. Energy* **2019**, *130*, 749–761. [CrossRef]
28. Van Houtven, J.; Hooyberghs, J.; Laukens, K.; Valkenborg, D. CONSTANd: An efficient normalization method for relative quantification in small-and large-scale omics experiments in R BioConductor and Python. *J. Proteome Res.* **2021**, *20*, 2151–2156. [CrossRef]
29. Jian, R.; Hu, X.; Han, T.; Wan, J.; Gan, W.; Chen, Z.; Zhang, Y.; Cao, C. Optimization of induction heating parameters for improving Self-healing performance of asphalt mixture through partial least square model. *Constr. Build. Mater.* **2023**, *365*, 130019. [CrossRef]
30. Wang, B.; Hu, X.; Cao, C.; Wan, J.; Gan, W.; Chen, Z.; Cai, C. Development and characterization of permeative anti-rutting agent for asphalt mixture enhancement. *Constr. Build. Mater.* **2023**, *364*, 129937. [CrossRef]
31. Tian, Y.; Sun, L.; Li, H.; Zhang, H.; Harvey, J.; Yang, B.; Zhu, Y.; Yu, B.; Fu, K. Laboratory investigation on effects of solid waste filler on mechanical properties of porous asphalt mixture. *Constr. Build. Mater.* **2021**, *279*, 122436. [CrossRef]

materials

Article

Influence of Filler Type and Rheological Properties of Asphalt Mastic on the Asphalt Mastic–Aggregate Interaction

Guangxun E [1], Jizhe Zhang [2,*], Quanjun Shen [1], Ping Ji [3], Jing Wang [2] and Yushuai Xiao [2]

[1] Shandong Key Laboratory of Highway Technology and Safety Assessment, Shandong Hi-Speed Group Co., Ltd., Jinan 250101, China
[2] School of Qilu Transportation, Shandong University, Jinan 250100, China
[3] Shandong Hi-Speed Engineering Consulting Group Co., Ltd., Jinan 250100, China
* Correspondence: jizhe.zhang@sdu.edu.cn

Abstract: The asphalt mastic–aggregate interaction plays an important role in the overall properties of asphalt mixtures and their durability in service in flexible pavements. This paper aims to study the influence of the physico-chemical features of fillers and the rheological properties of asphalt mastics on the bonding behavior between asphalt and aggregate, and the interfacial deterioration mechanism when subjected to static water immersion and pressured water immersion. It was found that the filler type (limestone powder, basalt powder, and granite powder) had a certain influence on the complex modulus of asphalt mastics, and its pore volume and specific surface area had significant effects on the phase angles and permeability of asphalt mastics. The effect of water pressure can accelerate the deterioration of bond strength of the asphalt mastic–aggregate interface in the short term, indicating that the dynamic water pressure generated by the driving load promotes the water damage process in asphalt pavements. In comparison, the residual bond strength ratio of the granite–asphalt mastic aggregate was the highest, while its bond strength was lower than that of the interface between limestone–asphalt mastics and limestone aggregate. This demonstrated that a low asphalt mastic complex modulus and a high phase angle are helpful in improving the durability of asphalt mixtures subjected to static and pressured water immersion conditions.

Keywords: filler; asphalt mastic; interfacial bond strength; asphalt–aggregate interaction; moisture damage

Citation: E, G.; Zhang, J.; Shen, Q.; Ji, P.; Wang, J.; Xiao, Y. Influence of Filler Type and Rheological Properties of Asphalt Mastic on the Asphalt Mastic–Aggregate Interaction. *Materials* **2023**, *16*, 574. https://doi.org/10.3390/ma16020574

Academic Editor: Simon Hesp

Received: 18 November 2022
Revised: 19 December 2022
Accepted: 28 December 2022
Published: 6 January 2023

1. Introduction

Owing to the advantages of smooth surfaces, driving comfort, low noise, and easy maintenance, flexible asphalt pavements have become the main pavement type for high-grade highways in most countries [1,2]. The asphalt mixture used in flexible asphalt pavements is a multiphase composite, which is composed of asphalt binder, filler, voids, aggregates of different sizes, and the asphalt–aggregate interface [3]. Therefore, it is normally considered to be a heterogeneous material and its properties can be strongly influenced by different material variables and their proportions.

In previous research, pavement engineers and researchers recognized that the asphalt mastic and the asphalt mastic–aggregate interface strongly determine the overall performance of the asphalt mixtures and their durability in service life [4]. The asphalt mastic in the asphalt mixture is the cementing component and consists of bitumen and fine aggregate, and its composition and properties directly affect the road performance of asphalt mixtures. The physico-chemical properties of the mineral fillers incorporated have a significant impact on the performance of asphalt mastics. Zhang et al. used oxide analytical reagents to represent mineral aggregate fillers and to study their effects on the properties of asphalt mastics and the interfaces. It was found that the effects of calcium oxide (CaO) were greater than those of silicon dioxide (SiO_2) due to the stronger interaction between asphalt binder and CaO [5]. Based on the research on the mesoscopic properties of mineral fillers, Lv et al. found that physical properties such as mineral filler fineness and mesoscopic gradation had a certain correlation with the mesoscopic strength of asphalt mastics [6]. Barra et al. analyzed the influence of the properties of limestone mineral powder and granite mineral powder on the softening point, penetration, and adhesion of asphalt mastic [7]. White studied the shear creep response of an airport asphalt mastic and the results indicated that two types of asphalt mastics had different properties due to different dusts [8,9]. Compared with the size of the filler, its shape, surface texture, specific surface area, and mineral components had more significant effects on the properties. Based on the microstructure of fillers, Geber et al. analyzed the effects of the particle size distribution, microscopic morphology, mercury intrusion porosity, specific surface area, and hydrophobicity of limestone powder and dolomite powder on the rheological properties of asphalt mastics [10]. The particle size, hydrophobicity, and content of fillers have significant effects on the rheological properties of asphalt mastics.

In recent years, more studies have also been carried out to investigate the adhesion of the asphalt mastic–aggregate interface. In general, four mechanisms have been adopted to explain the adhesion between asphalt mastic and mineral aggregate: surface energy, chemical reaction, molecular orientation, and mechanical contact [11]. Yi et al. studied the influence of different factors on the adhesion of the asphalt–aggregate interaction based on the surface energy theory and showed that the measured surface energy using an AFM method can represent the surface characteristics of materials [12]. Based on the molecular orientation theory, Huang et al. found that the fundamental cause of adsorption of asphalt on the aggregate surface was the polarity of the asphalt and aggregate and that the compositions of asphaltene and gelatine greatly determined adhesion [13]. Using digital imaging techniques and an abrasion method, Kuang et al. studied the macroscopic changes on the particle surface of limestone and granite before and after treatment and their influences on interfacial adhesion between asphalt and limestone/granite. It was found that the interfacial bonding between asphalt/mastic and aggregate in asphalt mixtures is mainly attributed to the micro/meso-physical and chemical actions in the vicinity of the interface [14].

Most importantly, factors such as water, temperature, environmental conditions, and other incorporated recycled waste, such as waste oil and industrial waste powder, also have important effects on the rheology and durability of asphalt mixtures [15–17]. Because, in practice, asphalt pavements have a long service life, they suffer from different degrees of defects. Water/moisture damage is a common early defect that can lead to the loosening and peeling of aggregates, which seriously affects the pavements service performance and shortens its service life [17–19]. Under the dual action of water and driving traffic load, the asphalt mastic–aggregate interface may become the weakest zone of the asphalt mixture, even after initially exhibiting a good interfacial bond strength.

The stripping potential of asphalt mixtures is usually evaluated based on the cohesion bond between the binder and aggregates. It is often caused by the loss of the mastic–aggregate bond and results in poor durability of asphalt mixtures. This study, therefore, aimed to experimentally assess the correlation between the physico-chemical features of fillers, the rheological properties of asphalt mastics, and the asphalt mastic–aggregate interaction before and after static and dynamic water attack. Three types of fillers and aggregates that are normally used in asphalt mixtures (limestone, basalt, and granite) were chosen and characterized. The influence of fillers on the rheological properties of asphalt mastics was characterized using a dynamic shear rheometer (DSR). The direct bond strength of the asphalt–aggregate interface as a key property indicator was measured and evaluated. The water attack testing was conducted by increasing the water pressure. Understanding the bond strength development and the deterioration of the asphalt mastic–aggregate interface will help to provide scientific guidelines for the design of asphalt mixtures.

2. Materials and Experimental

2.1. Materials

2.1.1. Bitumen

The bitumen used as a binder in the asphalt mixtures was a type of AH-70# road petroleum (produced by a local company in Shandong, China). It was an unmodified type with a penetration value of 60–70 (according to ASTM D5, it was 60/70 grade). Its properties, which were in accordance with Chinese test methods, are listed in Table 1.

Table 1. Properties of AH-70 bitumen.

Items	Results	Unit	Requirement
Needle penetration (25 °C)	68.3	0.1 mm	60–80
Ductility (10 °C)	40.1	cm	no less than 25
Ductility (15 °C)	>150	cm	no less than 100
Softening point	48.2	°C	no less than 45
Viscosity (135 °C)	0.450	Pa·s	-
Density (15 °C)	1.035	g/cm^3	-
Solubility	99.6	%	no less than 99.5

2.1.2. Filler

The mineral powders used as fillers in the asphalt mixtures in this study were limestone mineral powder (LP), basalt mineral powder (BP), and granite mineral powder (GP), which were sourced from these three types of minerals. These mineral particles, which were less than 4.75 mm in size, were ground in a laboratory mill for 2 min and then passed through a 0.075 mm sieve to obtain powder. The densities of the three kinds of mineral powder were measured using a pycnometer (Chinese standard, T0352) and the results are listed in Table 2.

Table 2. Relative densities of mineral powders.

Property	LP	BP	GP
Density/g/cm^3	2.67	2.87	2.69

The particle size distributions of these three fillers, measured using a laser diffraction technique (Tester of LS230), are shown in Table 3. D_{10}, D_{50}, and D_{90} represent the minimum particle size with a pass rate of 10%, 50%, and 90%, respectively. It was found that the D_{10}, D_{50}, and D_{90} of LP exhibited a relatively smaller size than those of the other BP and GP fillers. This indicates that LP was finer than the others, and BP and GP had similar sizes.

Table 3. Particle size distributions of three mineral powders.

Items	LP	BP	GP
D_{10}	2.75	3.27	3.89
D_{50}	13.08	26.17	26.16
D_{90}	52.33	62.33	62.23

The mineral composition of the mineral powder has an important influence on its chemical reaction with bitumen [20]. Table 4 shows the oxide chemical components of the three mineral fillers measured by X-ray fluorescence (XRF) (Rigaku, Tokyo, Japan, Supermini200). It can be seen that the chemical composition of the three mineral powders was significantly different. The main component of LP was CaO and its content was about 83% by mass, while the content of SiO_2 was the lowest (5.8%) compared to the others. The LP is, therefore, considered as an alkaline mineral. As regards the basalt powder, its main chemical components were SiO_2, Al_2O_3, Fe_2O_3, and CaO, and the SiO_2 content was 46% (mass fraction), which makes it a neutral mineral powder. The main components of the granite powder were SiO_2 and Al_2O_3, and its SiO_2 content was 64% (by mass). Thus, it is regarded as an acidic mineral powder. These chemical components cause the powders to have noticeably different chemical reactions with bitumen.

Table 4. Oxide composition of mineral fillers (%).

Oxide	Filler Type		
	LP	BP	GP
CaO	82.828	9.879	2.877
SiO_2	5.843	45.812	63.87
Al_2O_3	4.15	18.552	16.854
Fe_2O_3	0.573	11.656	3.116
MgO	4.786	5.954	0.692
K_2O	0.351	1.893	5.252
Na_2O	/	2.581	6.023
Other	1.469	3.673	1.316

In order to detect the mineral crystals in the three kinds of fillers and then to identify their influence on the performance of asphalt mastics and interfaces, their diffraction patterns were obtained by XRD (Rigaku SmartLab), and the mineral crystal analysis was carried out using the HighScore Plus software. The results are shown in Figure 1. It was found that (1) the main mineral crystalline components of LP were calcite ($CaCO_3$), dolomite ($CaMg(CO_3)_2$), and a small amount of quartz (SiO_2). Its acidic polar group forms a stable double bond structure, and the chemical bond is not easily destroyed when exposed to water. (2) The main mineral components of BP were olivine ($Mg_2(SiO_4)$) and pyroxene ($Ca(Mg,Fe)Si_3O_6$). (3) The main mineral components of GP were quartz (SiO_2), albite ($Na[AlSi_3O_8]$), and potassium feldspar ($K[AlSi_3O_8]$). The acidic quartz is normally weakly bonded with bitumen and does not easily form a strong bonding interface.

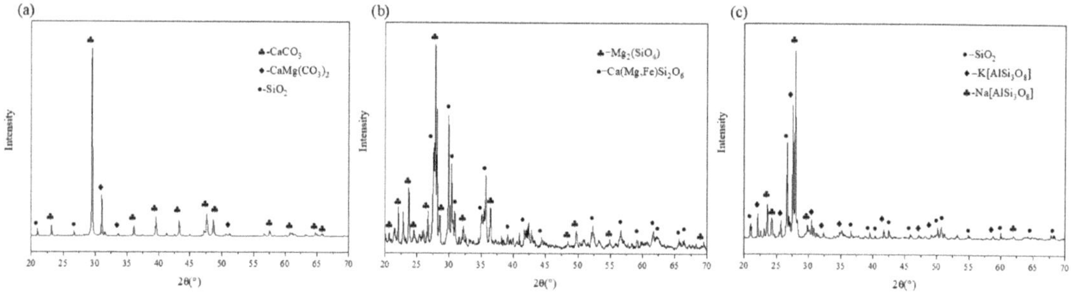

Figure 1. XRD diffractions of mineral fillers: (**a**) limestone powder; (**b**) basalt powder; (**c**) granite powder.

2.1.3. Aggregates

The cylindrical aggregate samples, made of limestone, basalt, and granite, were prepared to a size of Φ20 × 20 mm. First, a drill with an inner diameter of 20 mm was used to take a core of a large volume of the aggregate mineral, and then a high-precision double-sided cutting machine was used to cut the cylindrical core sample, ensuring a height of 20 mm. Thereafter, both surfaces of the sample were polished with sandpaper to ensure a similar texture. Finally, the aggregate samples were washed in boiling water at 100 °C and dried. Their physical properties are listed in Table 5. The technical indicators of the three aggregates met the requirements of the Chinese Specification JTG F40-2004 "Technical Specification for Highway Asphalt Pavement Construction".

Table 5. Basic performance index of aggregates.

Properties	Limestone	Basalt	Granite	Requirement
Density/g/cm^3	2.816	3.111	3.070	no less than 2.6
Crushing value/%	23.1	10.2	19.8	no less than 26
Abrasion value/%	20.2	16.6	14.3	no less than 28
Adhesion grade	5	5	3	-

2.2. Preparation of the Specimens

2.2.1. Design of Asphalt Mastics

In the study, asphalt mastic was defined and designed as a mixture of bitumen and calcareous or siliceous fillers. In order to minimize the effect of the mineral powder volume on the asphalt mastics, they were designed with a ratio of bitumen to filler = 1:1 by volume [21–23]. The filler was first dried at 150 °C for 3 h. Then, the bitumen was heated in a furnace at 150 °C for 1 h. After placing the filler in a vessel with a temperature of 150 °C, an enhanced stirring process was conducted to mix the bitumen and filler homogenously at a speed of 1000 rpm for 30 min. The three types of asphalt mastics (LM, BM, and GM) were then obtained in order to carry out the following tests.

2.2.2. Design of the Asphalt Mastic–Aggregate Interface Specimens

In order to quantitatively measure the adhesion of the asphalt mastic–aggregate interface, the asphalt mastic–aggregate interface specimens were prepared as shown in Figure 2. The mechanical failure tests were conducted by controlling the asphalt mastic thickness within 1 mm. The sample preparation processes included the following: (1) Installing a mold with the aggregates; (2) aligning the cylindrical aggregates (Φ20 × 20 mm) in self-made fixtures; (3) loading the asphalt mastic in between two pieces of aggregates; (4) controlling the thickness of the asphalt mastic within 1 mm; (5) demolding the asphalt–aggregate interface specimen; (6) removing the extra asphalt mastic around the interface. Then, the bond strength of the interface was tested using a testing machine. In this study, a 1 mm spacer was used to control the thickness of the asphalt mastic, and three types of mastics and three aggregates were used to prepare nine groups of asphalt mastic–aggregate interface specimens. The specific specimen designs and names are shown in Table 6.

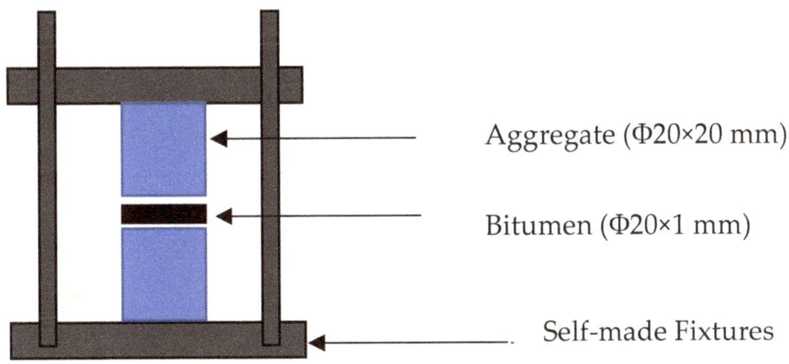

Figure 2. Schematic of the preparation of asphalt mastic–aggregate interface specimens.

Table 6. Specimen designs and names of asphalt mastic–aggregate interfaces.

Type of Interface	Name
Limestone mastic–Limestone	LM–L
Limestone mastic–Basalt	LM–B
Limestone mastic–Granite	LM–G
Basalt mastic–Limestone	BM–L
Basalt mastic–Basalt	BM–B
Basalt mastic–Granite	BM–G
Granite mastic–Limestone	GM–L
Granite mastic–Basalt	GM–B
Granite mastic–Granite	GM–G

2.3. *Experimental Tests*

2.3.1. Rheological Testing of Asphalt Mastics

- A dynamic shear rheometer (DSR) was employed in this study to evaluate the rheological properties of the asphalt mastics, focusing on the influence of fillers. In this test, mastic specimens with a 2 mm thickness were prepared and sandwiched between two flat plates with a diameter of 8 mm. One of the two plates was fixed and the other one was oscillated back and forth around the central axis at a certain angular velocity. The mastic specimens were tested at a temperature interval of 10 °C in the range of 30–60 °C and the scanning rate was 10 rad/s. According to the curves, the complex shear modulus G*, phase angle δ, and rutting factor (G*/sin δ) of the asphalt mastics at different temperatures were obtained. The results were employed to evaluate high-temperature performance of the asphalt mastics at moderate temperatures;
- The bending-beam rheometer (BBR) test was employed to characterize the low-temperature cracking resistance of the asphalt mastics. In this research, the BBR test was performed at −6 °C, −12 °C, and −18 °C. During testing, a constant load of 980 ± 50 mN was added in the middle of the mastic beam for 240 s. The deflection was automatically recorded in order to calculate the creep stiffness (S) and m-value (m). In order to resist thermal cracking at low temperatures, the creep stiffness (S) and the m-value must meet certain requirements.

2.3.2. Bond Strength Testing of the Asphalt Mastic–Aggregate Interface

As shown in Figure 3a–e, a universal testing machine with an accuracy of 1 N was employed to measure the maximum force using a deformation-controlled model. Its loading speed was 0.01 mm/s. The bond strength of the interface was then calculated by:

$$f_t = \frac{F}{S} \tag{1}$$

where F is the maximum failure force; S is the interfacial area; f_t is the bond strength of the interface.

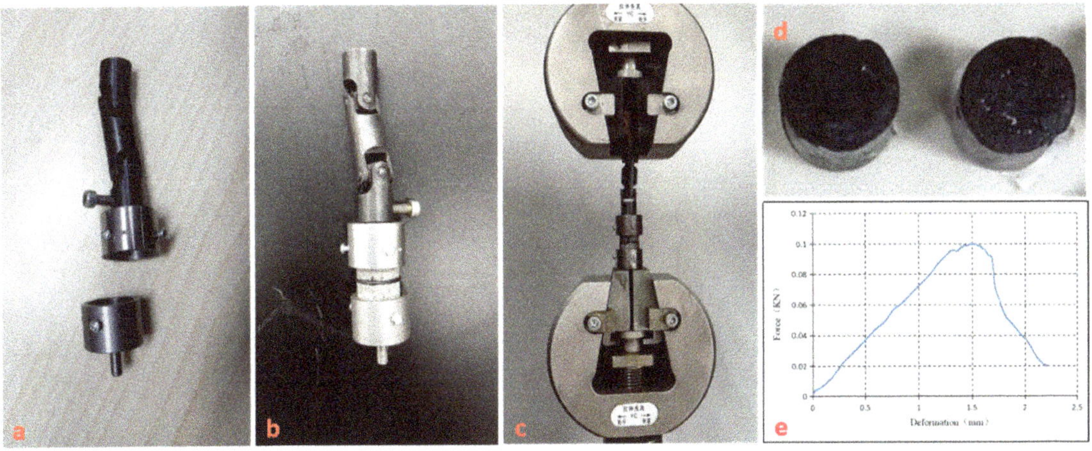

Figure 3. Testing of asphalt mastic–aggregate interface specimens: (**a**) holding fixture; (**b**) specimen fixed by holding fixture; (**c**) specimen fixed into universal testing machine; (**d**) specimen failure surface; (**e**) loading curve.

2.3.3. Water Absorption of Asphalt Mastics

In order to understand the diffusion behavior of water in the asphalt mastics under normal temperature and pressure, the moisture absorption rates of the asphalt mastics subjected to different water immersion periods were measured using a gravimetric method, and the change principle of the water contents in the asphalt mastics was analyzed by measuring the change in its mass against time [24]. The specific testing processes were given as follows: (1) A customized aluminum plate mold (the mass is m_0) was used to prepare an asphalt mastic film of 50 mm × 50 mm × 0.3 mm; (2) an analytical balance was used to weigh the aluminum plate and the asphalt mastic before water immersion (the mass is m_1); (3) the samples were immersed in distilled water and removed at regular intervals; (4) the water was wiped off with filter paper and the mass m_t of the asphalt mastic plus the aluminum plate was measured. The test results were recorded after soaking times of 1 h, 4 h, 12 h, and 24 h . . .

The moisture absorption rate (M_t) of the asphalt mastic at the time of immersion t was then calculated by:

$$M_t = \frac{m_t - m_1}{m_1 - m_0} \times 100\% \quad (2)$$

2.3.4. Water Attack Testing of the Asphalt Mastic–Aggregate Interface

In order to investigate the influence of different water conditions on the bond strength of the asphalt mastic–aggregate interface, the temperature (10~40 °C), static water immersion time (7 d, 14 d), and water pressure action time (12 h, 24 h) were assessed to test their influence on the mechanical properties of the interface. As calculated by the effect of standard axle load and average velocity, the variation range of the pore water pressure of the asphalt pavement surface layer is generally 0.20~0.57 MPa [25,26]. In this study, a self-designed pressure device was used to simulate the water pressure. Its pressure was 0.5 MPa.

3. Results and Discussion

3.1. Physical Features of Three Types of Filler

The Brunauer–Emmett–Teller (BET) method was applied to calculate the specific surface area of the fillers according to nitrogen adsorption isotherm measurements. Using the Barrett–Joyner–Halenda (BJH) model, the pore size distributions of the fillers were derived from the adsorption branches of the isotherms. The pore volume distributions of the mineral fillers as assessed by physical adsorption under high vacuum conditions (measured using Micromeritics ASAP2020 PLUS) are shown in Figure 4. The pore size distribution ranges of LP and GP were roughly similar, ranging from 4 to 20 nm, while the pore size distribution of BP was relatively wider, ranging from 4 to 40 nm. The BJH pore volume, the average pore size, and BET specific surface area of the fillers are listed in Table 7. It can be observed that the pore volumes of LP and GP were 0.007 cm^3/g and 0.009 cm^3/g, respectively, which were obviously lower than that of BP with 0.034 cm^3/g. The average pore sizes of BP and GP were similar, i.e., 3.810 nm and 3.819 nm, respectively, and the average pore size of LP was slightly smaller at 3.059 nm. According to the specific surface area calculated by BET, it can be seen that the specific surface area of LP was the smallest (1.899 m^2/g), while the specific surface area of BP was the largest (9.008 m^2/g). Moreover, the specific surface area of GP (3.42 m^2/g) was between LP and BP. As previously reported, the differences in the pore volume and specific surface area of the mineral powders certainly affect the selective absorption of the asphalt component according to the filler, and then affect the performance of the asphalt mastic and mixture [27].

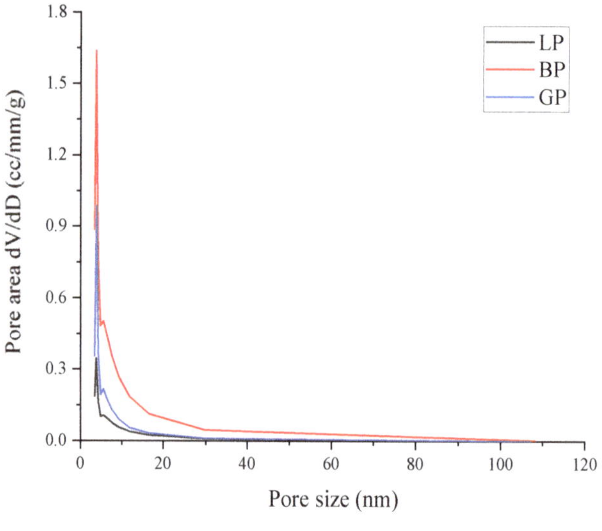

Figure 4. Pore size distribution of mineral powders by BET.

Table 7. Physical parameters of mineral powders.

Pore Features	LP	BP	GP
Pore volume/cm^3/g	0.007	0.034	0.009
Average pore size/nm	3.059	3.810	3.819
Surface area/m^2/g	1.899	9.008	3.42

The microscopic morphology of the mineral fillers directly affects the selective absorption of asphalt, resulting in changes in the rheological properties of asphalt mastics. The micromorphology of the fillers as assessed using SEM (Tescan, Brno, Czech Republic, Vega 3) is shown in Figure 5. It can be observed that the particle morphology and surface texture of the three mineral fillers were different. The LP particles were relatively smooth, without obvious edges and corners, and the particle size was relatively fine, with a small amount of fine flocculent particles attached to the surface of the coarse ones. BP had a relatively complex surface texture, a rough texture, a large number of holes, and small channels. The GP particles had a clear outline, more polygonal particles, obvious edges and corners, and no obvious holes in the particles. It can be hypothesized that BP absorbs light components of bitumen into the particles and this influences the rheological response.

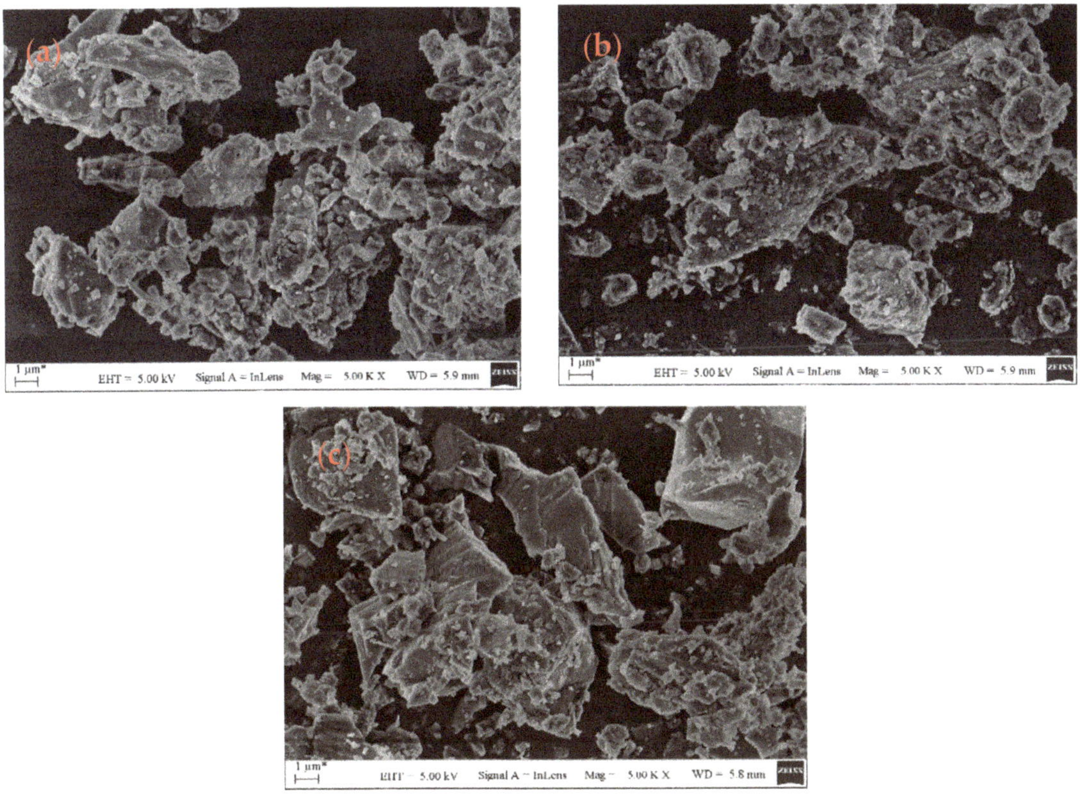

Figure 5. Morphology of mineral filler surface: (**a**) LP; (**b**) BP; and (**c**) GP.

3.2. Rheological Properties of Asphalt Mastics at High Temperature

A dynamic shear rheology tester (DSR) was employed to conduct a temperature sweep test in order to evaluate the rheological properties of the asphalt mastics. Their complex modulus and phase angle curves are shown in Figure 6. The complex modulus is a measure of the total resistance of a material when it is repeatedly sheared and deformed, and the higher its value, the stronger the ability of the asphalt mastic to resist deformation [28]. At the same temperature, the complex modulus of LM was the highest, indicating that its high-temperature deformation resistance is strong, which may be related to the particle size of LP and its chemical reaction with bitumen. The complex modulus of BM was second and that of GM was the lowest, which may be due to the difference in the pore volume and specific surface area. With the increase in temperature, the asphalt mastic changed from a viscoelastic state to a viscous fluid state, which increased the viscous components in the asphalt mastics and increased the phase angle. Under the same temperature, the phase angle of LM was the lowest, the phase angle of GM was the highest, and the phase angle of BM was in between. The LP had the smallest particle size and better dispersibility in the bitumen, resulting in LM exhibiting the lowest phase angle. The particle sizes of BP and GP were similar, but their specific surface areas and pore volumes were significantly different, and the pore volume of BP was three times that of GP. BM had the most structural bitumen and the least free bitumen, resulting in a highly elastic composition. The phase angle of BM was lower than that of GM. From these results, it can be seen that both the chemical composition and morphology of the fillers determine the rheological properties. First, the alkaline filler, LP, exhibited a very good chemical behavior with bitumen and this resulted in its high complex modulus, low phase angle, and even, smooth morphology. The porous features of BP enhance its complex modulus.

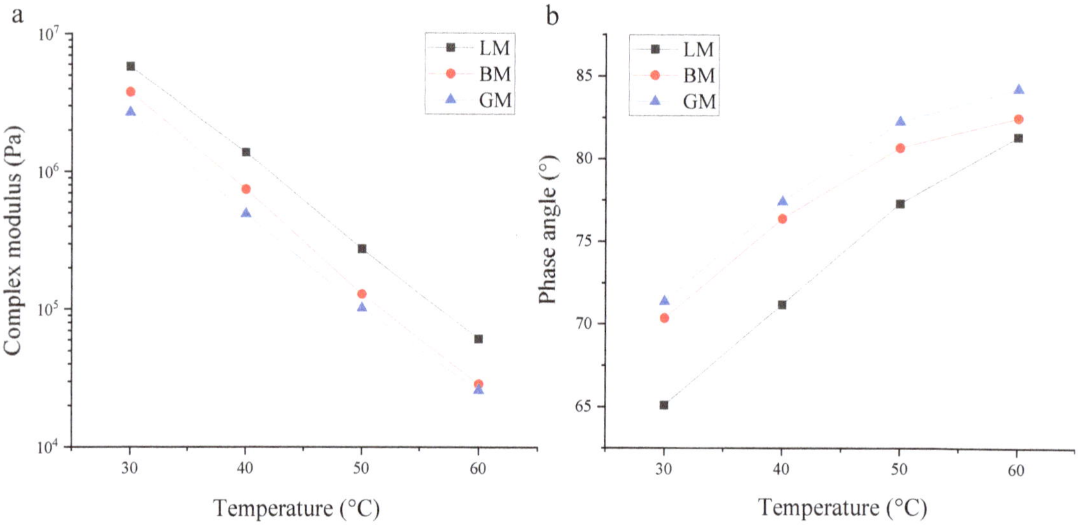

Figure 6. Rheological properties of asphalt mastics: (**a**) complex modulus and (**b**) phase angle.

The rutting factor G*/sinδ, which is commonly used in the Superpave specification, characterizes the long-term deformation ability of asphalt mastic, with a high value indicating a high ability to permanently deform. Figure 7 shows the rutting factors of the different types of asphalt mastics. At the same temperature, the order of the rutting factor was LM, BM, and GM, which was related to the particle size, pore volume, and surface area. The particle size of LP was the smallest, which made it more dispersed in bitumen, so LM had better resistance to permanent deformation. The specific surface area and pore volume of BP were three times those of the granite mineral powder, so that the permanent deformation resistance of BM was higher than that of GM.

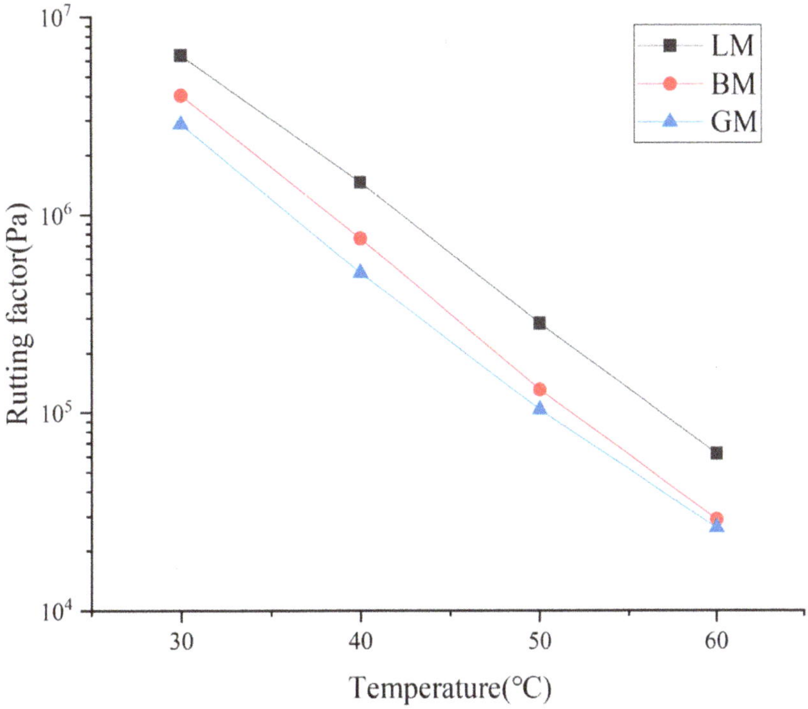

Figure 7. Rutting factors of asphalt mastics.

3.3. *Rheological Properties of Asphalt Mastics at Low Temperature*

The creep stiffness modulus S as assessed using the BBR characterizes the low-temperature performance of asphalt mastics. The higher the S value, the worse the low-temperature cracking resistance of the mastic. The creep rate m characterizes the change rate of the stiffness of the mastic with time, with high values indicating low deformation [29]. The results of the stiffness modulus S and m values of asphalt mastics are shown in Figure 8. It can be seen that the S of the three asphalt mastics decreased with the increase in the temperature, and the m increased with the increase in the temperature. This shows that, with the decrease in the temperature, the low-temperature cracking resistance of the three asphalt mastics reduced. The S and m values of the three mastics were not obviously different, indicating that the type of mineral filler had relatively less effect on the low-temperature performance of asphalt mastics. BP had the largest specific surface area, and it was shown to adsorb more light bitumen components. When the content of structural components increased, it resulted in the highest S value for BM at −18 °C and −12 °C. The slope m-value was introduced. A low slope value indicates a lower capacity to endure the stresses produced at low temperatures. In Figure 8b, BM at different temperatures (−6, −12, and −18) did not follow the same trend as LM and BM. This might be because of porous physical features of BP resulting in a different low-temperature resistance. The S value of LM was relatively low, the m value was high, and the cracking resistance was good, which was also related to its small specific surface area and pore volume.

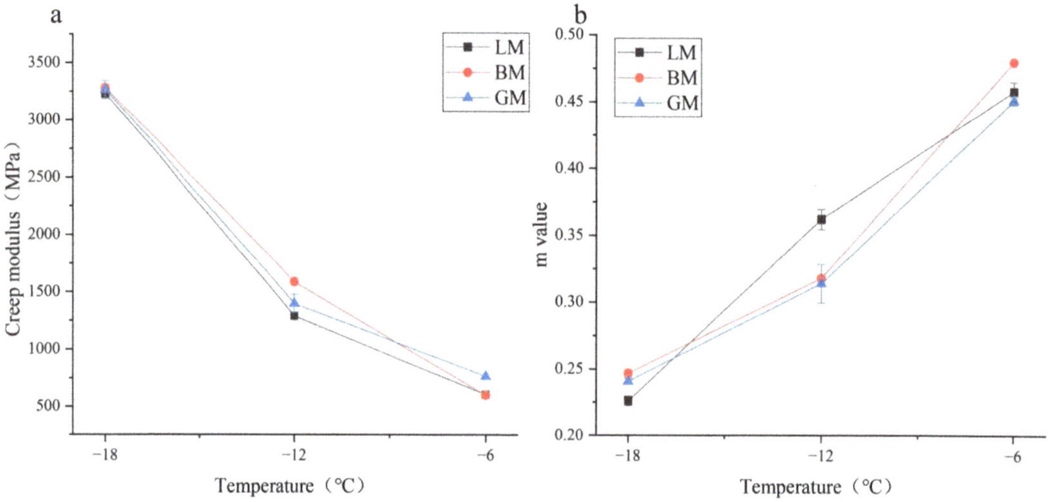

Figure 8. Rheological properties of asphalt mastics at low temperatures: (**a**) stiffness modulus and (**b**) creep rate.

3.4. Water Diffusion of Asphalt Mastics

In order to study the diffusion characteristics of water in asphalt mastics subjected to normal temperature and pressure, the moisture absorption curves of asphalt mastics under different immersion times were obtained using a gravimetric method. The results are shown in Figure 9. With the prolongation of the immersion time, water continuously diffused into the asphalt mastics and gradually became saturated. After soaking for 384 h, the change in the moisture absorption rate of the asphalt mastic tended to be gentle and essentially reached saturation. When exposed to the condition of water immersion, the moisture absorption rate of BM was the highest, followed by LM and GM. This could be related to the pore features of BP, i.e., a high pore volume and surface area.

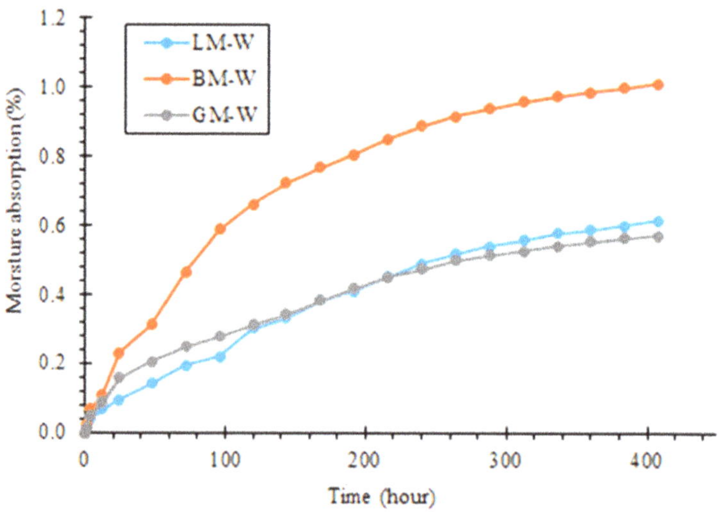

Figure 9. Moisture absorption curves of asphalt mastics.

The moisture absorption rate represents the moisture absorption characteristics of different asphalt mastics. To understand the diffusion properties of water in the asphalt mastic, the moisture absorption curves were fitted based on the Fick diffusion model using Equation (3) [30], and the diffusion coefficients of D are listed in Table 8. It can be seen from the fitting results under the same conditions that the order of diffusion coefficients of water in the three types of mastics was BM, LM, and GM. Because BP had the largest specific surface area and pore volume, and had a fluffy structure, many pores, and small channels in the particles, BM had the ability to absorb more water. LM and GM had similar diffusion coefficients.

$$\frac{M_t}{M_\infty} = 1 - \sum_{n=0}^{\infty} \frac{8}{(2n+1)^2\pi^2} e^{\frac{-D(2n+1)^2\pi^2 t}{l^2}} \tag{3}$$

where n is a natural number, D is the diffusion coefficient, l is the thickness of the sample, and $M\infty$ is the equilibrium moisture absorption rate.

Table 8. Diffusion coefficient of water in different asphalt mastics.

Type of Mastic	Condition	D/(cm^2/s)	R^2
LP mastic		1.054×10^{-10}	0.8396
BP mastic	Water Immersion	1.078×10^{-10}	0.8912
GP mastic		0.938×10^{-10}	0.8725

3.5. Bond Strength of the Asphalt Mastic–Aggregate Interface

3.5.1. Influence of Temperature

The bond strength results of the asphalt mastic–aggregate interfaces at different temperatures are shown in Figure 10. At 10 °C and 20 °C, the LM–L, LM–B, and LM–G interfaces had the highest bond strengths, compared to the other interfaces. For 30 °C and 40 °C, the BM–L, BM–B, and BM–G interfaces had the highest bond strengths. This shows that the interface bond strengths of the asphalt mastics were strongly determined by temperature and the mastic type. When increasing the temperature, the interface strength decreased gradually. The effect of temperature on the interface bond strength of the interfaces exhibited similar trends. Under 20 °C, the bond strength of the three interface combinations using LM was greater than that of BM and GM.

Fracture interfaces are illustrated in Figure 11. The failure modes of the asphalt mortar–aggregate interface were divided into two types: cohesion failure inside the asphalt mastic and adhesion failure at the asphalt mastic–aggregate interface. Under 20 °C, the failure mode was mainly the cohesion failure of the asphalt mastic–aggregate interface, indicating that the LM–aggregate interface had the best adhesion in the dry state and at a relatively low temperature. With the increase in temperature, the failure mode gradually changed from cohesion failure to adhesion failure. The bond strength of the specimen was mainly dominated by the adhesive strength of the asphalt mastic. The bond strength of the specimen prepared using BM was the highest, indicating that the adhesive strength of the BM system was higher, which was related to the larger pore volume, specific surface area, and surface morphology of BP. Moreover, the structural component of bitumen in the asphalt mastic can produce good adhesion with coarse aggregates.

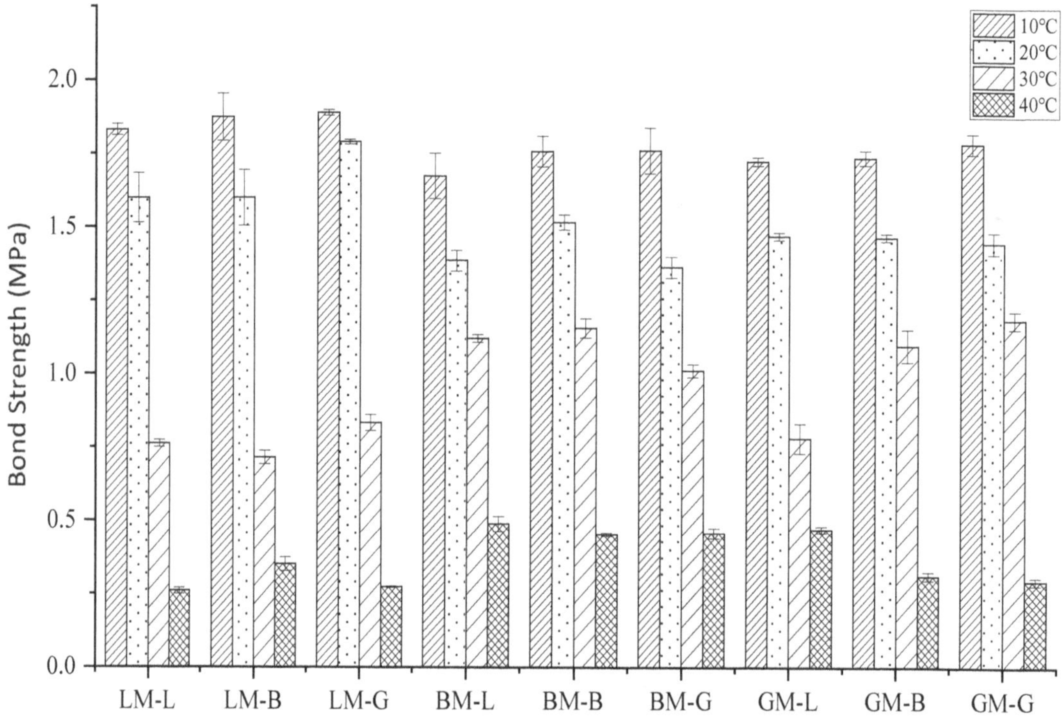

Figure 10. Bond strengths of asphalt mastic–aggregate interfaces at different temperatures.

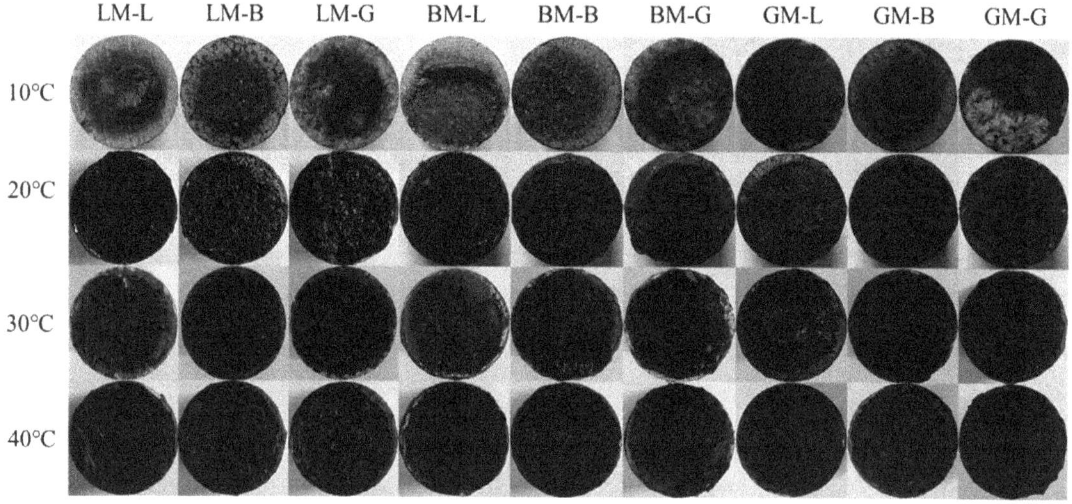

Figure 11. Failure modes of asphalt mortar–aggregate interfaces at different temperatures.

3.5.2. Influence of Water Immersion without Pressure

Figure 12 shows the residual bond strength ratios of asphalt mastic–aggregate interfaces under immersion in water for 7 days (W7) and 14 days (W14) without pressure. The residual bond strength was defined as the ratio between the bond strength before water immersion and after water exposure. It can be seen that when subjected to water immersion, all interfacial bond strengths decreased. This also gradually decreased with the extension of the immersion period from 7 days to 14 days. For the same immersion period, the residual bond strength ratios of BM–L, BM–B, and BM–G were between 62% and 76%, which were higher than those of the other mastic–aggregate interfaces. This may be attributed to the higher content of structural asphalt in BM, and the larger pore volume of basalt helps to produce stable mechanical interlocking with asphalt, while the difference in the diffusion coefficient between the aggregate and asphalt does not play a dominant role. For the same asphalt mastic, the mastic–granite aggregate interfaces had relatively lower residual bond strength ratios compared to other two types of aggregates. This is because the interface formed by the weakly acidic mineral components of the granite aggregate and the asphalt mastic peels off easily when exposed to water, resulting in a rapid attenuation of its strength. Moreover, it was noticed that after water immersion, the residual bond strength ratios of the LM–aggregate interfaces were lowest, indicating that the water attack was more serous in the LM–aggregate interfaces. The specimens prepared from LM did not obtain the ideal residual bond strength ratios, which may be affected by two factors: (1) The specimens prepared from LM (LM–L, LM–B, and LM–G) were in a dry state. Their bond strengths were the highest, resulting in the highest initial bond strength values; (2) the small pore volume and specific surface area of LP result in the mechanical interlocking with asphalt being weak, and thus, it can be eroded by water. Figure 13 shows the failure images of each asphalt mastic–aggregate interface when subjected to water immersion. It can be seen that the interface was completely damaged via adhesion failure, indicating that after immersion in water for 7 d and 14 d, the water can reach the asphalt mastic–aggregate interface and the interface is adhesively damaged.

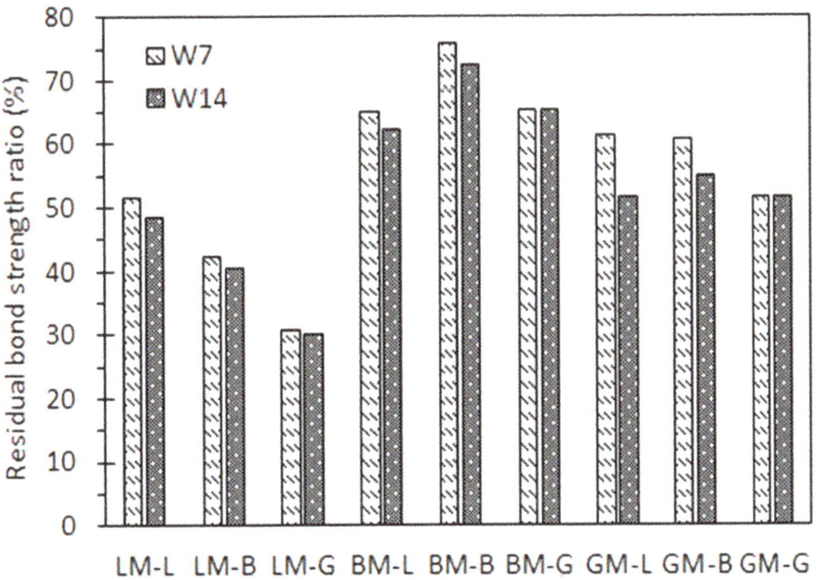

Figure 12. Residual bond strength ratios of asphalt mortar–aggregate interfaces under 7-day and 14-day water immersion.

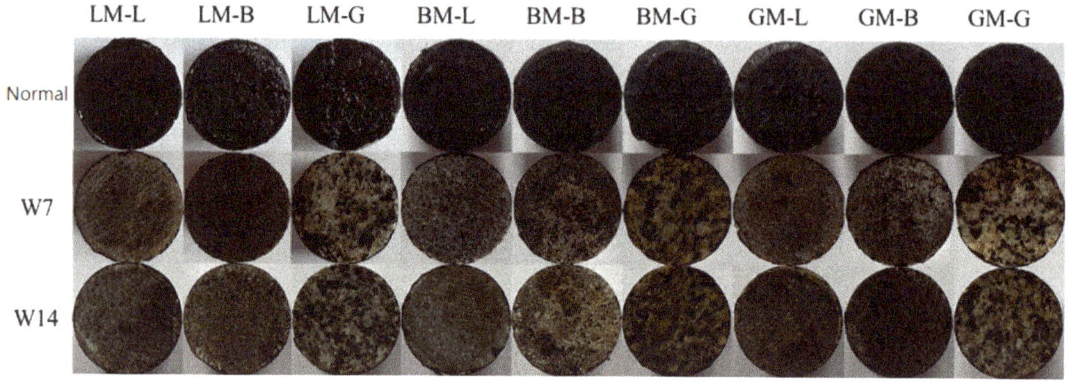

Figure 13. Failure of asphalt mastic–aggregate interface under water immersion.

3.5.3. Influence of Water Pressure

Figure 14 shows the residual bond strength ratios of the asphalt mastic–aggregate interfaces subjected to a water pressure of 0.5 MPa for 12 h and 24 h. It can be seen that with the prolongation of the water pressure action time, the residual bond strength ratio of each type of interface gradually decreased. For comparison, the residual bond strength of the GM–aggregate interfaces (GM–L, GM–B, and GM–G) was the highest, which may be due to the dominant effect of the low diffusion coefficient of GM under the action of water pressure. Figure 15 shows the interface failure images of each interface after water pressure for different times. It can be seen that, in the absence of water pressure (0 h), interface cohesion failure occurred. When water reached to the bonding interface with pressure (12 h and 24 h), interface failure developed from cohesion failure to adhesion failure.

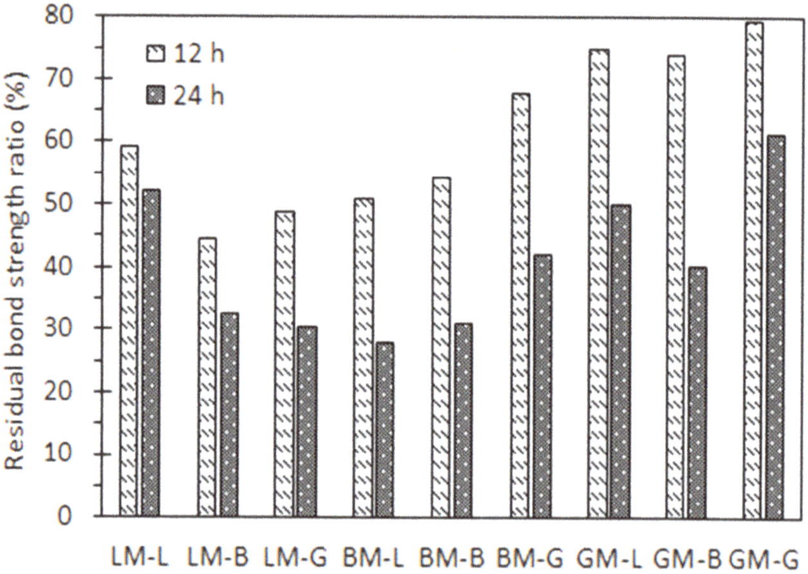

Figure 14. Residual bond strengths of asphalt mastic–aggregate interfaces under water pressure.

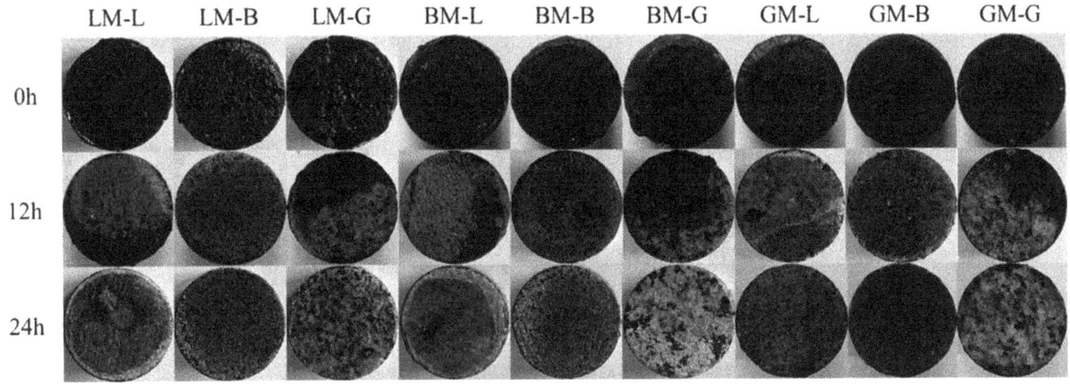

Figure 15. Failure of asphalt mortar–aggregate interfaces under water pressure.

It can be seen from Figures 12 and 14 that the residual bond strength ratios of the asphalt mastic–interfaces under water pressure for 24 h were even lower than those under static water immersion for 14 days. The water pressure can accelerate the attenuation of the asphalt mastic–aggregate interface bond strength, indicating that the hydrodynamic pressure generated by the traffic load can promote the water damage process of the asphalt pavement. This is also consistent with the conventional understanding that alkaline limestone has strong adhesion and acid granite has weak adhesion. Pressure immersion can accelerate the deterioration process of asphalt mastic–aggregate specimens as a result of water. However, by comparison, it was found that the attenuation law of the interfacial bond strength of the asphalt mastic–aggregate interface under water pressure was different under static water immersion. It can be seen that the water stability of the interface specimens was related to the properties of the aggregates and fillers and the diffusion behavior of the asphalt mastics.

4. Conclusions

The influence of the physico-chemical features of fillers and the rheological properties of asphalt mastics on the bonding behavior between asphalt and aggregate, and the interfacial deterioration mechanism subjected to static water immersion and pressured water immersion was tested experimentally and evaluated. The main findings are as follows:

- The mineral filler type influenced the complex modulus and low-temperature performance index of the asphalt mastic, and the difference in pore volume and specific surface area changed the content of the structural asphalt components in the asphalt mastics, thereby affecting the phase angle. The specific surface area of the basalt filler was the largest, resulting in a high content of structural asphalt components, and its mastic had a relatively higher stiffness modulus at −18 °C and −12 °C. The chemical composition of the filler was the primary factor in determining rheological behavior and the morphology was the secondary factor in influencing the properties of asphalt mastics;
- When exposed to the water immersion, the moisture absorption rate of the basalt mastic was the highest, followed by the limestone and granite mastics, and the granite mastic had the lowest diffusion coefficient. This may be related to the pore features of the basalt filler, which had a high pore volume and surface area;
- The alkaline limestone aggregate exhibited strong initial adhesion with bitumen and the acid granite aggregate exhibited weak adhesion. When the complex modulus of asphalt mastic was high and its phase angle was low, it resulted in a good initial bonding behavior with aggregate in the dry state. However, this does not indicate that such an interface between alkaline limestone aggregate and asphalt mastic exhibits good durability, e.g., against water attack;

- Static and pressured water immersion conditions can accelerate the deterioration process of asphalt mastic–aggregate interfaces. For the three asphalt mastics, when the complex modulus of the asphalt mastic was low and its phase angle was high, the durability of asphalt mixtures subjected to static and pressured water immersion conditions improved. The asphalt mastic with the acid granite filler initially exhibited relatively weak adhesion in the asphalt mastic, but it showed good water attack resistance between the asphalt mastic and coarse aggregate;
- The deterioration mechanism of the interfacial bond strength of the asphalt mastic–aggregate interface under static water immersion was different from pressured water immersion. It was found that the water stability of the asphalt mastic–aggregate interface was strongly related to the properties of the aggregates and fillers and the diffusion behavior of the asphalt mastics, which influenced the rheological properties of the asphalt mastics.

Author Contributions: Conceptualization, G.E. and Q.S.; methodology, P.J.; investigation, J.W.; writing—original draft preparation, Y.X. and J.Z.; writing—review and editing, J.Z. All authors have read and agreed to the published version of the manuscript.

Funding: This research was funded by Open Fund of Shandong Key Laboratory of Highway Technology and Safety Assessment.

Institutional Review Board Statement: Not applicable.

Informed Consent Statement: Not applicable.

Data Availability Statement: The data presented in this study are available on request from the corresponding author. The data are not publicly available due to confidentiality agreement.

Conflicts of Interest: The authors declare no conflict of interest.

References

1. Thives, L.P.; Ghisi, E. Asphalt mixtures emission and energy consumption: A review. *Renew. Sustain. Energy Rev.* **2017**, *72*, 473–484. [CrossRef]
2. Zhang, J.Z.; Liu, S.J.; Yao, Z.Y.; Wu, S.P.; Jiang, H.G.; Liang, M.; Qiao, Y.N. Environmental aspects and pavement properties of red mud waste as the replacement of mineral filler in asphalt mixture. *Constr. Build. Mater.* **2018**, *180*, 605–613. [CrossRef]
3. Zhu, X.Y.; Yang, Z.X.; Guo, X.M.; Chen, W.Q. Modulus prediction of asphalt concrete with imperfect bonding between aggregate-asphalt mastic. *Compos. Part B Eng.* **2011**, *42*, 1404–1411. [CrossRef]
4. Kakar, M.R.; Hamzah, M.O.; Valentin, J. A review on moisture damages of hot and warm mix asphalt and related investigations. *J. Clean. Prod.* **2015**, *99*, 39–58. [CrossRef]
5. Zhang, J.P.; Fan, Z.P.; Hu, D.L.; Hu, Z.; Pei, J.Z.; Kong, W.C. Evaluation of asphalt-aggregate interaction based on the rheological properties. *Int. J. Pavement Eng.* **2018**, *19*, 586–592. [CrossRef]
6. Lv, D.; Zheng, C.F.; Qin, Y.; Bi, H.; Li, K.Y.; Huang, J.F. Analysing the effects of the mesoscopic characteristics of mineral powder fillers on the cohesive strength of asphalt mortars at low temperatures. *Constr. Build. Mater.* **2014**, *65*, 330–337. [CrossRef]
7. Barra, B.; Breno, L.; Guerrero, Y.; Bernucci, L. Characterization of granite and limestone powders for use as fillers in bituminous mastics dosage. *An. Acad. Bras. Ciênc.* **2014**, *86*, 995–1002. [CrossRef]
8. White, G. Shear creep response of an airport asphalt mastic. *Int. J. Pavement Eng.* **2017**, *18*, 567–577. [CrossRef]
9. White, G. The contribution of asphalt mastic to shear resistance. In Proceedings of the 6th Eurasphalt & Eurobitume Congress, Prague, Czech Republic, 1–3 June 2016.
10. Geber, R.; Simon, A.; Kocserha, I.; Buzimov, A.; Buzimov, A.Y. Microstructural and rheological analysis of fillers and asphalt mastics. *J. Phys. Conf. Ser.* **2017**, *790*, 012009. [CrossRef]
11. Terrel, R.L.; Shutte, J.W. *Summary Report on Water Sensitivity*; Report No. SHRP-A/IR-89-003 for Strategic Highway Research Program; National Research Council: Washington, DC, USA, 1989.
12. Yi, J.Y.; Pang, X.Y.; Feng, D.C.; Pei, Z.S.; Xu, M.; Xie, S.N.; Huang, Y.D. Studies on surface energy of asphalt and aggregate at different scales and bonding property of asphalt-aggregate system. *Road Mater. Pavement Des.* **2018**, *19*, 1102–1125. [CrossRef]
13. Huang, M.; Zhang, H.L.; Gao, Y.; Wang, L. Study of diffusion characteristics of asphalt-aggregate interface with molecular dynamics simulation. *Int. J. Pavement Eng.* **2021**, *22*, 319–330. [CrossRef]
14. Little, D.N.; Allen, D.H.; Bhasin, A. *Modeling and Design of Flexible Pavements and Materials*; Springer: Berlin/Heidelberg, Germany, 2017.
15. Fakhri, M.; Javadi, S.; Sassani, A.; Torabi-Dizaji, M. Zinc Slag as a Partial or Total Replacement for Mineral Filler in Warm Mix Asphalt and Its Effects on Self-Healing Capacity and Performance Characteristics. *Materials* **2022**, *15*, 736. [CrossRef] [PubMed]

16. Fakhri, M.; Norouzi, M.A. Rheological and ageing properties of asphalt bio-binders containing lignin and waste engine oil. *Constr. Build. Mater.* **2022**, *321*, 126364. [CrossRef]
17. Mehrara, A.; Khodaii, A. A review of state of the art on stripping phenomenon in asphalt concrete. *Constr. Build. Mater.* **2013**, *38*, 423–442. [CrossRef]
18. Zhang, J.Z.; Sun, C.J.; Li, P.Z.; Liang, M.; Jiang, H.G.; Yao, Z.Y. Experimental study on rheological properties and moisture susceptibility of asphalt mastic containing red mud waste as a filler substitute. *Constr. Build. Mater.* **2019**, *211*, 159–166. [CrossRef]
19. Mansour, F.; Vahid, V. Effect of Liquid Nano material and hydrated lime in improving the moisture behaviour of HMA. *Transp. Res. Proc.* **2016**, *17*, 506–512. [CrossRef]
20. Chen, Y.; Xu, S.B.; Tebaldi, G.; Romeo, E. Role of mineral filler in asphalt mixture. *Road Mater. Pavement Des.* **2022**, *23*, 247–286. [CrossRef]
21. Antunes, V.; Freire, A.C.; Quaresma, L.; Micaelo, R. Influence of the geometrical and physical properties of filler in the filler-bitumen interaction. *Constr. Build. Mater.* **2015**, *76*, 322–329. [CrossRef]
22. Antunes, V.; Freire, A.C.; Quaresma, L.; Micaelo, R. Effect of the chemical composition of fillers in the filler-bitumen interaction. *Constr. Build. Mater.* **2016**, *104*, 85–91. [CrossRef]
23. Apeagyei, A.K.; Grenfell, J.R.A.; Airey, G.D. Evaluation of Moisture Sorption and Diffusion Characteristics of Asphalt Mastics Using Manual and Automated Gravimetric Sorption Techniques. *J. Mater. Civ. Eng.* **2014**, *26*, 04014045. [CrossRef]
24. Apeagyei, A.K.; Grenfell, J.R.A.; Airey, G.D. Influence of aggregate absorption and diffusion properties on moisture damage in asphalt mixtures. *Road Mater. Pavement Des.* **2015**, *16*, 404–422. [CrossRef]
25. Gao, J.Q.; Chen, H.; Ji, T.J.; Liu, H.Y. Measurement of dynamic hydraulic pressure in asphalt pavement using fiber Bragg grating. *Transducer Microsyst. Technol.* **2009**, *28*, 59–61.
26. Gao, J.Q.; Guo, C.C.; Liu, Y.T. Measurement of pore water pressure in asphalt pavement and its effects on permeability. *Measurement* **2015**, *62*, 81–87. [CrossRef]
27. Choudhary, J.; Kumar, B.; Gupta, A. Utilization of solid waste materials as alternative fillers in asphalt mixes: A review. *Constr. Build. Mater.* **2020**, *234*, 117271. [CrossRef]
28. Faheem, A.F.; Bahia, H.U. Modelling of Asphalt Mastic in Terms of Filler-Bitumen Interaction. *Road Mater. Pavement Des.* **2010**, *11*, 281–303. [CrossRef]
29. Zhang, J.Z.; Yao, Z.Y.; Wang, K.; Wang, F.; Jiang, H.G.; Liang, M.; Wei, J.C.; Airey, G. Sustainable utilization of bauxite residue (Red Mud) as a road material in pavements: A critical review. *Constr. Build. Mater.* **2021**, *270*, 121419. [CrossRef]
30. Vasconcelos, K.L.; Bhasin, A.; Little, D.N.; Lytton, R.L. Experimental Measurement of Water Diffusion through Fine Aggregate Mixtures. *J. Mater. Civ. Eng.* **2011**, *23*, 445–452. [CrossRef]

materials

MDPI

Article

Swelled Mechanism of Crumb Rubber and Technical Properties of Crumb Rubber Modified Bitumen

Hongbin Zhu [1], Min Zhang [2], Yuanyuan Li [1,*], Yingxue Zou [3], Anqi Chen [3,*], Fu Wang [1], Langrun Liu [1], Dengjun Gu [1] and Shaoyun Zhou [1]

1 School of Civil Engineering and Architecture, Wuhan Institute of Technology, Wuhan 430074, China
2 School of Civil and Architectural Engineering, Henan University, Kaifeng 475004, China
3 State Key Laboratory of Silicate Materials for Architectures, Wuhan University of Technology, Wuhan 430070, China
* Correspondence: liyy@wit.edu.cn (Y.L.); anqi.chen@whut.edu.cn (A.C.)

Abstract: Crumb rubber modified bitumen (CRMB) has excellent high-temperature performance and fatigue resistance, and is widely used in asphalt pavement to cope with increasing traffic axle load and changing climate. Under conventional preparation conditions, the swelling degree of CR can directly impact the comprehensive properties of CRMB; however, physical and chemical properties research on swelling crumb rubber (SCR) and crumb rubber recycled bitumen (CRRB) in CRMB is relatively lacking. In this paper, the working performance of CRMB and CRRB in high-temperature and low-temperature conditions were studied through physical and working performance testing of bitumen. The CR and SCR were tested by scanning electron microscope (SEM), Fourier transform infrared spectrometer (FTIR), gel permeation chromatography (GPC), and particle size distribution (PSD) tests to study the physicochemical behavior and microscopic effects before and after CR swelling. The results showed that CR dosage was in the range of 10%, 15%, and 20%, as well as that CR dosages have a positive effect on the high- and low-temperature performance, storage stability, and elastic recovery of bitumen. The high-temperature PG grades of bitumen were directly improved by four grades, and the elastic recovery rate increased by 339.9%. CR improved the ultra-low temperature crack resistance of bitumen. Due to the absorption of lighter components by CR, the relative content of the heavy component of bitumen increased; however, its low-temperature performance decreased significantly. After swelling, the CR particle size increased and the range became wider, the surface complexity of CR became higher, and the specific surface area was larger. At the same time, CR carried out the transformation process from large and medium molecules to small molecules. During the swelling process, a new benzene ring structure appeared in the CR, and the C–C bond and C–S bond of CR broke, forming part of the C=C bond.

Keywords: crumb rubber (CR); bitumen; swelled mechanism; light component

Citation: Zhu, H.; Zhang, M.; Li, Y.; Zou, Y.; Chen, A.; Wang, F.; Liu, L.; Gu, D.; Zhou, S. Swelled Mechanism of Crumb Rubber and Technical Properties of Crumb Rubber Modified Bitumen. *Materials* **2022**, *15*, 7987. https://doi.org/10.3390/ma15227987

Academic Editor: Giovanni Polacco

Received: 18 October 2022
Accepted: 9 November 2022
Published: 11 November 2022

1. Introduction

Crumb rubber (CR) is prepared from waste tires after preprocessing, grinding, separation, and screening [1]. It can be seen from the tire manufacturing process that the basic components of CR are natural rubber or synthetic rubber, sulfur, carbon black, metal reinforcement materials, plastic fibers, and other additives [2]. The use of CR in asphalt mixture pavement has greatly improved the overall performance of pavement; on the one hand, CR can extend the service life of the pavement and reduce the noise of vehicle driving, while on the other hand more waste tires can be consumed, helping to promote the use of resources while reducing the demand for natural bitumen [2,3]. The application of CR on the road provides a safe, friendly, and green solution for the green resource utilization of waste tires [4].

Based on the excellent performance of CRMB, the preparation process of CRMB has attracted great attention in the study of roadways. Improving the preparation process, adding stabilizer or compatibilizer, and pretreating CR are the three main ways of improving the high- and low-temperature performance and storage stability of CRMB [5]. In terms of improving the preparation process, Flanigan [6] concluded that when CR and bitumen were prepared at 260 °C, it was possible to produce modified bitumen with excellent storage stability and uniformity. Zanzotto [7] found that the preparation temperature was proportional to the solubility of CR in modified bitumen when preparing CRMB. Dong [8–10] summarized that when CRMB was prepared at a temperature below 230 °C, its low-temperature performance increased first and then decreased with increased mixing time, while the low-temperature performance of the modified bitumen prepared at 250 °C to 270 °C decreased with increased stirring time. In addition, they pointed out that the low-temperature performance of the modified bitumen sheared at 250 °C for 1 h was the best. In terms of adding modifiers, styrene–butadiene–styrene (SBS) and CR are usually used to react and cross-link bitumen in order to improve its comprehensive performance [11–13]; polyethylene terephthalate (PET) is another available bitumen modifier [14]. Of course, it would be better deepen the reemployment of plastic into bitumen [15]. Wang [16] performed photothermal aging and freeze–thaw aging tests on SBS/CR-modified bitumen, and showed its excellent high and low-temperature performance and environmental durability. Huang [17] used SBS and CR to double modify bitumen, with ordinary CRMB as the control group, and conducted multiple stress creep recovery tests (MSCR) and freeze–thaw split tests. The results showed that the composite modification could improve the water stability and low-temperature crack resistance of bitumen. In terms of the pre-activation of CR, common pretreatment technologies include microwave radiation [18,19], biochemical treatment [20,21], and strong acid and strong alkali solutions [22,23]. Liang [5] mixed the aromatic oil with CR and heated it by microwave; the results showed that the high-temperature stability of the dry mix asphalt mixture was significantly improved and that resistance to permanent deformation was improved after the CR was activated by microwave. After microwave activation, the surface activity and specific surface area of CR were increased. Kabir [21] used microorganisms to desulfurize CR, which improved the interaction between CR and bitumen and reduced the degree of segregation. Ma [1] concluded that a modifier composed of bio-oil or its derivatives has the advantages of safety, environmental friendliness, low cost, and convenience of materials, and is feasible to apply it to CRMB. Li [24] used NaOH solution to treat CR and concluded that NaOH solution could improve the compatibility of bitumen with CR by removing zinc stearate impurities from the CR surface. It can be seen that the pre-activation treatment of CR is of great significance to the performance of bitumen, and the mechanism of pretreatment and direct treatment are consistent.

Mastering the reaction mechanism of CR in bitumen is the theoretical basis for improving the CR modification effect and optimizing bitumen performance. The elasticity, hardness, strength, and aging resistance of natural CR are poor, and cannot meet the axle load requirements of automobiles; thus, the rubber needs to be vulcanized when manufacturing tires [25]. During vulcanization, CR changes from plastic compounded rubber to highly elastic and hard cross-linked CR, and linear macromolecules are cross-linked into three-dimensional network macromolecules, greatly improving the physical, mechanical, and chemical properties of CR [2]. It can be seen that while vulcanization has a significant and positive impact on the performance of tires, the vulcanization reaction is not conducive to the interaction between CR and bitumen. The reaction of CR in bitumen involves swelling and degradation [26]; its theoretical basis is swelling degradation theory, which is mostly accepted by researchers. This theory explains the physical swelling and chemical desulfurization of CR. First, CR absorbs the light components of bitumen and forms a gel film, and the volume of the CR expands, which is known as the swelling reaction. High temperature causes the cross-linked sulfur bond of CR to be broken, and achieving desulfurization and degradation [27]. The compatibility between CR and bitumen can be enhanced by activated CR. The swelling degree of CR is related to the swelling time, the swelling temperature, and the composition of the bitumen [28]. CR absorbs the light components of bitumen, and the surface of the CR particle produces new active groups and forms a layer of highly asphaltene gel film [29]. The distance between CR particles is reduced and the particles are more closely connected, forming a high-viscosity semi-solid continuum system; as such, CRMB has excellent rutting resistance, crack resistance, and water damage resistance [30]. During swelling, the structure of bitumen changes from a colloid to a sol–gel, which leads to the improved adhesion and durability of the bitumen, while its temperature sensitivity decreases. The light components of bitumen absorbed by CR reduce the cracking resistance and workability of bitumen [1], and can improve the degradation effect of CR to promote the activation of CR. High temperatures break the cross-linked sulfur bonds (the C–S bond and S–S bond) of CR, forming active groups [18]. The fracture of the cross-linked sulfur bond changes the original stable three-dimensional network macromolecular structure of CR into chain-like CR hydrocarbon molecules with a linear structure, meaning that CR has more interfaces for bitumen to bind or attach [2,9].

The swelling and degradation theory has provided later researchers with a better understanding of the reaction of CR in bitumen. Researchers have invested more research into the reaction of CR in bitumen and the performance of CRMB, achieving excellent research results. For example, when CRMB is prepared at above 220 °C, CRMB has good uniformity and low-temperature performance [8–10]. However, the high temperature of 220 °C has an irreversible negative impact on bitumen, especially the aging of bitumen caused by the volatilization of light components of bitumen [31]. At the same time, swelling and degradation reactions exist at this temperature, meaning that it is not possible to only study CRMB with swelling reaction. There are few comprehensive research cases investigating swelling crumb rubber (SCR) and crumb rubber recycled bitumen (CRRB) in CRMB, leading to limited research on the swelling mechanism of CR. Therefore, a stronger theoretical basis is urgently needed for research on SCR and CRRB.

To further study the swelling mechanism of CR and the performance of CRMB, we used CR as a bitumen modifier to prepare bitumen with good high-temperature performance; CR can produce a swelling reaction in bitumen, making it possible to obtain CRMB with good storage stability. In this study, CRMB with different dosages was prepared by the swelling–shear–swelling process. Separate CR and bitumen with 15% CRMB were used to obtain SCR and CRRB. The elastic recovery and storage stability of the bitumen, the modification effect of CR on the bitumen, and the influence of CR dosage on the working performance of CRMB were studied through the three parameters. Scanning electron microscopy (SEM), Fourier transform infrared spectrometry (FTIR), gel permeation chromatography (GPC), and the particle size distribution (PSD) test were used to study the physicochemical behavior and microscopic effect of CR before and after swelling. The ultimate goal of the study was to determine the physical and chemical properties of CR before and after swelling, the swelling mechanism of CR, and the physical and working properties of CRRB.

2. Materials and Experimental Method

2.1. Bitumen and CR

In this study, the petroleum bitumen involved (including the bitumen before modification) was 70# base bitumen, the technical properties of which are shown in Table 1. The CR used was 40-mesh with a particle size of 0.425 mm, which was supplied by Jiangsu Zhonghong Environmental Protection Technology Co., Ltd. (Wuxi, China).

Table 1. Physical performance of 70# base bitumen.

Testing Parameters	25 °C Penetration	5 °C Ductility	Softening Point (R & B)
70# base bitumen	69.40 dmm	11.70 cm	49.3 °C

2.2. Preparation of Test Samples

2.2.1. Preparation of CRMB

The 70# base bitumen was preheated to a flowing state at 135 °C. As the CR used was a 40-mesh powder, it was necessary to pre-mix the designed amount of CR (10%, 15%, and 20%) with the bitumen at low speed to prevent splashing during the high-speed shearing process. The roughly mixed modified bitumen was then placed in an oil bath at 180 °C to allow the CR to swell in the bitumen. After 0.5 h, the high-speed shear instrument was switched on ana high-speed shearing was performed at 4000 rpm for 0.5 h. After shearing, the CR was left in an oil bath at 180 °C for 0.5 h, allowing the CR to continue to swell in the bitumen. Finally, high-speed shearing was carried out at 4000 rpm for 5 min to produce CRMB.

2.2.2. Preparation of SCR and CRRB

The CR and bitumen of CRMB were separated in order to test the physicochemical differences before and after CR swelling and the differences in the properties of bitumen after CR action. The 15% CRMB with a temperature of 180 °C was filtered through a 400-mesh sieve, the bitumen attached to the CR was washed with trichloroethylene solution, and the CR was left for 24 h to allow the trichloroethylene to evaporate. Then, the CR was placed in an oven at 60 °C for 5 min to remove excess water, and the CR was removed and cooled to room temperature to obtain SCR. After the extracted bitumen was heated to 180 °C, CRRB was obtained by three filtrations of 15% CRMB at 180 °C with a 400-mesh filter. The purpose of using a 300-mesh filter was to retain the original size distribution of swelling CR to a greater extent and obtain CRRB with higher purity. Figure 1 shows the picture of CR and SCR; compared with the loose state of CR, SCR showed different degrees of agglomeration, and CR particles were sticky and bonded to each other.

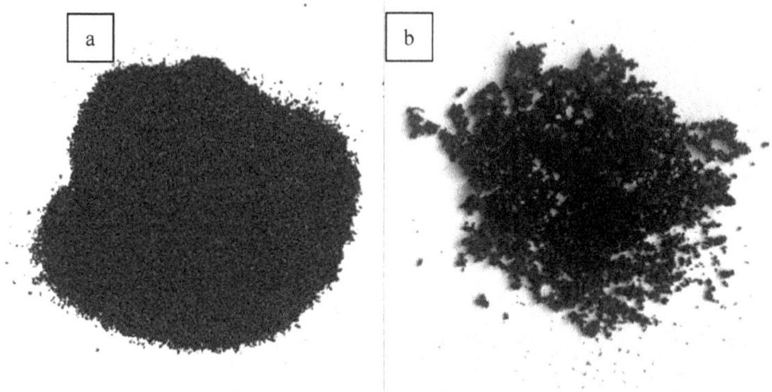

Figure 1. Macroimage of (**a**) CR and (**b**) SCR.

2.3. Testing of Bitumen

2.3.1. Physical Property and Viscoelastic Property Test

The physical, viscoelastic, and working properties of 70# base bitumen, CRMB, and CRRB were tested via the softening point test, the ductility test, the penetration test, the viscosity test, the elastic recovery test, and the segregation test, using two samples for each test. Among them, CRRB did not need the segregation test. The specific tests respectively reference standard test methods of bitumen and bituminous mixtures for highway engineering (JTG E20-2011): T0606-2011, T0605-2011, T0604-2011, T0625-2011, T 0662-2000, and T0661-2011.

2.3.2. High-Temperature Rheological Test

The high-temperature rheological properties of bitumen were detected using a dynamic shear rheometer (DSR), Smartpave102, Germany. Before the start of the test, the γ of the instrument was set to 12%, ω was 10 rad/s, and the test temperature ranged from 52 °C to 76 °C (the equipment limit, using linear fitting to find the failure temperature) by one point per degree. During testing, 1 g of 70# base bitumen, 10% CRMB, 15% CRMB, 20% CRMB, and CRRB were prepared and kneaded into spheres, and the instrument was started for testing; refer to T0628-2011 in JTG E20-2011 for the specific test process. The composite modulus (G*) and phase angle (δ)of bitumen can be used to evaluate the deformation resistance and viscoelasticity of bitumen at high temperatures. The rutting factor $G^*/\sin\delta$ can be calculated to further evaluate the rutting resistance of bitumen [32]. The failure temperature of bitumen is the value of the rutting factor $G^*/\sin\delta = 1.0$ kPa, through which the PG high-temperature grade of bitumen can be determined [33,34].

2.3.3. Low-Temperature Rheological Test

The creep stiffness (S value) and creep rate (m value) of 70# base bitumen, CRMB, and CRRB were measured by bending beam rheometer (BBR). Before the test, the test sample needs to be prepared and a mold is assembled after coating with the isolating agent. After cooling to room temperature, the sample was scraped flat with a hot scraper and frozen for 15 min before demolding. Two test samples were made for each bitumen; the size of the test samples was 125 mm × 12.7 mm × 6.35 mm. Refer to T0627-2011 in JTG E20-2011 for the specific test process. The samples were loaded with 980 mN at temperatures of −6 °C, −12 °C, and −18 °C successively. The S value and m value obtained by the BBR test were used to evaluate the deformation adaptability and low-temperature crack resistance, respectively. Because of thermal cracking, the S value must be less than or equal to 300 MPa and the m value must be greater than or equal to 0.300 [35].

2.4. *Testing of CR and SCR*

2.4.1. Particle Size Distribution Test

The particle size distribution of CR and SCR was measured by a laser particle size analyzer (Malvern Mastersizer 2000) and the difference in the particle size distribution before and after CR swelling was studied. The optical parameters were set before the test when the dispersing medium water was circulating normally, then the agitator was turned on and the CR sample was added to start the test. Absolute ethanol was used as a dispersant. Each sample was tested three times, and the particle size distributions of CR and SCR were tested in the range of 1 nm to 10,000 nm.

2.4.2. Micromorphology Test

Scanning electron microscopy (SEM) was used to analyze the micromorphology of CR and SCR. The instrument model was a Czech TESCAN MIRA LMS. Before the test, CR and SCR were glued to the conductive adhesive. To obtain the electrical conductivity of the samples, they were sprayed with 10 mA gold using an Oxford Quorum SC7620 gold-spraying instrument. CR and SCR with electrical conductivity were placed into the sample bin of the SEM and vacuuming steps were carried out. When the vacuum degree that could be tested was reached, vacuuming was stopped and the microscopic morphology of the CR and SCR samples was taken. The accelerating voltage was 3 kV, and images with multiple of 1000× and 5000× were taken.

2.4.3. Relative Molecular Weight Test

The relative molecular weights of CR and SCR were measured using gel permeation chromatography (GPC), on an Agilent PL-GPC50 consisting of two parts: a Waters 1515 high-pressure liquid chromatography (HPLC) pump and a Waters 2414 refractive index (RI) detector. Before the test, about 20 mg of CR sample was placed in a 10 mL volumetric flask and CR was dissolved in 10 mL mobile phase solvents, specifically, tetrahydrofuran (THF), for 24 h. The sample was filtered with a 0.45 mm Polytetrafluoroethylene (PTFE) filter; the concentration of the test sample was required to be 2.0 mg/mL. During the test, the mobile phase sample was passed through an HPLC pump and pumped into the column at a certain flow rate. The column was kept at 35 °C and the mobile phase flow rate was 1.0 mL/min. The GPC curves of CR and SCR were divided into thirteen regions according to the retention time, in which the combined five left regions are macromolecular regions (LMS), the combined middle four regions are middle molecular regions (MMS), and the combined right four regions are small molecular regions (SMS) [36].

2.4.4. Chemical Structure Test

The surface functional groups of CR and SCR were analyzed by Fourier transform infrared spectrometry (FTIR), and the chemical structure of CR before and after swelling was studied. The instrument was a Thermo Scientific Nicolet iS20. During the test, CR and potassium bromide were mixed and ground in a mortar. The resolution of the instrument was set to 4 cm^{-1}, the scanning times was 32, and the transmittance of the wavenumbers between 400 cm^{-1} and 4000 cm^{-1} was detected.

2.5. Technical Map

Figure 2 shows the technical map of this study.

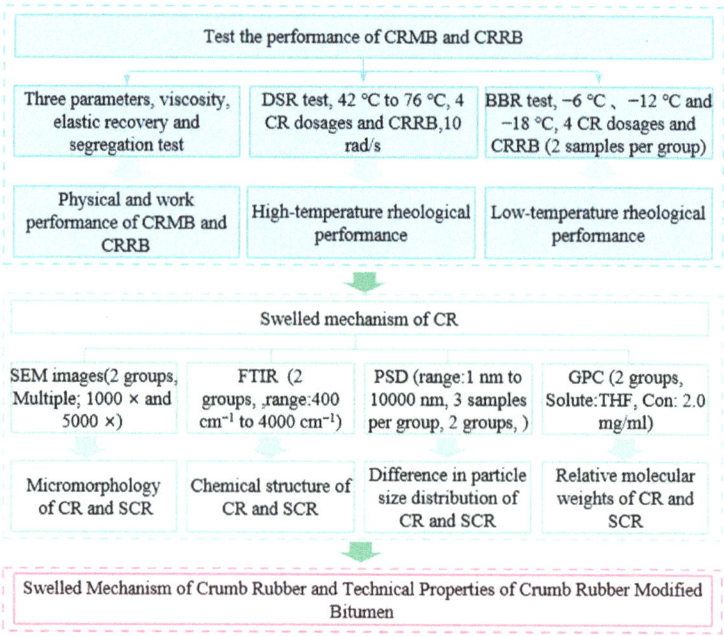

Figure 2. Technical map.

3. Results and Discussion

3.1. Technical Performance of CRMB and CRRB

3.1.1. Physical Property

Figure 3 shows the results of the three parameters for 70# base bitumen, 10% CRMB, 15% CRMB, 20% CRMB, and CRRB. Figure 3a shows that the mixing of bitumen and CR led to a decrease in the penetration of bitumen. With the continuous increase of CR dosage, the penetration of bitumen decreased continuously. Figure 3b shows that the mixture of bitumen and CR led to an increase in the softening point of bitumen. With continuous increase of the dosage of CR, the softening point increased continuously. Figure 3c shows that the mixing of bitumen and 10% CR led to a decrease in the ductility of bitumen; however, with the continuous increase of CR dosage (10% to 20%), the ductility of bitumen increased. The mixing of bitumen and CR had a great influence on the penetration of bitumen, and a relatively small effect on their respective softening points and ductility. The penetration of bitumen decreased and the softening point increased, indicating that its high-temperature stability improved and its ductility increased, which means that its low-temperature crack resistance was improved [37]. It can be seen that CR promoted the high-temperature performance of bitumen, and that the dosage of CR was positively correlated with the low-temperature performance of bitumen. While this has disadvantages in terms of the low-temperature performance of bitumen, the basic parameters of bitumen conform to the corresponding specifications. The experimental results with respect to the physical properties show that even after removing the particle effect of CR, CRRB had good high-temperature performance. Its high-temperature performance was close to that of 10% CRMB; however, its low-temperature crack resistance was poor, and there were obvious faults with the other four kinds of bitumen. There are two main reasons for this. One is that the light components of bitumen were volatilized, meaning that the relative content of the heavy component of bitumen increased during high-temperature stirring [31]. Second, during the swelling process, CR absorbed the light components of bitumen, the free wax content of bitumen decreased, the oil content decreased, and the relative content of the heavy component of bitumen increased [1]. However, the content of light components of bitumen decreased and the content of heavy component increased, increasing the high-temperature performance and low-temperature performance of the bitumen [38,39]. Therefore, while CRRB had good high-temperature performance, it had poor low-temperature crack resistance.

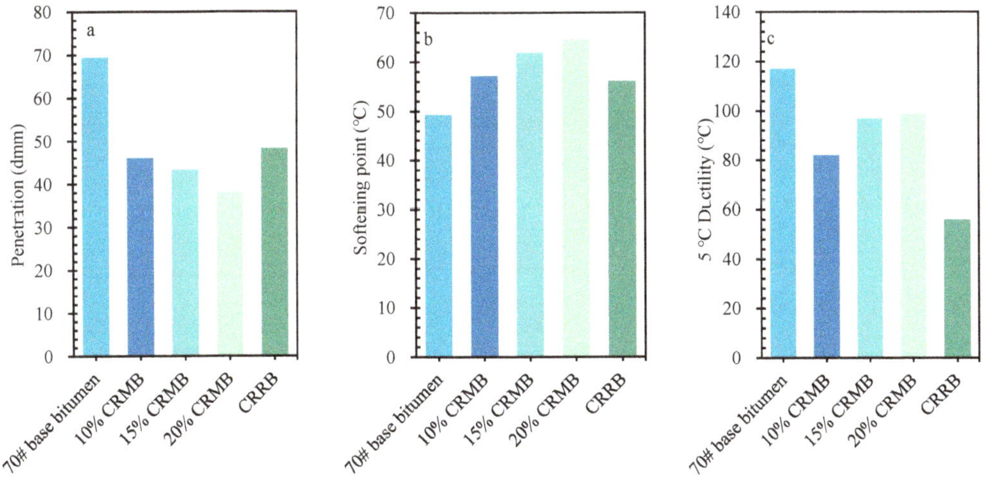

Figure 3. Physical properties of bitumen: (**a**) penetration; (**b**) softening point; (**c**) ductility.

3.1.2. High-Temperature Rheological Property

Figure 4 shows the G* and δ curves, which are used to study the shear deformation resistance and viscoelasticity of bitumen at high temperatures. Bitumen is a viscoelastic material; due to the relationship between stress and strain, it experiences a hysteresis effect. The closer the δ value is to 90°, the closer the material is to a viscous material, while the closer the δ value is to 0°, the closer the material is to an elastic material [40]. The characteristics of viscoelastic properties exhibited by bitumen at different temperature conditions are determined by the definitions of G* and δ. In the range of 52 °C to 76 °C, the G* values of all five kinds of bitumen decreased continuously with the increase in temperature, while the δ increased with the increase in temperature. This phenomenon indicates that during the process of heating the bitumen gradually becomes soft, its ability to resist shear deformation gradually decreases, and it changes from elastic to viscous. Relative to 70# base bitumen, the G* value of CRMB gradually increased and the δ value gradually decreased with increasing CR dosage, indicating that CR can improve the high-temperature shear resistance of bitumen, increasing the elasticity and decreasing the viscosity. CRRB had good high-temperature performance, close to that of 10% CRMB. These results show that the shear deformation resistance and elasticity of the five kinds of bitumen were negatively correlated with temperature, while the viscosity was positively correlated with temperature under the same CR dosage condition. At the same temperature, the shear deformation resistance and elasticity of bitumen were positively correlated with CR dosage while the viscosity was negatively correlated with CR dosage. Thus, at higher dosages it was easier for the bitumen to exhibit elastic properties, and the high-temperature shear deformation resistance was better.

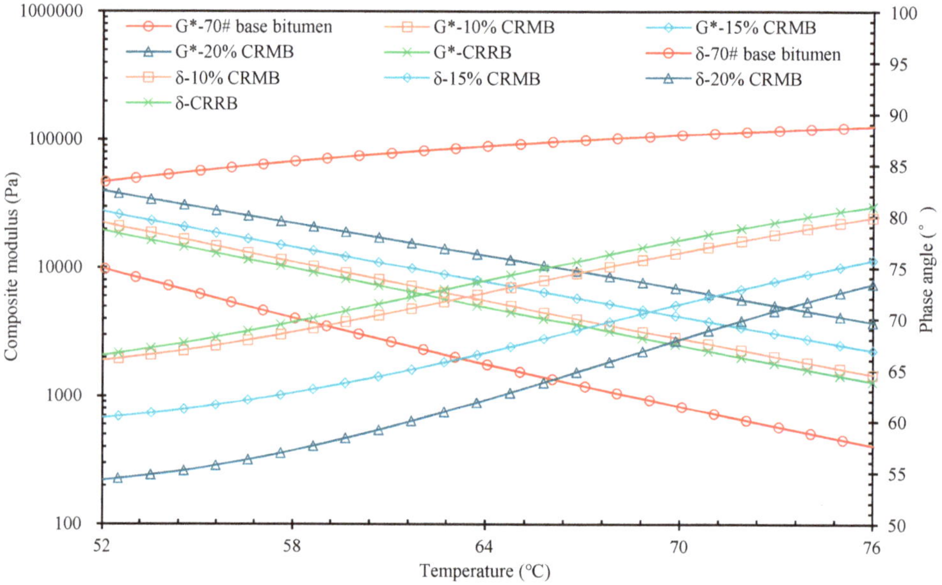

Figure 4. G* and δ curve.

Figure 5 shows the rutting factor (G*/sinδ) curve. It can be observed that the rutting factor of bitumen increased significantly after adding CR. Table 2 shows the corresponding temperature results when the rutting factor G*/sinδ = 1.0 kPa. After linear fitting, the failure temperatures of G*/sinδ of 70# base bitumen, 10% CRMB, 15% CRMB, 20% CRMB, and CRRB at 1.0 kPa were 68.8 °C, 88.6 °C, 96.6 °C, 104.1 °C, and 89.9 °C, respectively. The failure temperature of 10% CRMB was 30.62% higher than that of 70# base bitumen, which was close to that of CRRB. A higher the failure temperature indicates better rutting resistance. The corresponding high-temperature grades were PG 64, PG 88, PG 94, PG 100, and PG 88 respectively. When 10% CR was mixed with bitumen, the PG grade of bitumen was rapidly increased by four grades, and the PG grade of CRRB was increased by four grades as well. In comparison, the PG grades of 15% CRMB and 20% CRMB were only increased by one grade. It can be seen from these results that CR can rapidly improve the rutting resistance of bitumen at high temperatures. The high-temperature performance results of G* and δ curves are consistent with the results for the penetration degree and softening point.

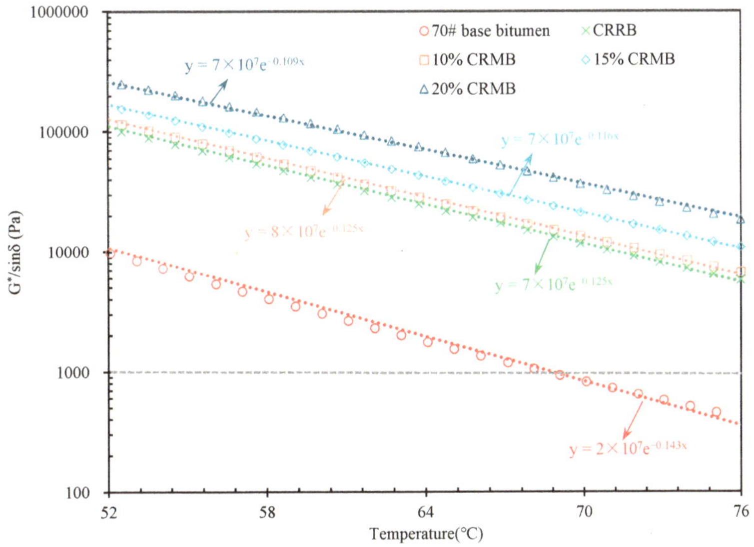

Figure 5. Rutting factor (G*/sinδ) curve.

Table 2. Failure temperature of bitumen.

Testing Parameters	70# Base Bitumen	10% CRMB	15% CRMB	20% CRMB	CRRB
Failure temperature (°C)	68.8	88.6	96.6	104.1	89.9

3.1.3. Low-Temperature Rheological Property

Figures 6 and 7 show the m value and S value of the five kinds of bitumen at −6 °C, −12 °C, and −18 °C. As shown in the figure, the m value of CRMB was between 70# base bitumen and CRRB, and the m value of CRRB was basically between 70# base bitumen and CRRB. In terms of m value and S value alone, the low-temperature performance of 70# base bitumen was better than that of CRMB, and that of CRMB was better than CRRB. A higher S value indicates worse low-temperature ductility, while the m value indicates the rate of change in the S value; thus, a larger m value indicates a higher relaxation rate and better low-temperature performance. At the same time, it can be seen that the regularity of the S value and m value was not very clear, and it was not possible to scientifically to evaluate the low-temperature rheological properties of bitumen by a single analysis of the S value or m value. The equation k = m/S was used to evaluate the rheological properties of bitumen at low temperatures. A higher value of k indicates better rheological properties of bitumen at low temperatures [35,41]. Table 3 shows the k value of bitumen at different temperatures; as the test temperature decreased, the k value of bitumen decreased as well. At −6 °C, the k value of 70# base bitumen was much greater than 10% CRMB; however, as the temperature decreased, the k value of 70# base bitumen began to approach 10% CRMB. This means that the low-temperature performance of 70# base bitumen was greatly affected by the temperature of bitumen; in addition, CR slowed the rate of decrease in the failure temperature under ultra-low temperature environmental conditions and enhanced its ultra-low temperature crack resistance. The test results for the three low temperatures show that within the range of 10% to 20% CR dosage, the greater the CR dosage, the greater the k value. The low-temperature performance of bitumen was positively correlated with the CR dosage. Compared with 70# base bitumen, 20%, 15%, and 10% CRMB had better low-temperature performance at −6 °C, −12 °C, and −18 °C, respectively, showing that CR incorporation improved the low-temperature performance of bitumen and could reduce the sensitivity of bitumen to temperature. With the increase in the dosage, the increase in the rate of the k value decreased significantly, indicating that the degree of improvement of CR on low-temperature bitumen performance decreased with increasing dosage. At the same time, the k value of CRRB was always the lowest among the five kinds of bitumen; CRRB had the worst low-temperature performance, and the low-temperature performance of CRRB had obvious faults compared with 70# base bitumen, for reasons consistent with the above analysis.

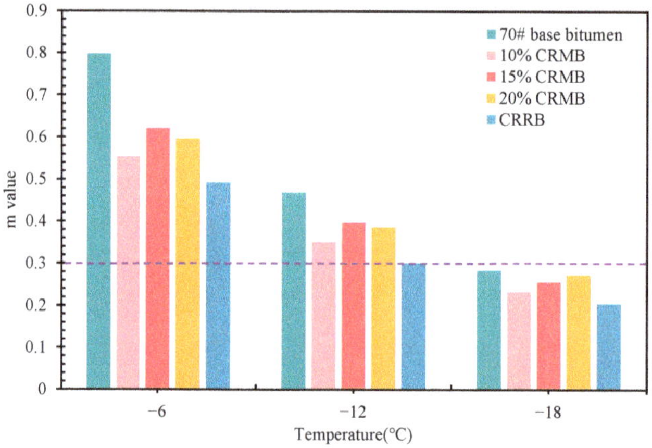

Figure 6. m value of bitumen.

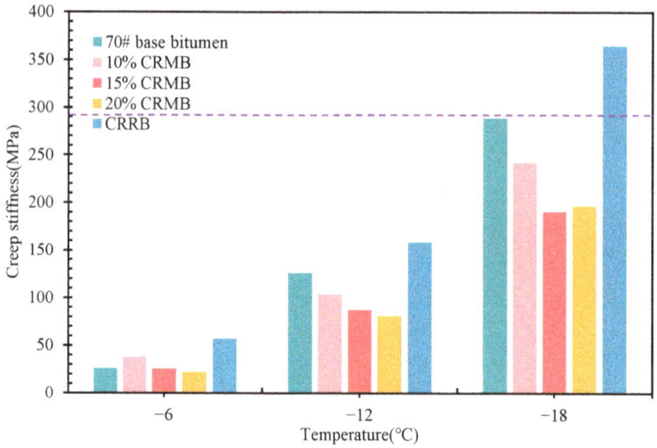

Figure 7. S value of bitumen.

Table 3. k value of bitumen at different temperatures.

Testing Parameters	−6 °C	−12 °C	−18 °C
70# base bitumen	0.0306	0.0037	0.0010
10% CRMB	0.0145	0.0034	0.0010
15% CRMB	0.0240	0.0045	0.0013
20% CRMB	0.0270	0.0048	0.0014
CRRB	0.0086	0.0019	0.0006

To study the low-temperature performance of bitumen, it was necessary to study the low-temperature failure temperature of bitumen. Table 4 shows the failure temperature and low-temperature PG grade of the bitumen samples. It can be seen that the failure temperature of 70# base bitumen was slightly lower than 10% CRMB. According to the above conclusion, the low-temperature performance of 70# base bitumen was greatly affected by the temperature of the bitumen. The lower the temperature of the bitumen, the faster the failure rate was reached with respect to its low-temperature performance. Thus, the failure temperature of 70# base bitumen was reached earlier than 10% CRMB failure. The failure temperature results show that within the range of 10% to 20% CR dosage, the larger the CR dosage, the smaller the failure temperature. The low-temperature performance of bitumen was positively correlated with the dosage of CR. With increasing dosage of CR, the decline in the rate of failure temperature decreased obviously. At the same time, the failure temperature of CRRB was the highest among the five kinds of bitumen; CRRB had the worst low-temperature performance, and its low-temperature performance had obvious faults compared with 70# base bitumen. The low-temperature PG grade of the five kinds of bitumen was −22 °C.

Table 4. Failure temperature and PG grade of bitumen at low temperature.

Testing Parameters	Failure Temperature (°C)	Low-Temperature PG Degree (°C)
70# base bitumen	−14.31	−22
10% CRMB	−14.59	−22
15% CRMB	−16.11	−22
20% CRMB	−16.25	−22
CRRB	−12.07	−22

In summary, within the range of 10% to 20% CR dosage, the higher the CR dosage, the better the low-temperature performance. The low-temperature performance was positively correlated with the CR dosage, and with increasing dosage, the failure rate of the low-temperature performance decreased. The low-temperature performance of 70# base bitumen was greatly affected by low temperature, and the lower the temperature, the faster the low-temperature failure. CR slowed the rate of failure temperature decrease in the ultra-low temperature environment, and the sensitivity of bitumen to temperature was reduced. The low-temperature performance of CRRB was far worse than that of 70# base bitumen. The low-temperature rheological change trend of bitumen was consistent with the ductility test.

3.1.4. Elastic Recovery Performance

Figure 8 shows the elastic recovery rate of bitumen at 25 °C. CR can significantly improve the elastic recovery of bitumen. With the continuous increase of CR dosage, the elastic recovery rate of bitumen increased. The elastic recovery of bitumen was increased by 339.89% when adding 10% CR. The elastic recovery rate of 15% CRMB was increased by 7.35% on average, and the same was true for 20% CRMB. Compared with 15% CRMB, the CRRB can be increased by 99.52% compared with 70# base bitumen. This may be due to the aromatics of bitumen being decreased and its elastic components increased, making the elastic recovery performance of bitumen better [42]. Thus, the elastic recovery performance of CRRB was better than that of 70# base bitumen.

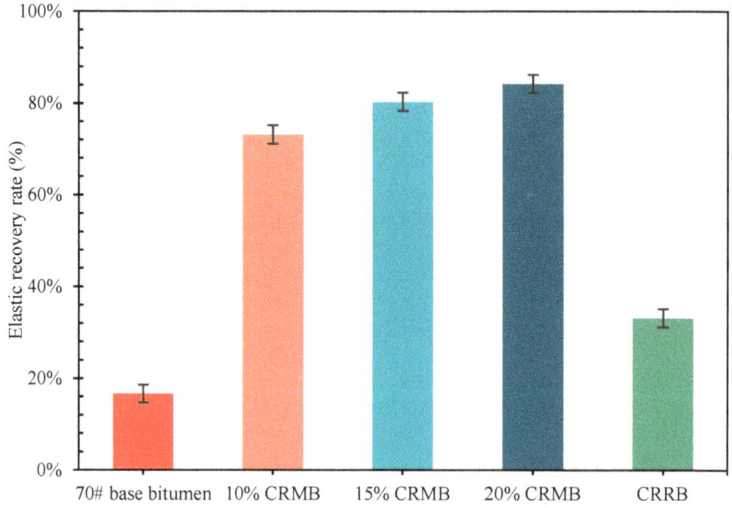

Figure 8. Elastic recovery of bitumen.

3.2. Working Performance of CRMB

3.2.1. Brookfield Viscosity

Figure 9 shows the viscosity and temperature curves drawn according to the viscosity results of 70# base bitumen, 10% CRMB, 15% CRMB, 20% CRMB, and CRRB. Due to CR being mixed with the bitumen, the viscosity of bitumen at 135 °C and 175 °C was greatly increased, and the temperature required for the preparation of the bitumen mixture was increased as well. The optimum compaction temperature range of asphalt mixture is the temperature value when the bitumen viscosity is 0.28 ± 0.03 Pa·s, and the optimum mixing temperature range is the temperature value when the bitumen viscosity is 0.17 ± 0.02 Pa·s [43,44]. Table 5 shows the median of the compaction temperature range and mixing temperature range of the five kinds of bitumen. After 10% CR was mixed with bitumen, the mixing temperature and compaction temperature increased significantly. The mixing temperature of bitumen increased by 10.74%, and the compaction temperature increased by 17.74%. The mixing temperature and compaction temperature of 15% CRMB and 20% CRMB were not greatly improved compared with the former. CRRB had a high viscosity compared with 70# base bitumen; its compaction temperature increased by 10.60%, and the mixing temperature increased by 6.28%.

Table 5. Median of compaction temperature range and mixing temperature range of bitumen.

Sample	70# Base Bitumen	10% CRMB	15% CRMB	20% CRMB	CRRB
Median of compaction temperatures range	152.7	179.8	182.0	185.4	168.9
Median of mixing temperature range	164.3	181.9	183.6	186.2	174.6

3.2.2. Storage Stability

Figure 10 shows the softening point difference of the three modified bitumen samples. The density of commonly used CR is about 1.13 g/cm^3, while that of commonly used 70# base bitumen is about 1.03 g/cm^3. The CR in CRMB sinks to the bottom of the bitumen over time [2]. Technical specifications for the construction of highway asphalt pavement require that the softening point difference of CRMB should be −5 °C to 10 °C [3]; the maximum softening point difference of the three modified bitumen samples was 6.85 °C, which obviously meets this requirement. With the continuous increase of CR dosage, the segregation softening point difference of bitumen decreased, which is consistent with Li [35,45]. The softening point difference of 10% CRMB was 42.77% lower than that of 15% CRMB, and the softening point difference of 15% CRMB was only 19.39% lower than that of 20% CRMB. According to Stokes' sedimentation theory, 10% of CR was continuously segregated in the aluminum tube and most of the CR failed to form a stable system at the bottom of the aluminum tube. CR always maintained a certain sedimentation rate, resulting in a large softening point difference. Furthermore, 15% and 20% of CR segregated constantly in the aluminum tube; however, at this time a part of the CR (within 10% to 15%) was able to form a stable system at the bottom of the aluminum tube and the CR sedimentation rate slowed down, meaning that the difference in the softening point was small. It can be seen that there was a dosage value that led the change rate of the CR softening point difference to drop sharply within the dosage of 10% to 15%.

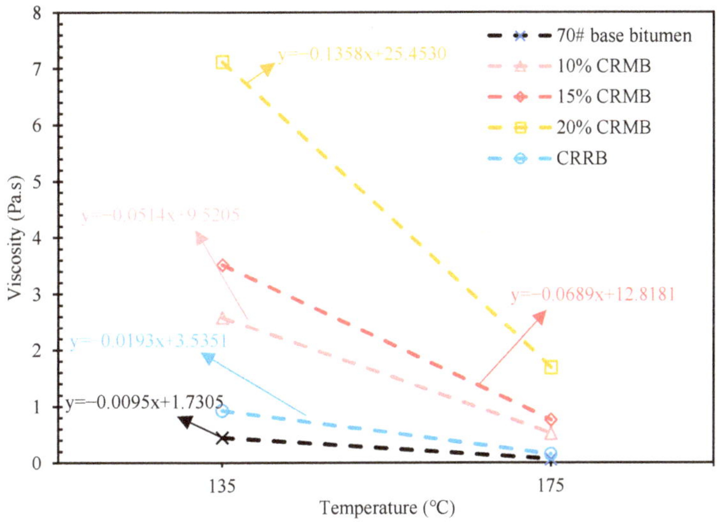

Figure 9. Viscosity temperature curves of bitumen.

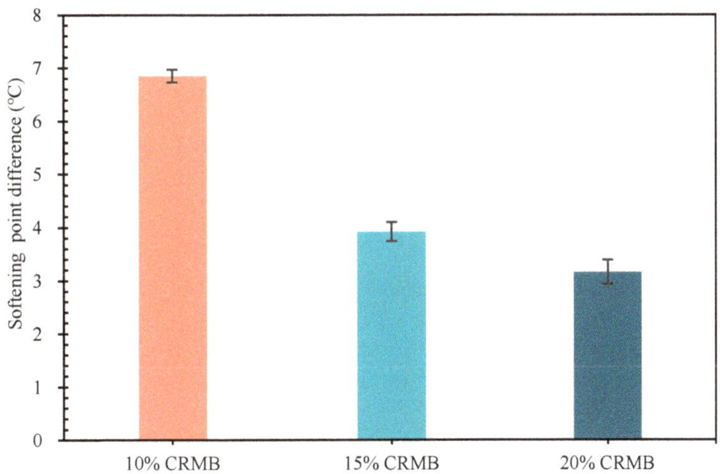

Figure 10. Softening Point Difference of CRMB.

3.3. Swelled Mechanism of CR

3.3.1. Particle Size Distribution of CR

Figure 11 shows the results of the particle size distribution of CR and SCR. It can be seen that the peak position of the particle size curve of SCR is to the right compared with the CR, the peak of CR is at 400 μm, and the peak of SCR is at 502 μm, while the particle size distribution range of SCR is larger than that of CR. The particle size curves of CR and SCR intersect at 532 μm; 66.2% of the total number of CR particles are from 0 μm to 532 μm, and only 35.9 % of the total number of SCR particles are from 0 μm to 532 μm, indicating that the particle size of CR increased after dissolution. Table 6 shows the average particle sizes of CR and SCR. It can be seen that the D10, D50, and D90 of SCR are larger than those of CR. The area average diameter (D[3,2]) and volume average diameter (D[4,3]) of SCR are larger than CR as we, with D[3,2] being increased by 52.3% and D[4,3] increased by 49.7%. In addition, it can be seen that the volume of CR expanded by about 50% under this preparation condition, while its mesh size decreased. The above conclusions prove that CR undergoes a swelling reaction and its volume increases after CR absorbs the light components of bitumen via the action of the bitumen [9,46].

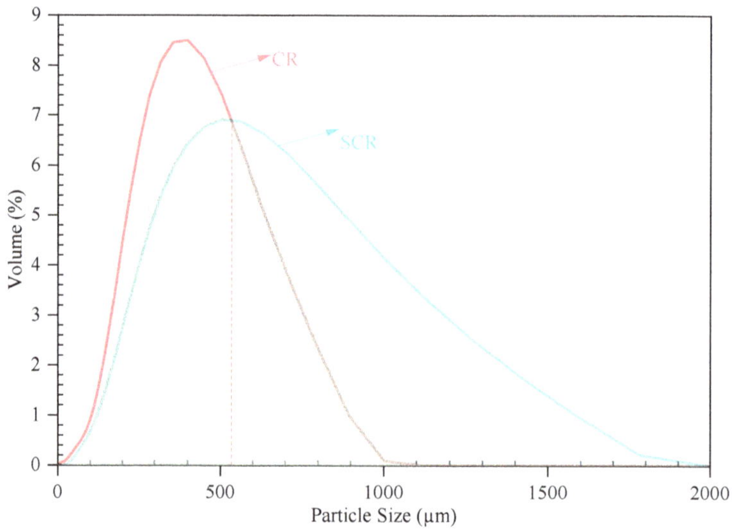

Figure 11. Particle size distribution curve of CR and SCR.

Table 6. Statistics of average particle size for CR and SCR.

Sample	D10 (μm)	D50 (μm)	D90 (μm)	D[3,2] (μm)	D[4,3] (μm)
CR	137.976	322.548	592.279	214.524	345.878
SCR	175.095	449.226	969.408	326.616	517.881

3.3.2. Micromorphology of CR

Figure 12a,b shows the microscopic morphology of CR. It can be seen that the surface smoothness of CR is better and angularity is prominent; this is the result of high-strength shear tires, the surface of which have a visual "hardness feel". Figure 12c,d shows the microscopic morphological results of SCR. It can be seen that after the swelling effect of bitumen, the swelling effect on the surface morphology of CR is more obvious; more holes and obvious grooves appear on the surface of SCR, the smoothness is reduced and the complexity is higher. The surface of SCR has a soft flocculent edge, in contrast to the hard angularity of CR, and intuitively the surface of SCR has a "delicate feel". Under the effect of swelling, the admixture in CR falls off, leading to more holes and grooves, and in turn led to the larger specific surface area and stronger interface sense of SCR, which make SCR more stable in the bitumen.

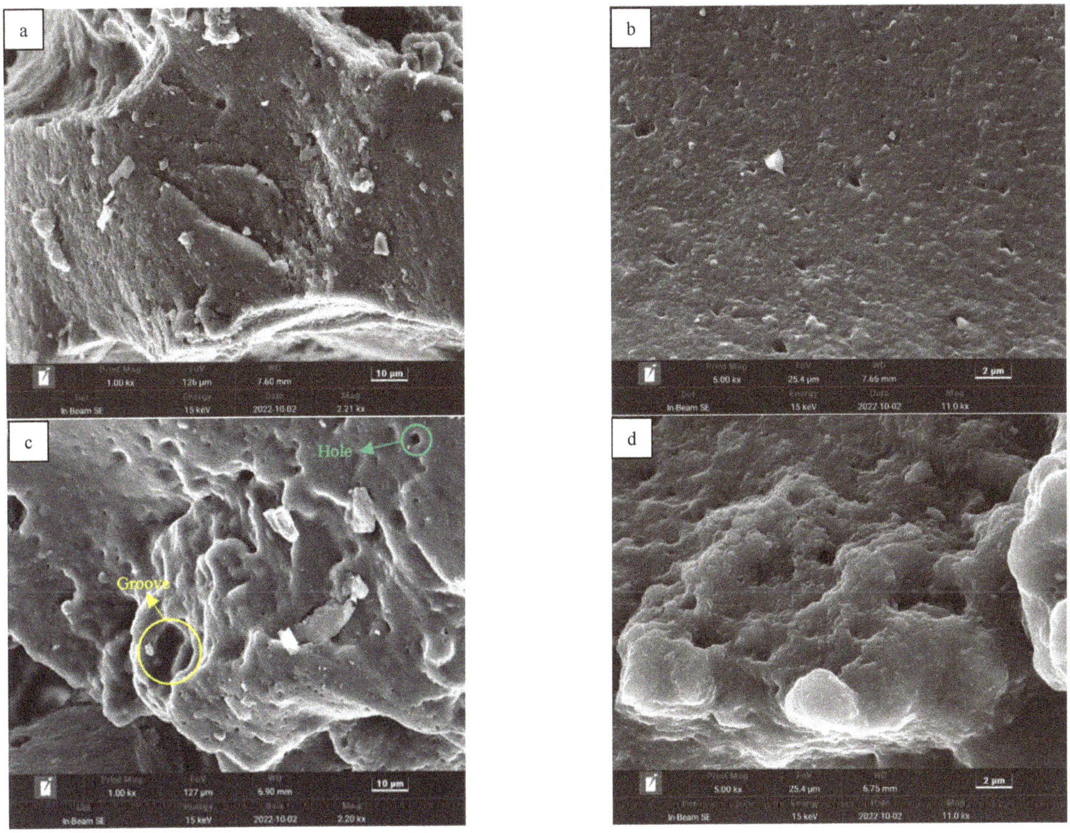

Figure 12. Micromorphology: (**a**) CR 1000×; (**b**) CR 5000×; (**c**) SCR 1000×; (**d**) SCR 5000×.

3.3.3. Relative Molecular Weight of CR

Figure 13 shows the molecular weight distribution of CR and SCR. After swelling, the molecular weight distribution curve of CR shifts to the left, indicating that CR mainly undergoes the transformation from large molecules to small molecules during the swelling reaction. Table 7 shows the average molecular weight (Mn), heavy average molecular weight (Mw), and polydispersity (pD) of CR and SCR. It can be seen that after CR swelling, the pD increased by 71.83%, indicating that CR swelling made the distribution of CR significantly wider and the concentricity of molecular weight distribution worse. It can be seen that Mn decreased by 35.80% and Mw increased by 10.25%. The decrease in the value of Mn was 3.5 times the increase in the value of Mw, indicating that the molecular weight of CR became smaller after swelling, and there was a trend of small molecules turning into large molecules [47].

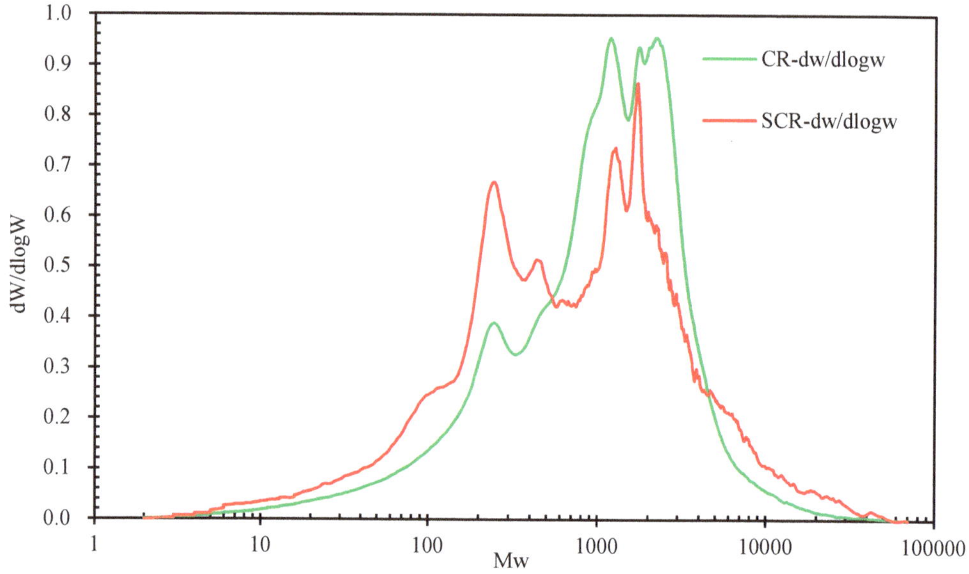

Figure 13. Molecular weight distribution of CR and SCR.

Table 7. Mn, Mw, and pD of CR and SCR.

Sample	Mn	Mw	pD
CR	634	2049	3.23
SCR	407	2259	5.55

Figure 14 show the chromatogram of CR and SCR. It can be seen that after swelling, the strong peak of CR at the combination of MMS and SMS region shifts to the small molecule region after the dissolution; only one peak of the original two strong peaks of CR remains and is completely attributed to the small molecule region. This indicates that breakage of the medium molecular main Chain of CR occurs, producing small molecular weight molecules. The macromolecular region did not change significantly, and only the lifting of the peaks was somewhat slowed down, which did not affect the macromolecules significantly under this preparation condition. Table 8 shows the relative proportions of LMS, MMS, and SMS for CR and SCR, respectively; it can be seen that the proportion of MMS decreases by 40%, while the proportion of SMS increases by 50.5%. The CR swelling process is characterized by the transition from large and medium molecules to small molecules. The small increase in the proportion of LMS is due to the transition from large and medium molecules to small molecules as well. A number of the original medium molecules can be classified as large molecules after lysis, i.e., "medium molecules" were medium molecules in CR and large molecules in SCR. Furthermore, when the CR was swollen, only a small amount of large molecules were transformed into small and medium molecules, which is consistent with the above conclusion and confirms that large molecules were less susceptible to CR interactions with bitumen than small molecules [48].

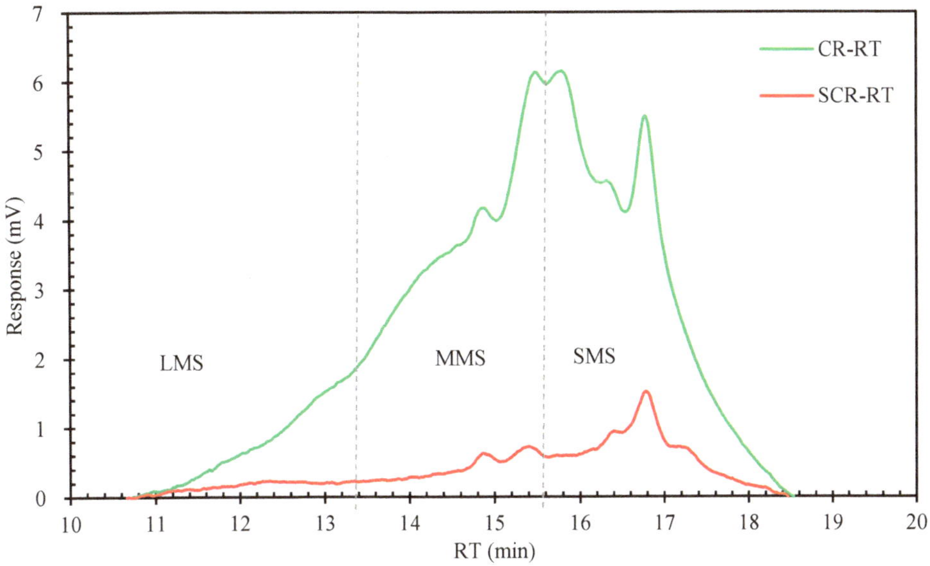

Figure 14. Chromatogram of CR and SCR.

Table 8. Relative proportion of three regions of CR and SCR.

Sample	LMS	MMS	SMS
CR	14.2%	54.7%	31.1%
SCR	16.0%	32.7%	46.8%

3.3.4. Chemical Structure of CR

Figure 15 shows the infrared spectra of CR and SCR. For convenience, Figure 15 only shows the infrared spectra from 3000 cm^{-1} to 500 cm^{-1}. It can be seen that the infrared spectra of CR have absorption peaks at 2923 cm^{-1}, 1427 cm^{-1}, 1253 cm^{-1}, etc. These absorption peaks are caused by the vibration of hydrocarbon bonds, and it is known that hydrocarbons are the main components of CR. The presence of absorption peaks at 2923 cm^{-1}, 970 cm^{-1}, 814 cm^{-1} to 657 cm^{-1}, 575 cm^{-1} to 540 cm^{-1}, and 521 cm^{-1} to 477 cm^{-1} indicates that the CR contained unsaturated bis- and benzene ring-conjugated olefins linked by sulfide bonds [8]. Comparing the infrared spectra of CR and SCR, the two are obviously different, and it can be concluded that the chemical structure of CR changes when it swells in the bitumen. Specifically, the original CR has an absorption peak at 1427 cm^{-1} and a strong absorption peak at 1297 cm^{-1} to 1037 cm^{-1}, indicating the presence of ester groups, while only a few ester groups are present in SCR after CR swelling. The original CR has an absorption peak at 1190 cm^{-1}, indicating the presence of amide, while this amide disappears after CR swelling. CR added a =C-H stretching vibration on the benzene ring at 2657 cm^{-1}, indicating the emergence of a new benzene ring structure in SCR, while SCR added a =C-NH stretching vibration on the benzene ring at 1956 cm^{-1}. After the swelling of CR, its characteristic peak at 1427 cm^{-1} is significantly weakened, and it is known that bitumen can break the methyl conjugate bond of CR. SCR has a C=C bond absorption peak at 1528 cm^{-1}, while the C–C bond of CR disappears at 873 cm^{-1} and its weak C–S bond disappears at 640 cm^{-1} and 574 cm^{-1}. It can be seen that the C–C and C–S of bonds of the CR were broken after swelling, promoting the formation of the C=C bond to a certain extent.

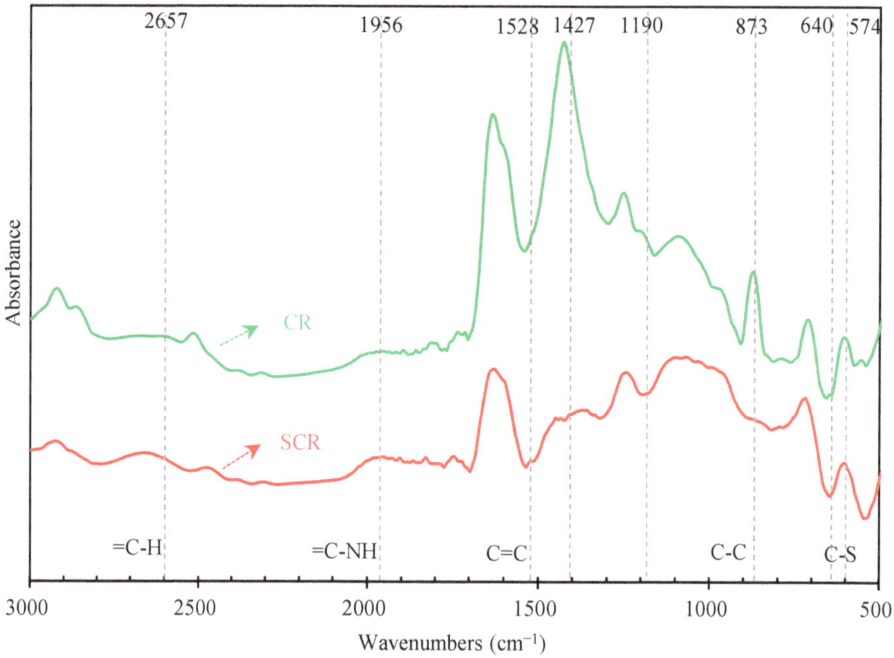

Figure 15. Infrared spectrum of CR and SCR.

4. Conclusions

To study the physicochemical properties of CR and SCR and the working properties of CRRB, we prepared CRMB, then SCR and CRRB were separated from CRMB and tested. The working properties and high and low-temperature properties of CRMB and CRRB were investigated to characterize the physicochemical properties of CR before and after swelling in order to study the swelling mechanism.

(1) The results of our high-temperature performance tests showed that CR significantly improved the high-temperature rutting resistance of bitumen. The high-temperature failure temperature of 10% CRMB was 30.62% higher than that of 70# base bitumen, and its high-temperature PG grade rapidly improved by four grades. Meanwhile, CR significantly improved the elastic recovery of bitumen; the elastic recovery rate of 10% CRMB was improved by 339.89% compared to 70# base bitumen. After modification by CR, the dosage of CR was positively correlated with the low-temperature performance of bitumen; CR decelerated the failure temperature decrease rate of bitumen in an ultra-low temperature environment, and its ultra-low temperature crack resistance was enhanced.

(2) CR raised the viscosity of bitumen, resulting in a significant increase in mixing and compaction temperatures, which can be mitigated by the addition of warm mixes. After modification with CR, the separation softening point difference of CRMB decreased with the increase in dosage, and its storage stability was better. The results in terms of softening point difference for the three dosages show that the softening point difference of 10% CRMB decreases more significantly than that of 15% CRMB, with a specific decrease of 42.8%.

(3) When CR was swollen it absorbed light components of bitumen, increasing the relative content of the heavy component of bitumen and decreasing the relative content of oil. Therefore, CRRB maintained good high-temperature performance; its high-temperature performance was close to that of 10% CRMB, although its low-temperature performance

was poor, and there were obvious faults with the other four kinds of bitumen. Furthermore, due to its increased elastic component, CRRB had an elastic recovery rate of 33.16%.

(4) After CR swelling, the size distribution range of SCR was larger than that of CR and the size of CR increased, as shown by the expansion of CR volume by about 50%. The surface of the swollen CR became more complex, with a larger specific surface area and a stronger sense of interface. After the swelling reaction, the molecular weight distribution of SCR increased by 71.8% compared to CR and the percentage of MMS decreased by 40%, while the percentage of SMS increased by 50.5%. In addition, CR carried out the transformation process from large and medium molecules to small molecules. After swelling, the ester group content of CR decreased significantly and a new benzene ring structure appeared. Finally, the C–C and C–S bonds of the CR were broken to generate partial C=C bonds.

At this stage, although the performance of CRMB and CRRB and the physicochemical properties of SCR were studied, CR in a bitumen settling system needs to be further explored in conjunction with Stokes' sedimentation theory. This study investigated only the performance of CRMB; the next step should be from the perspective of CRMB research for green construction.

Author Contributions: Conceptualization, writing—original draft preparation, methodology, H.Z.; Conceptualization, M.Z.; Software, project administration, funding acquisition, writing—review and editing, Y.L.; validation, writing—review and editing, A.C.; formal analysis, Y.Z. and L.L.; investigation, D.G.; data curation, S.Z.; validation, supervision, F.W. All authors have read and agreed to the published version of the manuscript.

Funding: This research received no external funding. Financial support was provided by the transportation technology project of the department of transport of Hubei province (No. 2022-11-1-10), the scientific research fund project of the Wuhan institute of technology (No. K2021032), the strength formation mechanism and application of self-compacting asphalt pavement materials at ambient temperature for urban roads of science and technology planning project of Hubei Provincial Department of Housing and Urban-Rural Development (No. 202171), and test help from Shiyanjia Lab (www.shiyanjia.com).

Institutional Review Board Statement: Not applicable.

Informed Consent Statement: Not applicable.

Data Availability Statement: Not applicable.

Conflicts of Interest: The authors declare no conflict of interest.

References

1. Ma, J.; Sun, G.; Sun, D.; Zhang, Y.; Falchetto, A.C.; Lu, T.; Hu, M.; Yuan, Y. Rubber asphalt modified with waste cooking oil residue: Optimized preparation, rheological property, storage stability and aging characteristic. *Constr. Build. Mater.* **2020**, *258*, 120372. [CrossRef]
2. Zheng, W.; Wang, H.; Chen, Y.; Ji, J.; You, Z.; Zhang, Y. A review on compatibility between crumb rubber and asphalt binder. *Constr. Build. Mater.* **2021**, *297*, 123820. [CrossRef]
3. Zhu, Y.; Xu, G.; Ma, T.; Fan, J.; Li, S. Performances of rubber asphalt with middle/high content of waste tire crumb rubber. *Constr. Build. Mater.* **2022**, *335*, 127488. [CrossRef]
4. Li, J.; Wang, J.; Xiao, F.; Amirkhanian, S.N. Characterizing Compatibility of Crumb Rubber Modified Asphalt by Customized Drainage Method. *J. Test. Eval.* **2020**, *49*, 20190856. [CrossRef]
5. Liang, M.; Qiu, Z.; Luan, X.; Qi, C.; Guo, N.; Liu, Z.; Su, L.; Yao, Z.; Zhang, J. The Effects of Activation Treatments for Crumb Rubber on the Compatibility and Mechanical Performance of Modified Asphalt Binder and Mixture by the Dry Method. *Front. Mater.* **2022**, *9*, 845718. [CrossRef]
6. Flanigan, T.P. Process for Liquefying Tire Rubber and Product Thereof. US Patent 5,492,561, 20 February 1996.
7. Zanzotto, L.; Kennepohl, G.J. Development of Rubber and Asphalt Binders by Depolymerization and Devulcanization of Scrap Tires in Asphalt. *Transp. Res. Rec. J. Transp. Res. Board* **1996**, *1530*, 51–58. [CrossRef]
8. Dong, R.; Zhao, M. Research on the pyrolysis process of crumb tire rubber in waste cooking oil. *Renew. Energy* **2018**, *125*, 557–567. [CrossRef]
9. Dong, R.; Zhao, M.; Tang, N. Characterization of crumb tire rubber lightly pyrolyzed in waste cooking oil and the properties of its modified bitumen. *Constr. Build. Mater.* **2019**, *195*, 10–18. [CrossRef]

10. Dong, R.; Zhao, M.; Xia, W.; Yi, X.; Dai, P.; Tang, N. Chemical and microscopic investigation of co-pyrolysis of crumb tire rubber with waste cooking oil at mild temperature. *Waste Manag.* **2018**, *79*, 516–525. [CrossRef]
11. Cq, A.; Wf, A.; Gy, B.; Ling, H.B.; Bx, A.; Xl, A. Influence of crumb rubber particle size and SBS structure on properties of CR/SBS composite modified asphalt. *Constr. Build. Mater.* **2020**, *235*, 117517.
12. Cheng, Y.; Chai, C.; Liang, C.; Chen, Y. Mechanical Performance of Warm-Mixed Porous Asphalt Mixture with Steel Slag and Crumb-Rubber–SBS Modified Bitumen for Seasonal Frozen Regions. *Materials* **2019**, *12*, 857. [CrossRef] [PubMed]
13. Manosalvas, M.; Gallego, J.; Saiz, L.; Bermejo, J.M. Rubber modified binders as an alternative to cellulose fiber—SBS polymers in Stone Matrix Asphalt. *Constr. Build. Mater.* **2016**, *121*, 727–732. [CrossRef]
14. Leng, Z.; Padhan, R.K.; Sreeram, A. Production of a sustainable paving material through chemical recycling of waste PET into crumb rubber modified asphalt. *J. Clean. Prod.* **2018**, *180*, 682–688. [CrossRef]
15. Veropalumbo, R.; Russo, F.; Oreto, C.; Biancardo, S.A.; Zhang, W.; Viscione, N. Verifying laboratory measurement of the performance of hot asphalt mastics containing plastic waste. *Measurement* **2021**, *180*, 109587. [CrossRef]
16. Cw, A.; Kd, A.; Liang, S.B.; Xj, A.; Cheng, S. Stability improvement technology of SBS/crumb rubber composite modified asphalt from Xinjiang China. *J. Clean. Prod.* **2022**, *359*, 132003.
17. Huang, W.; Liu, X.; Zhang, S.; Zheng, Y.; Ding, Q.; Tong, B. Performance-Guided Design of Permeable Asphalt Concrete with Modified Asphalt Binder Using Crumb Rubber and SBS Modifier for Sponge Cities. *Materials* **2021**, *14*, 1266. [CrossRef]
18. Yang, X.; Shen, A.; Li, B.; Wu, H.; Lyu, Z.; Wang, H.; Lyu, Z. Effect of microwave-activated crumb rubber on reaction mechanism, rheological properties, thermal stability, and released volatiles of asphalt binder. *J. Clean. Prod.* **2019**, *248*, 119230. [CrossRef]
19. Liang, M.; Xin, X.; Fan, W.; Ren, S.; Shi, J.; Luo, H. Thermo-stability and aging performance of modified asphalt with crumb rubber activated by microwave and TOR. *Mater. Des.* **2017**, *127*, 84–96. [CrossRef]
20. Jm, A.; Mh, A.; Ds, A.; Tong, L.A.; Gs, B.; Sl, A.; Lei, X.A. Understanding the role of waste cooking oil residue during the preparation of rubber asphalt. *Resour. Conserv. Recycl.* **2020**, *167*, 105235.
21. Kabir, S.F.; Zheng, R.; Delgado, A.G.; Fini, E.H. Use of microbially desulfurized rubber to produce sustainable rubberized bitumen. *Resour. Conserv. Recycl.* **2020**, *164*, 105144. [CrossRef]
22. Li, J.; Chen, Z.; Xiao, F.; Amirkhanian, S.N. Surface activation of scrap tire crumb rubber to improve compatibility of rubberized asphalt. *Resour. Conserv. Recycl.* **2021**, *169*, 105518. [CrossRef]
23. Segre, N.; Monteiro, P.J.; Sposito, G. Surface Characterization of Recycled Tire Rubber to Be Used in Cement Paste Matrix. *J. Colloid Interface Sci.* **2002**, *248*, 521–523. [CrossRef] [PubMed]
24. Liu, L.; Liu, Z.; Yang, C.; Huang, Y.; Li, W. Effect of the binary compound method of TOR promoting dissolution and NaOH solution surface treatment on the performance of rubber asphalt. *Constr. Build. Mater.* **2021**, *305*, 124697. [CrossRef]
25. Zedler, Ł.; Colom, X.; Cañavate, J.; Saeb, M.; Haponiuk, J.T.; Formela, K. Investigating the Impact of Curing System on Structure-Property Relationship of Natural Rubber Modified with Brewery By-Product and Ground Tire Rubber. *Polymers* **2020**, *12*, 545. [CrossRef]
26. Yang, X.; Shen, A.; Guo, Y.; Lyu, Z. Effect of process parameters on the high temperature performance and reaction mechanism of CRMA. *Pet. Sci. Technol.* **2018**, *36*, 1537–1543. [CrossRef]
27. Dong, D.; Huang, X.; Li, X.; Zhang, L. Swelling process of rubber in asphalt and its effect on the structure and properties of rubber and asphalt. *Constr. Build. Mater.* **2012**, *29*, 316–322. [CrossRef]
28. Wang, H.; Liu, X.; Apostolidis, P.; Erkens, S.; Scarpas, T. Numerical investigation of rubber swelling in bitumen. *Constr. Build. Mater.* **2019**, *214*, 506–515. [CrossRef]
29. Abdelrahman, M.A.; Carpenter, S.H. Mechanism of Interaction of Asphalt Cement with Crumb Rubber Modifier. *Transp. Res. Rec. J. Transp. Res. Board* **1999**, *1661*, 106–113. [CrossRef]
30. Liu, S.; Zhou, S.B.; Peng, A.; Xuan, W.; Li, W. Analysis of the performance and mechanism of desulfurized rubber and low-density polyethylene compound-modified asphalt. *J. Appl. Polym. Sci.* **2019**, *136*, 48194. [CrossRef]
31. Wang, S.; Huang, W.; Liu, X.; Lin, P. Influence of high content crumb rubber and different preparation methods on properties of asphalt under different aging conditions: Chemical properties, rheological properties, and fatigue performance. *Constr. Build. Mater.* **2022**, *327*, 126937. [CrossRef]
32. Xie, J.; Yang, Y.; Lv, S.; Zhang, Y.; Zhu, X.; Zheng, C. Investigation on Rheological Properties and Storage Stability of Modified Asphalt Based on the Grafting Activation of Crumb Rubber. *Polymers* **2019**, *11*, 1563. [CrossRef] [PubMed]
33. Li, J.; Xiao, F.; Amirkhanian, S. High temperature rheological characteristics of plasma-treated crumb rubber modified binders. *Constr. Build. Mater.* **2019**, *236*, 117614. [CrossRef]
34. Chen, A.; Airey, G.D.; Thom, N.; Li, Y. Characterisation of fatigue damage in asphalt mixtures using X-ray computed tomography. *Road Mater. Pavement Des.* **2022**, 1–19. [CrossRef]
35. Li, J.; Xiao, F.; Amirkhanian, S.N. Storage, fatigue and low temperature characteristics of plasma treated rubberized binders. *Constr. Build. Mater.* **2019**, *209*, 454–462. [CrossRef]
36. Chen, A.; Qiu, Y.; Wang, X.; Li, Y.; Wu, S.; Liu, Q.; Wu, F.; Feng, J.; Lin, Z. Mechanism and Performance of Bituminous Mixture Using 100% Content RAP with Bio-Rejuvenated Additive (BRA). *Materials* **2022**, *15*, 723. [CrossRef]
37. Ye, S.; Ge, M.R.; Guo, B. Study on the Influence of Asphalt Performance for Ultraviolet Aging. *Adv. Mater. Res.* **2015**, *1065–1069*, 798–801. [CrossRef]

38. Wei, C.; Zhang, H.; Duan, H. Effect of catalytic-reactive rejuvenator on structure and properties of aged SBS modified asphalt binders. *Constr. Build. Mater.* **2020**, *246*, 118531. [CrossRef]
39. Wei, C.; Duan, H.; Zhang, H.; Chen, Z. Influence of SBS modifier on aging behaviors of SBS-modified asphalt. *J. Mater. Civ. Eng.* **2019**, *31*, 04019184. [CrossRef]
40. Chen, A.; Deng, Q.; Li, Y.; Bai, T.; Chen, Z.; Li, J.; Feng, J.; Wu, F.; Wu, S.; Liu, Q.; et al. Harmless treatment and environmentally friendly application of waste tires—TPCB/TPO composite-modified bitumen. *Constr. Build. Mater.* **2022**, *325*, 126785. [CrossRef]
41. Liu, S.; Cao, W.; Shang, S.; Qi, H.; Fang, J. Analysis and application of relationships between low-temperature rheological performance parameters of asphalt binders. *Constr. Build. Mater.* **2010**, *24*, 471–478. [CrossRef]
42. Liu, H.; Fu, L.; Jiao, Y.; Tao, J.; Wang, X. Short-Term Aging Effect on Properties of Sustainable Pavement Asphalts Modified by Waste Rubber and Diatomite. *Sustainability* **2017**, *9*, 996. [CrossRef]
43. Wang, F.; Zhu, H.; Shu, B.; Li, Y.; Gu, D.; Gao, Y.; Chen, A.; Feng, J.; Wu, S.; Liu, Q.; et al. Microwave heating mechanism and self-healing performance of asphalt mixture with basalt and limestone aggregates. *Constr. Build. Mater.* **2022**, *342*, 127973. [CrossRef]
44. Wang, F.; Zhu, H.; Li, Y.; Gu, D.; Gao, Y.; Feng, J.; Shu, B.; Li, C.; Wu, S.; Liu, Q.; et al. Microwave heating mechanism and Self-healing performance of scrap tire pyrolysis carbon black modified bitumen. *Constr. Build. Mater.* **2022**, *341*, 127873. [CrossRef]
45. Li, J.; Yao, S.; Xiao, F.; Amirkhanian, S.N. Surface modification of ground tire rubber particles by cold plasma to improve compatibility in rubberised asphalt. *Int. J. Pavement Eng.* **2020**, *23*, 651–662. [CrossRef]
46. Ren, S.; Liu, X.; Lin, P.; Erkens, S. Influence of swelling-degradation degree on rheological properties, thermal pyrolysis kinetics, and emission components of waste crumb rubber modified bitumen. *Constr. Build. Mater.* **2022**, *337*, 127555. [CrossRef]
47. Gao, Y.; Gu, F.; Zhao, Y. Thermal oxidative aging characterization of SBS modified asphalt. *J. Wuhan Univ. Technol. Sci. Ed.* **2013**, *28*, 88–91. [CrossRef]
48. Yu, H.; Leng, Z.; Zhang, Z.; Li, D.; Zhang, J. Selective absorption of swelling rubber in hot and warm asphalt binder fractions. *Constr. Build. Mater.* **2020**, *238*, 117727. [CrossRef]

materials

MDPI

Article

Bionic Synthesis of Mussel-like Adhesive L-DMA and Its Effects on Asphalt Properties

Jinyi Wu, Quantao Liu * and Shaopeng Wu

State Key Laboratory of Silicate Materials for Architectures, Wuhan University of Technology, Wuhan 430070, China; jinyi_wu@whut.edu.cn (J.W.); wusp@whut.edu.cn (S.W.)
* Correspondence: liuqt@whut.edu.cn

Abstract: Cracks are inevitable during the service life of asphalt pavement and the water at the fracture surfaces tends to cause the grouting materials to fail. Studies have shown that the catechol groups in adhesion proteins secreted by mussels can produce strong adhesion performance in the water. In this paper, the mussel-like adhesive L-Dopa Methacrylic anhydride (L-DMA) was prepared based on the concept of bionic design and used to improve the properties of asphalt. By using Fourier-transform infrared spectroscopy (FTIR) and Thermogravimetric analysis (TGA), the thermal stability and structural composition of L-DMA were investigated. Then, the rheological and low-temperature properties of L-DMA-modified asphalt were investigated using the dynamic shear rheological (DSR) test and bending beam rheological (BBR) test. Moreover, the modification mechanism was explored by FTIR. It was found that L-DMA can be effectively synthesized and has good thermal stability. The incorporation of L-DMA increases the composite modulus, viscosity, creep recovery rate and rutting factor of asphalt binder, resulting in an enhancement of its high-temperature performance. At a high L-DMA content of 10%, the low-temperature performance of the modified asphalt was enhanced. The modification of L-DMA to asphalt is mainly a physical process. Hydrogen bonds and conjugated systems generated by the introduction of catechol groups enhance the adhesion properties of asphalt. In general, L-DMA improves the properties of asphalt and theoretically can improve the water resistance of asphalt, which will be explored in future research.

Keywords: L-DMA; mussel bionic materials; modified asphalt; modification mechanism; rheological properties

Citation: Wu, J.; Liu, Q.; Wu, S. Bionic Synthesis of Mussel-like Adhesive L-DMA and Its Effects on Asphalt Properties. *Materials* **2022**, *15*, 5351. https://doi.org/10.3390/ma15155351

Academic Editor: Simon Hesp

Received: 29 June 2022
Accepted: 2 August 2022
Published: 3 August 2022

1. Introduction

The road network in China has significantly improved in recent years. According to the latest statistical bulletin of the Ministry of Transport of China, the total mileage of roadways in China has occupied first place globally, reaching 5.198 million kilometers. Asphalt pavement provides outstanding performance and ride comfort, accounting for more than 90% of the highways [1,2]. Nevertheless, as the mileage of asphalt pavements that must be maintained grows, the disease and disposal issues of asphalt pavement have drawn people's attention. During the service life, the pavement will be impacted by loading, temperature, water, and other factors [3], causing the asphalt to undergo physical and chemical changes such as volatilization, oxidation, decomposition, and polymerization [4]. The performance of asphalt will deteriorate as a result of these consequences, the stress relaxation capability of the asphalt pavement declines as stiffness rises, and the asphalt pavement is susceptible to shrinkage fractures at low temperatures. The pavement quality will worsen if the cracks are not fixed or treated promptly, resulting in mesh cracks, cracks, rutting, and other more significant pavement diseases [5]. When the load interval or temperature rises, some microcracks will mend spontaneously (self-healing performance of asphalt), but other cracks that cannot be healed will spread due to vehicle load and form larger cracks. Water in the road environment, such as rainwater, will seep into the cracks and then gradually infiltrate into the bottom of the road under vehicle load, which will increase the water content of the road base and even the subgrade and diminish the strength.

There are several treatments for different kinds of cracks [6], such as slurry seal, micro-surfacing, overlay, joint grouting, seam sticking, etc. Among these treatment approaches, joint grouting is the most extensively utilized due to its easy operation and high cost efficiency. At present, the asphalt pavement crack repair materials utilized in China generally comprise thermal asphalt, cold asphalt, and organic chemical specialist materials [7–9]. Polymer-modified asphalt, polymer-modified emulsified asphalt, and resin are widely employed as joint grouting materials in practical engineering applications [10]. However, from the standpoint of practical use, they have the faults of weak adhesion between grouting materials and fracture interface under water circumstances. To tackle this issue, a novel class of asphalt-based bionic material with good water resistance and strong adhesion was developed in this research.

Bionic design is an effective technique to emulate certain life traits and accomplish the intelligent design of materials. Mussels are a form of a marine creature with excellent adhesive capacity. The exceptional waterproof adhesion and universal adhesive qualities of mussels are particularly appealing. It is observed that mussels secrete various proteins via the Byssus, which are engaged in the adhesion process and are connected to the adhesion capacity and are solidified quickly after contact with saltwater to create an adhesive plate and securely bond to the substrate. The research discovered that the primary component of mucus is mussel adhesion protein (MAP) [11,12]. One of its most distinctive structural properties is the amino acid containing DOPA (dihydroxyphenylalanine). The catechol group (also known as catechol) in DOPA possesses chemical adaptability, versatility, and affinity diversity, which is the key to the super-strong adhesion property of mussels [13,14].

Due to the chemical versatility and affinity diversity of catechol functional groups, DOPA exhibits unique chemical properties. Thus, DOPA can boost the adhesion and cohesion of mussel adhesive proteins [15–17]. Through its chemical versatility, DOPA can undergo different reactions to achieve bonding and cross-link curing [18,19]. As shown in Figure 1, DOPA can be oxidized to Dopaquinone under certain conditions. After that, Dopaquinone can be rearranged and dehydrogenated to form Dehydro DOPA, which results in further cross-linking; DOPA quinone can also undergo Michael addition and Schiff base formation with amino and thiol groups; it can also form Carboxylated dihydroxyindole through intramolecular cyclization, and finally form cross-links. At the same time, Dopaquinone and DOPA can also be disproportionated to form free radicals, and finally coupled to tannins (Quinone tanning).

The magical adhesion properties of mussel adhesive proteins make them have unlimited potential applications in biomedical, national defense, marine engineering, and other fields. However, the ultra-low preparation quantity and expensive price limit the application of this excellent adhesive material. In order to obtain the adhesion ability equivalent to that of mussel adhesion protein in a wet state, people began to develop adhesion materials based on mussel adhesion protein. The study of adhesion materials using mussel adhesion protein was a gradual transition from natural mussel extraction to artificial synthesis. The first made use of the traditional protein extraction method and recombinant gene cloning method to prepare mussel adhesive materials. In the late 1980s, BioPolymers. Ltc successfully prepared mussel adhesive protein from mussel byssus glands by traditional extraction methods and developed a 'Cell-Tak' superadhesive [20,21]. 'Cell-Tak' has also been successfully used to enhance the adhesion of biomedical slides, plastics, or metal surfaces to cellular tissue samples. Subsequently, Hwang et al. recombined, expressed and purified the genes of Mgfp-5 and Mgfp-3A mussel adhesion proteins by recombination technology to obtain a superadhesive like 'Cell-Tak' [22,23]. The recombinant proteins with adhesion properties similar to those of 'Cell-Tak' were obtained. However, the cost of recombinant gene technology is also surprising due to the fact that the regulatory mechanism is still unclear. Lv et al. also tried to perform some work on the extraction of natural protein and genetic engineering preparation of mussel adhesive materials [24]. Although mussel adhesion protein can be obtained by direct extraction or gene recombination, the separation and purification of mussel adhesion protein from mussels is an extremely complex process with a meager yield. This makes it very challenging to directly apply this adhesive to commercial applications, however, it also further highlights the urgency of developing synthetic mussel bionic adhesives with strong waterproof bonding properties. According to the intrinsic adhesion mechanism of mussel adhesive protein (DOPA, the key factor of adhesion performance), this paper aims to prepare polymeric adhesives to mimic the effect of natural mussel adhesion proteins.

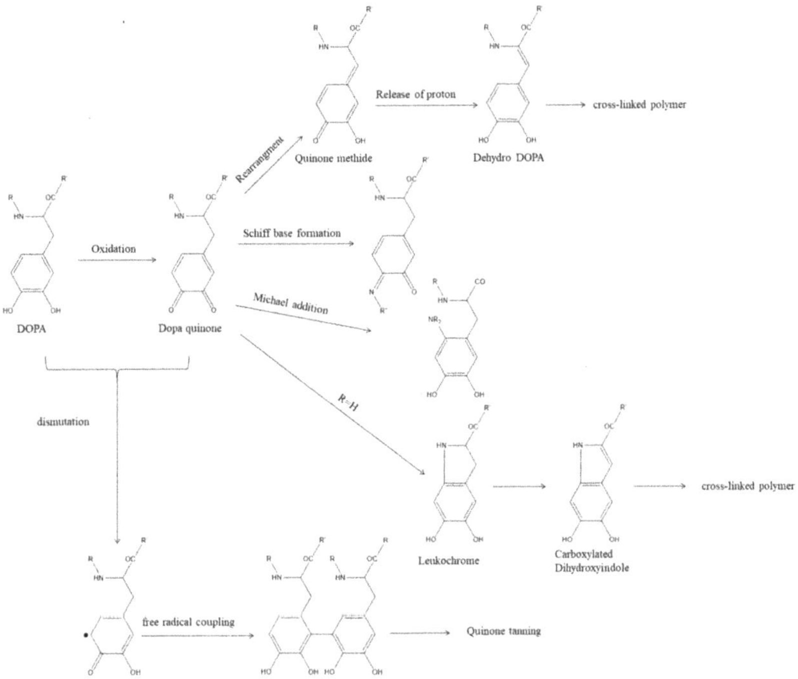

Figure 1. Possible cross-linking pathways of DOPA.

Although biomimetic materials have been used in asphalt materials, this work marks the first time the bionic mussel adhesion protein has been applied to asphalt materials. By analyzing its impact on the overall performance of asphalt, this study aims to investigate the viability of using bionic mussel adhesion protein in asphalt. L-Dopa Methacrylic anhydride (L-DMA) was synthesized based on the features of bionic mussel adhesion protein in this study and used to modify 70# base asphalt with varied quantities to prepare L-DMA-modified asphalt (L-DMA-A). The thermal stability and structural composition of L-DMA were investigated. In addition, the high-temperature rutting resistance and low-temperature cracking resistance of L-DMA-modified asphalt were examined by DSR and BBR. Finally, the modification mechanism of L-DMA-modified asphalt was analyzed by FTIR. The flowchart of the experiment design is shown in Figure 2.

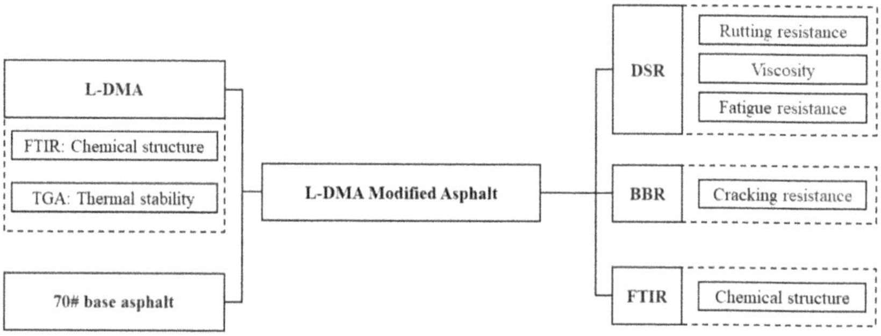

Figure 2. Flowchart of the experiment design.

2. Materials and Experiments

2.1. *Raw Materials*

In this study, the chemicals used to prepare L-DMA included sodium tetraborate ($Na_2B_4O_7$•$10H_2O$), sodium bicarbonate ($NaHCO_3$), Levodopa (L-DOPA), methacrylic anhydride (MAA), magnesium sulfate ($MgSO_4$), ethyl acetate (EAc), tetrahydrofuran (THF), and cyclohexane. The first four chemicals were purchased from Aladdin Ltd. (Shanghai Aladdin Biochemical Technology Co. Ltd., Shanghai, China) and the latter three were purchased from Sinopharm Chemical Reagent Co. Ltd., Shanghai, China. All chemicals were chemically pure and used without further purification. The asphalt used, 70# base asphalt, was from Hubei Guochuang Hi-tech Materials Co. Ltd., of China (Wuhan), and the basic properties are shown in Table 1.

Table 1. Properties of the 70# base asphalt.

Properties	Test Value	Requirements
Penetration (25 °C, 0.1 mm)	68.9	60–80
Ductility (15 °C, cm)	184	≥100
Softening point (°C)	48.4	≥46
Density (g/cm^3)	1.034	-

2.2. *Synthesis of L-DMA*

The mechanism of the synthesis process of L-DMA [25] is shown in Figure 3. An amount of 24 g of $Na_2B_4O_7$•$10H_2O$ and 10 g of $NaHCO_3$ were accurately weighed and dissolved in 200 mL deionized water and aerated with Ar for 20 min, then 10.4 g of L-DOPA (52.8 mmol) was added and continued to be bubbled with Ar. After 5 min, 9.5 g of MAA (58.1 mmol) was dissolved in 50 mL of THF and slowly dropped to the above solution. During the dropwise addition, 1 mol/L NaOH solution was used to keep the pH of the solution above 8. The reaction mixture continued to be stirred for 18 h at room temperature and protected from light.

The reaction mixture was filtered by extraction to remove the precipitate from the bottom of the flask, and then the pH was adjusted to less than 2 with 4 mol/L HCl solution, followed by extraction with 100 mL EAc. The organic phase was collected after three extractions with 100 mL of EAc, and the aqueous phase was removed. An amount of 4 g of anhydrous magnesium sulfate was added to the organic phase, stirred for 12 h, and filtered to remove the residual water. Finally, the organic phase was concentrated by a rotary evaporator to 50 mL.

Finally, the concentrated ethyl acetate solution was slowly dropped into 400 mL of cyclohexane with high-speed stirring (at 2000 rpm). After the dropwise addition, the solution was stirred for 30 min, and then placed at 4 °C for 18 h to allow complete recrystallization. Finally, the solid was filtered out and dried in a vacuum oven. The synthesis process is shown in Figure 4.

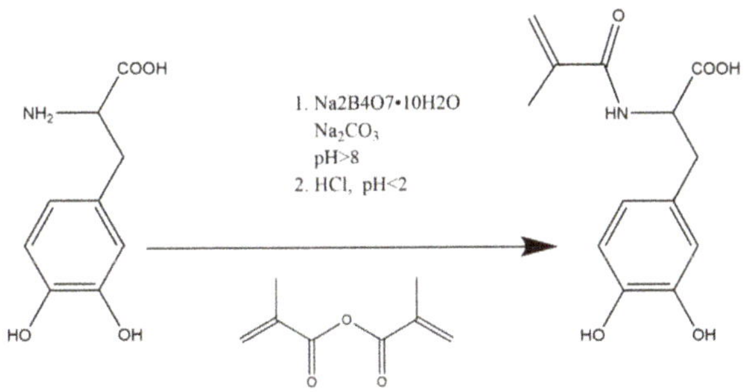

Figure 3. Mechanism of the synthesis of L-DMA.

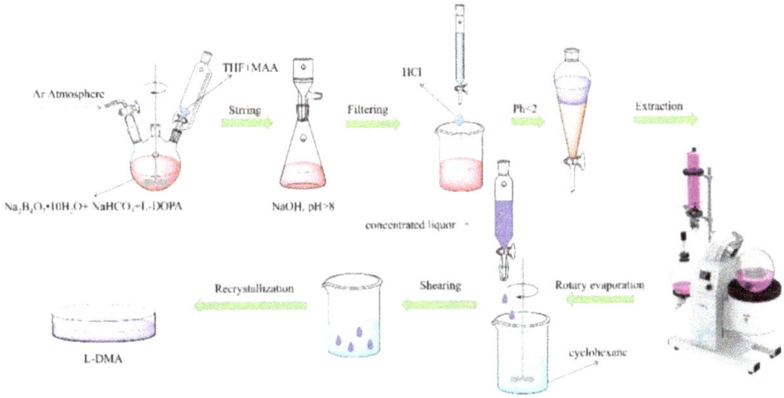

Figure 4. Preparation procedure of L-DMA.

2.3. Chemical Structure and Thermal Stability Characterizations of L-DMA

FTIR spectra of MAA, L-DOPA, and L-DMA were recorded on a Thermo Nicolet 6700 Fourier-transform spectrophotometer (Waltham, MA, USA). In all cases, 16 scans were collected at a resolution of 2 cm^{-1} for the spectrum in the range of 4000–400 cm^{-1}.

TGA was used to determine the thermal stability of L-DMA, which was carried out using the synchronous thermal analyzer (TA TGA55, New Castle, DE, USA) The L-DMA was protected in an N_2 atmosphere at a flow rate of 20 mL/min, in the temperature range from room temperature to 800 °C at a heating rate of 10 °C/min.

2.4. Preparation of L-DMA-Modified Asphalt

We added a certain mass ratio of L-DMA into the heated 70# base asphalt and mixed manually for ten minutes. To guarantee a good dispersion of L-DMA, a shear apparatus was used to disperse L-DMA at a rate of 4000 r/min for 40 min at a temperature of 170 °C. The dosage of L-DMA was 1%, 2.5%, 5%, 7.5%, and 10% of asphalt, respectively.

2.5. Property Characterizations of L-DMA-Modified Asphalt

2.5.1. DSR Test

DSR (MCR-101, Anton Paar, Graz, Austria) examined the rheological properties of L-DMA-A. Asphalt's viscoelasticity is significantly impacted by exposure temperature changes as a typical viscoelastic material [26,27]. Specific test parameter settings for the temperature sweep are shown in the Table 2.

Table 2. Test parameters of temperature sweep.

Temperature	Plate Diameter	Plate Gap	Shear Frequency	Heating Rate
−10–30 °C	25 mm	1 mm	10 rad/s	2 °C/min
30–80 °C	8 mm	2 mm		

Asphalt pavement will be affected by vehicle load during the service process. Loading with different frequencies will cause different deformations of asphalt pavement, and the deformation size determines the high-temperature performance of asphalt pavement. Viscosity is an essential parameter of asphalt and the addition of L-DMA will affect the viscosity of asphalt binder. Therefore, it is necessary to study the variation trend of the viscosity of 70# asphalt and L-DMA-modified asphalt under different load frequencies. The specimens were subjected to a frequency sweep in the range of 0.01–10 Hz. In order to meet the practical application of pavement, the temperature was set to 58 °C.

Additionally, L-DMA-modified asphalt with different L-DMA contents were tested by loading–unloading mode in the Multiple Stress Creep and Recovery (MSCR) test, which can be used to characterize the high-temperature performance of asphalt [28]. In this paper, the MSCR tests under stresses of 0.1 kPa and 3.2 kPa were carried out at 58 °C. The whole test contained 20 cycles of creep and recovery. Specifically, the first 10 cycles were conducted under the stress of 0.1 kPa, and the following 10-cycle test under the constant stress of 3.2 kPa was further conducted to finish the whole test. Figure 5 shows the creep and recovery (loading–unloading) curves of asphalt in one cycle.

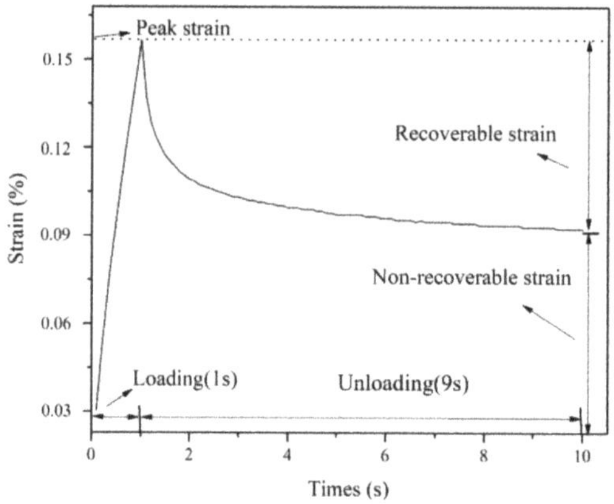

Figure 5. Change of strain with time during one creep—recovery cycle.

According to AASHTO M 332 Standard [29], the two evaluation indexes obtained by the MSCR test are the creep recovery rate R (%) and nonrecoverable creep compliance J_{nr}, which are used to evaluate the viscoelastic characteristics of asphalt under high-temperature conditions. The nonrecoverable creep compliance J_{nr} is the ratio of ultimate unrecoverable strain and applied stress in each loading cycle, which describes the contribution of asphalt to high-temperature deformation of asphalt mixtures. The creep recovery rate R (%) is the percentage of recoverable strain and total creep strain in the loading cycle, which characterizes the high-temperature elastic recovery performance of asphalt and reflects the influence of delayed elastic deformation of asphalt to a certain extent. According to the Equations (1)–(3) below, R (%) and J_{nr} under different stresses can be calculated.

$$J_{nr100}\left(\frac{1}{kPa}\right) = \frac{1}{10}\left\{\sum_{1}^{10} \frac{non\text{-}recoverable\ strain}{0.1}\right\} \quad (1)$$

$$J_{nr3200}\left(\frac{1}{kPa}\right) = \frac{1}{10}\left\{\sum_{1}^{10}\frac{non\text{-}recoverable\ strain}{3.2}\right\} \tag{2}$$

$$R(\%) = \frac{1}{10}\left\{\sum_{1}^{10}\frac{recoverable\ strain}{peakstrain} \times 100\%\right\} \tag{3}$$

2.5.2. BBR Test

Low-temperature cracks are inevitable in the use of asphalt pavement. Improving the low-temperature performance of asphalt pavement is the top priority of road workers. In general, BBR is most commonly used for the low-temperature performance evaluation of asphalt. According to the specification AASHTO T313-12 [30], 70# asphalt and L-DMA-modified asphalt beams were prepared, respectively, then held 1 Dh at the test temperature and were subjected to a 4 min loading test to measure the creep stiffness modulus S and creep rate m. S and m were selected at 60 s, if $S \leq 300$ MPa and $m \geq 0.3$ at this condition, then the asphalt is applicable at this temperature condition.

2.5.3. FTIR Spectroscopy Analysis

To study the chemical structure changes of L-DMA after modified asphalt and the modification mechanism, the FTIR Spectroscopy test was carried out. The sample was dissolved in carbon disulfide and then dropped onto a clear KBr lens. After drying for a moment, the sample was prepared and put into the instrument for testing.

3. Results and Discussion

3.1. Structural Analysis and Thermal Stability Analysis of L-DMA

The infrared spectra of L-DMA are shown in Figure 6. The absorption peak at 3370 cm^{-1} was the vibration peak of catechol phenolic hydroxyl, and the formation of hydrogen bonds widened the peak area. The N-H stretching vibration peak appeared at 3202 cm^{-1}. The C=O stretching vibration peak at 1730 cm^{-1} originated from the amide group, and the hydrogen bond made the electron shift outward and the bond force weaken. The bending vibration and deformation vibration of sec-amide appeared at 1655 cm^{-1} and 1520 cm^{-1}, respectively. However, the stretching vibration of C=C overlapped with the bending vibration of N-H, so the absorption peak appeared at 1655 cm^{-1}. At the same time, the absorption peaks at 1605 cm^{-1} and 1450 cm^{-1} were also observed, which were the stretching vibration of the L-DOPA aromatic ring. The single peak at 1260 cm^{-1} is the C-N stretching vibration.

Compared with the spectra of the reaction monomers MAA and DOPA, in the FTIR spectra of L-DMA, the absorption peaks of methacrylic anhydride in MAA at 1787 and 1725 cm^{-1} had obviously disappeared and changed into a single peak of C=O at 1730 cm^{-1}, and two amino peaks at 3210 and 3075 cm^{-1} at L-DOPA became one, indicating that the primary amine in L-DOPA reacted to form the secondary amide. From the above analysis, MAA and L-DOPA reacted successfully to form L-DMA.

Figure 7 shows the thermal weight loss curve of the L-DMA monomer powder, the initial thermal decomposition temperature of L-DMA is 221.5 °C with a mass loss of 5%, which is mainly due to the volatilization of bound water. There was a sharp mass loss between 224 °C and 478.8 °C, which corresponded to the decomposition of the C-N bond and methacrylate. Subsequently, the slow mass loss of the sample from 480 °C to 800 °C was the decomposition of the L-DMA main structure. The maximum temperature during asphalt mixing and paving is about 180 °C. Therefore, L-DMA can be used for asphalt modification.

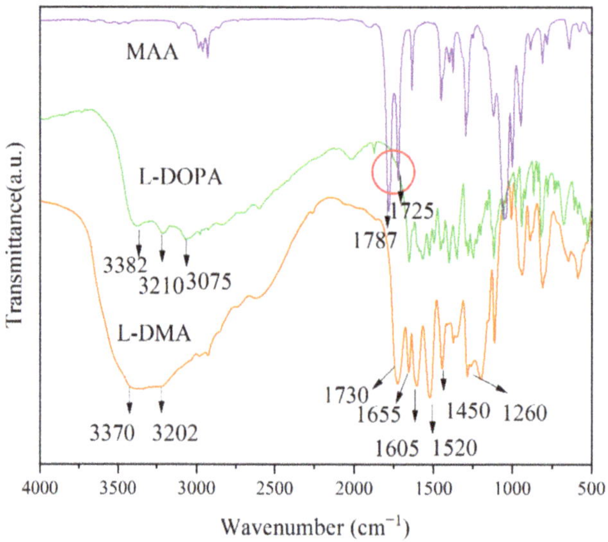

Figure 6. FT-IR spectra results of MAA, L-DOPA, and L-DMA.

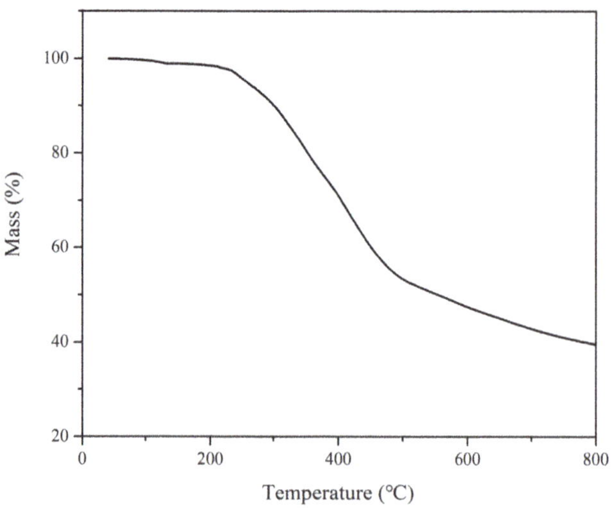

Figure 7. The TG curve of L-DMA.

3.2. Modification Mechanism of L-DMA

The modification mechanism of L-DMA-modified asphalt was analyzed by FTIR. Figure 8 compared the infrared spectra of L-DMA, 70# asphalt, and L-DMA-modified asphalt. It can be found that the characteristic absorption peaks of 70# base asphalt show asymmetric and symmetric stretching vibration of methylene at 2925 and 2854 cm^{-1}. C=C and C=O stretching vibrations appear at 1600 cm^{-1}; the absorption peaks of the benzene ring stretching vibration are 866 and 809 cm^{-1}. Combined with the analysis of the infrared spectrum of L-DMA in Figure 6, it can be inferred that the modification process of L-DMA on asphalt is mainly a physical modification process. With the addition of L-DMA, the conjugated effect between catechol and the aromatic benzene ring in asphalt resulted in the redshift of the vibration absorption peak of the benzene ring. The peak area widened at about 3370 cm^{-1}, suggesting that L-DMA can generate many hydrogen bonds with free hydroxyl and amino groups in asphalt through catechol groups, thus forming a strong hydrogen bond and promoting its combination with asphalt. The number of hydrogen bonds increased significantly with an increase in L-DMA content. The hydroxyl characteristic peak area of modified asphalt with 10% L-DMA content was larger than that of other contents. Moreover, due to the existence of the benzene ring, L-DMA can interact with the aromatic ring in asphalt through the Π-Π accumulation [31–33]. These can improve the adhesion performance of modified asphalt. At the same time, dense hydrogen bonds can also enhance the self-healing performance of asphalt [34–37]. Infrared spectra also show that L-DMA does not decompose and exists stably in asphalt.

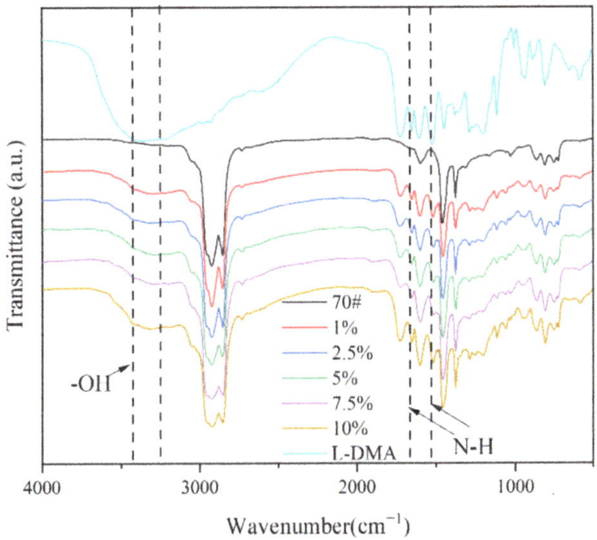

Figure 8. FT-IR spectra results of L-DMA, 70# base asphalt, and L-DMA-A.

3.3. *Rheological Properties of L-DMA-Modified Asphalt*

When L-DMA is added to asphalt as a modifier, physical and chemical modifications will inevitably occur, thus affecting the original structure and the viscoelastic properties of asphalt. In general, the composite shear modulus G* is the measurement of the total resistance of the material during repeated shear deformation [38,39]; the phase angle δ is the relative index of the number of recoverable and irreversible deformations.

The composite modulus and phase angle curves of L-DMA-modified asphalts with different contents of L-DMA in the temperature range of 30–80 °C are shown in Figure 9. It is clear that with an increase in L-DMA content, the composite modulus G* of modified asphalt increases and the phase angle δ decreases, indicating that the addition of L-DMA improves the shear deformation resistance of asphalt, and the elasticity of asphalt is obviously enhanced. The modified asphalt with 10% content has the most excellent high-temperature performance, which is mainly caused by the good adhesion role of L-DMA in asphalt. The catechol group in L-DMA generated a large number of hydrogen bonds with the free hydroxyl and amino groups in asphalt, thus forming a strong hydrogen bond and promoting the integration with asphalt [40,41]. Moreover, catechol and the asphalt aromatic ring accumulated to produce cohesion [42,43], increasing the proportion of elastic components in the modified asphalt and improving the shear deformation resistance of asphalt.

The evaluation index of anti-rutting performance of asphalt is quantified by the rutting factor, which is expressed as G*/sinδ. In general, the asphalt binder with a larger rutting factor shows stronger anti-rutting ability [44]. According to Superpave specification, AASHTO: MP1, at the highest pavement design temperature, the G*/sinδ value of matrix asphalt should be at least 1.0 kPa [45]. Figure 10 shows the rutting factors of base and different modified asphalts at the temperature range of 30–80 °C. The results show that the rutting factors of modified asphalts rise with an increase in L-DMA content. The modified asphalt with 10% content has the largest rutting factor because the incorporation of L-DMA will introduce phenolic hydroxyl groups, which will form hydrogen bonds with the aromatic components in the light component of the asphalt, thereby increasing the adhesion of the material. At high temperatures, the catechol group in L-DMA will be oxidized into semiquinone and quinone structures, which improves the overall cohesive energy [46,47]. In addition, the benzene ring can also produce Π-Π conjugate with the aromatic ring in asphalt, which increases the cohesive energy. These effects increase the cohesion and adhesion of modified asphalt. These enhanced bonding structures will hinder the free flow of modified asphalt, enhance the elastic recovery ability, and improve the rutting resistance.

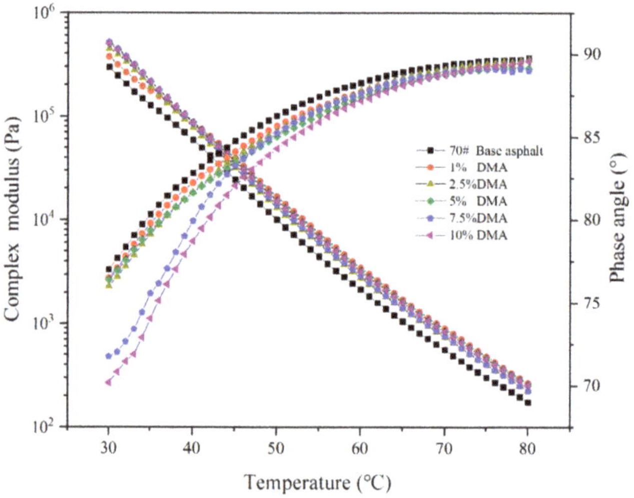

Figure 9. The result of the temperature sweep test (30–80 °C).

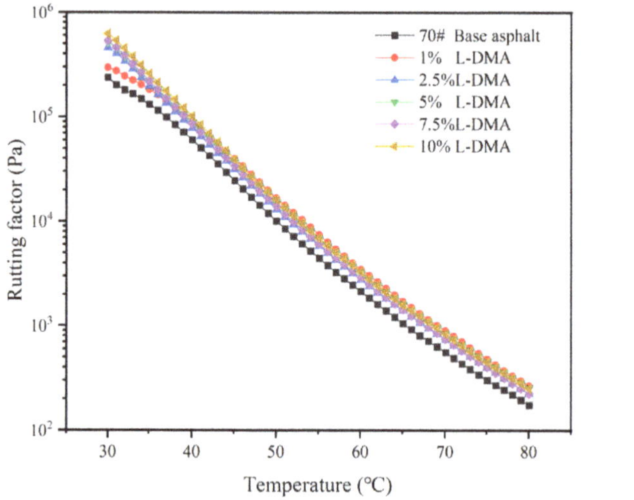

Figure 10. Rutting factor and temperature curves of different asphalts.

The low-temperature rheological property of asphalt binder is one of the important factors for its cracking resistance. The SHRP standard defines G*sinδ as the evaluation index of fatigue cracking. The smaller the value of G*sinδ, the better the ability to resist fatigue damage [48–50]. In this paper, the low-temperature rheological properties of 70# base asphalt and modified asphalt with different contents of L-DMA were tested in the temperature range of −10–30 °C and the test results are shown in Figure 11. The composite modulus of modified asphalt decreases with an increase in temperature, while the phase angle increases. When the content of L-DMA is 1%, the modified asphalt has the highest composite modulus. Then, with an increase in L-DMA content, the composite modulus of modified asphalt shows a downward trend. When the content of L-DMA is 10%, the composite modulus of modified asphalt is the lowest and is lower than that of 70# asphalt. The reason is the change of L-DMA's effect, and when the content is low, it mainly plays a role of filling in asphalt, which makes the low-temperature performance of modified asphalt worse. However, with an increase in L-DMA content, unsaturated rings such as benzene, quinone, and pyridine in aromatic components of asphalt will conjugate with the benzene ring in L-DMA, and the rings overlap with each other to form a layered structure, which reduces the composite modulus G* of modified asphalt.

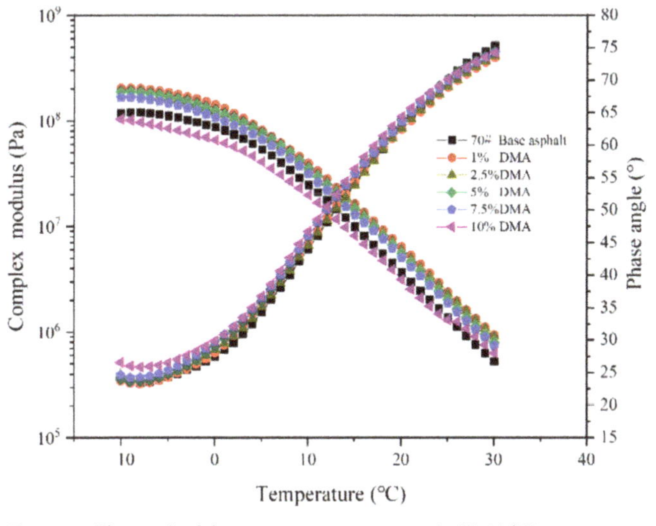

Figure 11. The result of the temperature sweep test (−10–30 °C).

Figure 12 shows the changing trend of the fatigue factors of 70# asphalt and modified asphalts with different contents of L-DMA at different temperatures. The results show that the fatigue factor of modified asphalt increases first and then decreases with an increase in L-DMA content. At a high content of 10%, the addition of L-DMA decreases the fatigue factor and enhances the fatigue resistance of asphalt due to the conjugation between the unsaturated rings of L-DMA and asphalt.

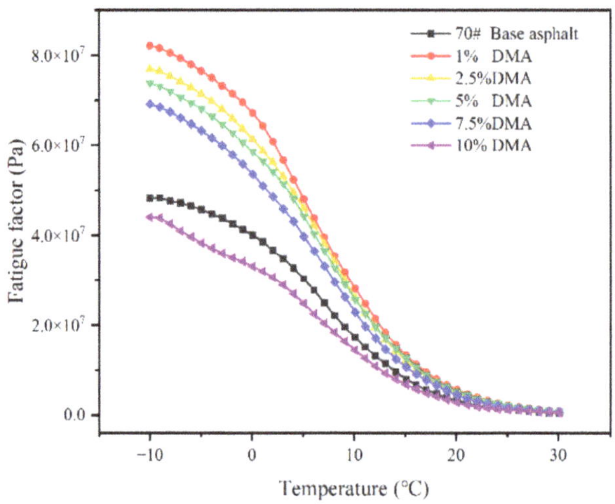

Figure 12. Fatigue factor and temperature curves of different asphalts.

The viscosities of 70# asphalt and L-DMA-modified asphalts mixed with different contents of L-DMA are shown in the Figure 13. It can be seen from the figure that with an increase in loading frequency, the viscosity of asphalt at 58 °C shows a downward trend. The viscosity of 70# asphalt is the lowest, while 10% L-DMA-modified asphalt has the highest viscosity. The viscosity can reflect the deformation resistance of the material to a certain extent, and the viscosity of 10% L-DMA-modified asphalt can still maintain a high level at 58 °C, which indicates that the addition of L-DMA contributes greatly to the improvement of the high-temperature performance of asphalt.

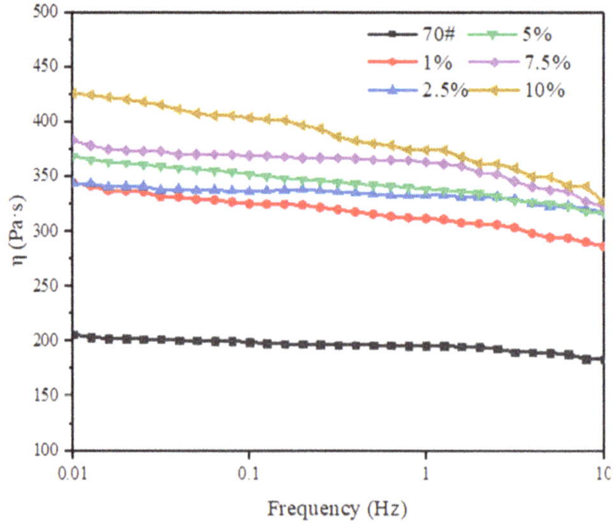

Figure 13. Frequency—Viscosity Curve at 58 °C.

The strain changes of 70# base asphalt and L-DMA-modified asphalt under different stress levels at 58 °C are shown in Figure 14. The result shows that the 70# base asphalt has the largest cumulative strain. With an increase in L-DMA content, the cumulative strain of the modified asphalt gradually decreases. This is because the cohesion between

L-DMA and asphalt improves the elasticity of asphalt materials, significantly improving the recoverable deformation ability of asphalt and thus improving the high-temperature performance of asphalt.

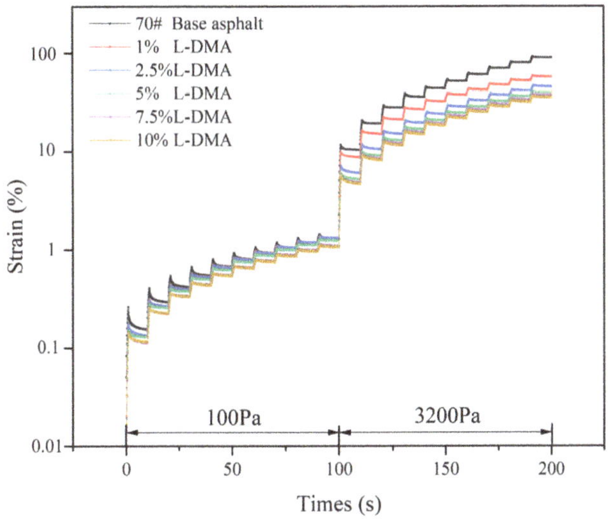

Figure 14. Strain during the MSCR test at 58 °C.

Table 3 shows the experimental results of creep recovery rate *R* (%) and unrecoverable creep compliance (J_{nr}) of base asphalt and modified asphalts at different stress levels. The results show that under the action of 100 Pa and 3200 Pa, the creep recovery rate R (%) of L-DMA-modified asphalt increases with an increase in L-DMA content, indicating that the addition of L-DMA enhances the recovery deformation ability of asphalt. The unrecoverable creep compliance (J_{nr}) decreases with an increase in L-DMA content, indicating that the incorporation of L-DMA improves the permanent deformation resistance of asphalt, which is consistent with the results of the rutting factor test. Comparing the unrecoverable creep compliance (J_{nr}) of asphalt and L-DMA-modified asphalts under different stress levels, it can be found that the unrecoverable creep compliance (J_{nr}) of 70# base asphalt changes greatly under 100 Pa and 3200 Pa, while the unrecoverable creep compliance (J_{nr}) of L-DMA-modified asphalts does not change significantly under different stress levels. It indicates that the addition of L-DMA reduces the dependence of asphalt on shear stress and enhances the shear deformation resistance of asphalt. The hydrogen bonds generated by the addition of L-DMA have certain reversibility, so the creep recovery rate *R* (%) of L-DMA-modified asphalt has greatly improved.

Table 3. The MSCR parameters for asphalt samples.

Samples	*R*-100	*R*-3200	J_{nr}-100	J_{nr}-3200
70# Base asphalt	4.05	2.55	2.66	3.56
1% L-DMA	5.67	2.95	2.62	2.75
2.5% L-DMA	6.29	3.58	2.48	2.43
5% L-DMA	6.32	3.62	2.18	2.23
7.5% L-DMA	6.49	3.91	2.14	2.17
10% L-DMA	7.32	4.11	1.78	1.83

3.4. Low-Temperature Crack Resistance of L-DMA Modified Asphalt

It can be seen from Figure 15 that with a decrease in temperature, the S modulus increases but the m value decreases, indicating that the lower the temperature, the worse the low-temperature performance of asphalt. When the temperature is lower than −12 °C, the m of 70# asphalt and L-DMA-modified asphalt will be far less than 0.3, and the S is much higher than 300 MPa, indicating that when the temperature is lower than −12 °C, the low-temperature performance of 70# asphalt and L-DMA-modified asphalt cannot meet the standard. At −12 °C, the S and m of 70# asphalt basically meet the requirements of the specification. It can be seen that with an increase in L-DMA content, the S modulus increases first and then decreases gradually, indicating that less L-DMA content will affect the deformation capacity of asphalt. However, with an increase in L-DMA content, the deformation capacity of modified asphalt at low temperatures is becoming better and better, and the deformation capacity of modified asphalt is the best when the content of L-DMA is 10%, meeting the standard. The m decreases first and then increases, suggesting that with the incorporation of L-DMA, the stress relaxation ability of asphalt to external load decreases first and then increases. When the content of L-DMA is 10%, the m is the largest, and the modified asphalt has the best stress relaxation ability to deal with external loads.

Combined with the low-temperature rheological test results of DSR, it is found that the low content of L-DMA has an adverse effect on the low-temperature performance of asphalt, but with an increase in L-DMA content, it has a significant improvement on the low-temperature performance of asphalt. This is mainly because the low content of L-DMA plays a filling role in asphalt, increasing the composite modulus G* and creep stiffness modulus S of L-DMA-modified asphalt, reducing the creep rate m, and worsening the low-temperature performance of modified asphalt. However, with an increase in L-DMA content, the unsaturated rings such as benzene, quinone, and pyridine in the aromatic components of asphalt will conjugate with the benzene ring in L-DMA, and the rings overlap with each other to form a layered structure. The interlayer friction is small, which reduces the G* value of modified asphalt, decreases the S value, and increases the m value, thereby improving the low-temperature performance of asphalt. This is consistent with the results of the low-temperature sweep test in DSR.

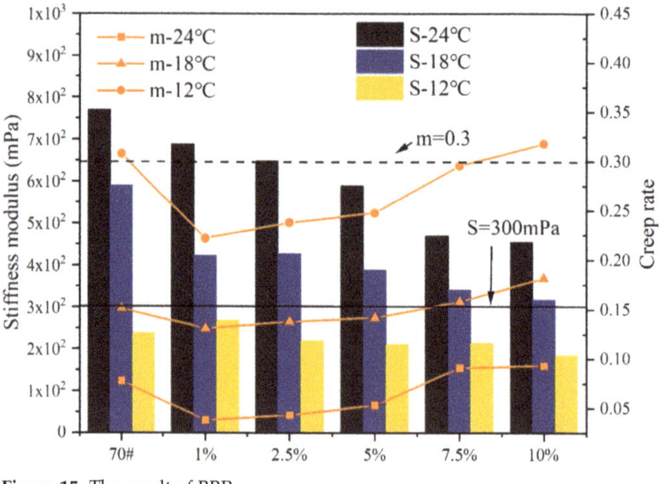

Figure 15. The result of BBR.

4. Conclusions

In this study, the mussel adhesion protein material L-DMA was synthesized and used to modify asphalt. The deformation and cracking resistances of L-DMA-modified asphalt were studied by DSR temperature sweep, frequency sweep, multi-stress creep tests, and BBR tests. In addition, the modification mechanism of L-DMA-modified asphalt was analyzed by FTIR test. According to the test results, the following conclusions can be drawn.

- Comparing with MAA and L-DOPA, the infrared spectrum of L-DMA has new characteristic peaks at 3370 cm^{-1}, 3202 cm^{-1}, 1730 cm^{-1}, 1655 cm^{-1}, 1520 cm^{-1}, 1605 cm^{-1}, 1450 cm^{-1}, and 1260 cm^{-1}, which proves that L-DMA was synthesized successfully. Moreover, L-DMA has good thermal stability and can be used in asphalt materials.
- The modification of L-DMA to asphalt is mainly a physical process. Hydrogen bond and aromatic ring conjugation introduced by catechol can improve the adhesion of asphalt.
- L-DMA has a noticeable improvement impact on the high-temperature performance of asphalt due to the strong hydrogen bonding, conjugation of benzene rings, and transformation of quinone and semiquinone by catechol at high temperatures. Compared with 70# asphalt, L-DMA-modified asphalt has a higher composite modulus, rutting factor, viscosity, and creep recovery rate.
- The impact of LDMA on the low-temperature fracture resistance of asphalt is complicated. With the rise of L-DMA content, the low-temperature performance of modified asphalt drops initially and later increases. When the L-DMA content reaches 10%, the low-temperature fracture resistance of L-DMA modified asphalt will be higher than that of 70# asphalt due to the layered structure formed by aromatic ring conjugation.

Overall, the addition of L-DMA has a positive impact on the high-temperature rutting resistance and a high content of L-DMA will have some improvement on the low-temperature cracking resistance of asphalt. The usage of L-DMA as a kind of mussel adhesion protein material improves the performance index of asphalt, and, theoretically, there is a possibility to improve water resistance and self-healing properties of asphalt, which will be explored in future research. Furthermore, due to the high cost and complex preparation process of L-DMA, further research is needed to decrease the cost and increase the preparation efficiency.

Author Contributions: Conceptualization, Q.L.; Investigation, J.W.; Methodology, J.W. and S.W.; Writing—original draft, J.W.; Writing—review and editing, Q.L. and S.W. All authors have read and agreed to the published version of the manuscript.

Funding: This research was funded by the National Natural Science Foundation of China (No.51978547), Technological Innovation Major Project of Hubei Province (2019AEE023) and Key R&D Program of Hubei Province (2020BCB064).

Institutional Review Board Statement: Not applicable. Studies do not involve humans or animals.

Informed Consent Statement: Not applicable.

Acknowledgments: The work presented in this paper is financially supported by the National Natural Science Foundation of China (No.51978547), Technological Innovation Major Project of Hubei Province (2019AEE023) and Key R&D Program of Hubei Province (2020BCB064).

Conflicts of Interest: The authors declare no conflict of interest.

References

1. Wang, C.; Wang, M.; Chen, Q.; Zhang, L. Basic performance and asphalt smoke absorption effect of environment-friendly asphalt to improve pavement construction environment. *J. Clean. Prod.* **2022**, *333*, 130142. [CrossRef]
2. Behnood, A.; Gharehveran, M.M. Morphology, rheology, and physical properties of polymer-modified asphalt binders. *Eur. Polym. J.* **2019**, *112*, 766–791. [CrossRef]
3. Li, Y.; Feng, J.; Yang, F.; Wu, S.; Liu, Q.; Liu, Z.; Li, C.; Gu, D.; Chen, A.; Jin, Y. Gradient aging behaviors of asphalt aged by ultraviolet lights with various intensities. *Constr. Build. Mater.* **2021**, *295*, 123618. [CrossRef]
4. Li, J.; Yu, J.; Wu, S.; Xie, J. The mechanical resistance of asphalt mixture with steel slag to deformation and skid degradation based on laboratory accelerated heavy loading test. *Materials* **2022**, *15*, 911. [CrossRef]
5. Moriyoshi, A.; Jin, T.; Nakai, T.; Ishikawa, H.; Tokumitsu, K.; Kasahara, A. Construction and pavement properties after seven years in porous asphalt with long life. *Constr. Build. Mater.* **2014**, *50*, 401–413. [CrossRef]
6. Hu, D.; Gu, X.; Yang, F.; Zhou, Z.; Bo, W.; Cui, B.; Pei, J. Atomic mechanisms of separation failure at the asphalt–aggregate interface and its dependence on aging and rejuvenation: Insights from molecular dynamics simulations and DFT calculations. *Appl. Surf. Sci.* **2022**, *598*, 153775. [CrossRef]

7. Gong, Y.; Wu, S.; Zhang, Y.; Pang, Y.; Ma, Y. Investigation of the High-Temperature and Rheological Properties for Asphalt Sealant Modified by SBS and Rubber Crumb. *Polymers* **2022**, *14*, 2558. [CrossRef] [PubMed]
8. Guo, M.; Tan, Y.; Du, X.; Lv, Z. Study on the cohesion and adhesion of hot-poured crack sealants. *Front. Struct. Civ. Eng.* **2017**, *11*, 353–359. [CrossRef]
9. Rubio, M.C.; Martínez, G.; Baena, L.; Moreno, F. Warm mix asphalt: An overview. *J. Clean. Prod.* **2012**, *24*, 76–84. [CrossRef]
10. Mirzaiyanrajeh, D.; Dave, E.V.; Sias, J.E.; McKay, Z.D.; Blankenship, P.P. Comprehensive Evaluation of Properties and Performance of Asphalt Mixtures with Reactive Isocyanate and Styrene-Butadiene-Styrene-Modified Binders. *J. Mater. Civ. Eng.* **2022**, *34*, 04022228. [CrossRef]
11. Deming, T.; Yu, M.; Hwang, J. Mechanical studies of adhesion and crosslinking in marine adhesive protein analogs. *Polym. Mater. Sci. Eng.* **1999**, *80*, 471–472.
12. Fant, C.; Elwing, H.; Höök, F. The influence of cross-linking on protein–protein interactions in a marine adhesive: The case of two byssus plaque proteins from the blue mussel. *Biomacromolecules* **2002**, *3*, 732–741. [CrossRef] [PubMed]
13. Ryu, J.H.; Messersmith, P.B.; Lee, H. Polydopamine surface chemistry: A decade of discovery. *ACS Appl. Mater. Interfaces* **2018**, *10*, 7523–7540. [CrossRef]
14. Chung, H.; Glass, P.; Pothen, J.M.; Sitti, M.; Washburn, N.R. Enhanced adhesion of dopamine methacrylamide elastomers via viscoelasticity tuning. *Biomacromolecules* **2011**, *12*, 342–347. [CrossRef]
15. Waite, J.H. Evidence for a repeating 3,4-dihydroxyphenylalanine- and hydroxyproline-containing decapeptide in the adhesive protein of the mussel, *Mytilus edulis* L. *J. Biol. Chem.* **1983**, *258*, 2911–2915. [CrossRef]
16. Waite, J.H.; Tanzer, M.L. Polyphenolic substance of Mytilus edulis: Novel adhesive containing L-dopa and hydroxyproline. *Science* **1981**, *212*, 1038–1040. [CrossRef] [PubMed]
17. Yu, M. Role of L-3, 4-dihydroxyphenylalanine in mussel adhesive proteins. *J. Am. Chem. Soc.* **1999**, *121*, 5825–5826. [CrossRef]
18. Zhang, C.; Li, K.; Simonsen, J. A novel wood-binding domain of a wood–plastic coupling agent: Development and characterization. *J. Appl. Polym. Sci.* **2003**, *89*, 1078–1084. [CrossRef]
19. Lee, B.P.; Dalsin, J.L.; Messersmith, P.B. Synthesis and gelation of DOPA-modified poly (ethylene glycol) hydrogels. *Biomacromolecules* **2002**, *3*, 1038–1047. [CrossRef] [PubMed]
20. Strausberg, R.L.; Link, R.P. Protein-based medical adhesives. *Trends Biotechnol.* **1990**, *8*, 53–57. [CrossRef]
21. Hwang, D.S.; Gim, Y.; Yoo, H.J.; Cha, H.J. Practical recombinant hybrid mussel bioadhesive fp-151. *Biomaterials* **2007**, *28*, 3560–3568. [CrossRef] [PubMed]
22. Hwang, D.S.; Yoo, H.J.; Jun, J.H.; Moon, W.K.; Cha, H.J. Expression of functional recombinant mussel adhesive protein Mgfp-5 in *Escherichia coli*. *Appl. Environ. Microbiol.* **2004**, *70*, 3352–3359. [CrossRef] [PubMed]
23. Doraiswamy, A.; Dinu, C.; Cristescu, R.; Messersmith, P.; Chisholm, B.; Stafslien, S.; Chrisey, D.; Narayan, R. Matrix-assisted pulsed-laser evaporation of DOPA-modified poly (ethylene glycol) thin films. *J. Adhes. Sci. Technol.* **2007**, *21*, 287–299. [CrossRef]
24. Lv, Y.; Zhang, Y.; Gao, W.; Wang, Y. Expression and functional evaluation of *Mytilus galloprovincialis* foot protein type 5 (Mgfp-5), the recombinant mussel adhesive protein. *Afr. J. Biotechnol.* **2017**, *16*, 1840–1848.
25. Zhang, X.; Liu, H.; Yue, L.; Bai, Y.; He, J. Mussel-mimetic polymer underwater adhesives with l-Dopa functionality: Influencing adhesion properties and simplified operation procedures. *J. Mater. Sci.* **2020**, *55*, 7981–7997. [CrossRef]
26. Krishnan, J.M.; Rajagopal, K. On the mechanical behavior of asphalt. *Mech. Mater.* **2005**, *37*, 1085–1100. [CrossRef]
27. Zhang, H.; Chen, Z.; Xu, G.; Shi, C. Evaluation of aging behaviors of asphalt binders through different rheological indices. *Fuel* **2018**, *221*, 78–88. [CrossRef]
28. Zhao, Z.; Wu, S.; Liu, Q.; Xie, J.; Yang, C.; Wang, F.; Wan, P. Recycling waste disposable medical masks in improving the performance of asphalt and asphalt mixtures. *Constr. Build. Mater.* **2022**, *337*, 127621. [CrossRef]
29. *AASHTO M 332*; Standard Specification for Performance-Graded Asphalt Binder Using Multiple Stress Creep Recovery (MSCR) Test. American Association of State Highway and Transportation Officials: Washington, DC, USA, 2020.
30. *AASHTO T 313*; Standard Method of Test for Determining the Flexural Creep Stiffness of Asphalt Binder Using the Bending Beam Rheometer (BBR). American Association of State Highway and Transportation Officials: Washington, DC, USA, 2012.
31. Bandara, N.; Zeng, H.; Wu, J. Marine mussel adhesion: Biochemistry, mechanisms, and biomimetics. *J. Adhes. Sci. Technol.* **2013**, *27*, 2139–2162. [CrossRef]
32. Li, A.; Mu, Y.; Jiang, W.; Wan, X. A mussel-inspired adhesive with stronger bonding strength under underwater conditions than under dry conditions. *Chem. Commun.* **2015**, *51*, 9117–9120. [CrossRef] [PubMed]
33. Xiong, X.; Liu, Y.; Shi, F.; Zhang, G.; Weng, J.; Qu, S. Enhanced Adhesion of Mussel-inspired Adhesive through Manipulating Contents of Dopamine Methacrylamide and Molecular Weight of Polymer. *J. Bionic Eng.* **2018**, *15*, 461–470. [CrossRef]
34. Yan, M. Phase Structure and Self-healing Behavior of Liquid Crystalline Block Copolymer Elastomer Cross-Linked by Hydrogen Bonds. Ph.D. Thesis, Xiangtan University, Xiangtan, China, 2015.
35. Xie, Z.; Hu, B.-L.; Li, R.-W.; Zhang, Q. Hydrogen bonding in self-healing elastomers. *ACS Omega* **2021**, *6*, 9319–9333. [CrossRef] [PubMed]
36. Jian, X.; Hu, Y.; Zhou, W.; Xiao, L. Self-healing polyurethane based on disulfide bond and hydrogen bond. *Polym. Adv. Technol.* **2018**, *29*, 463–469. [CrossRef]

37. Shao, C.; Chang, H.; Wang, M.; Xu, F.; Yang, J. High-strength, tough, and self-healing nanocomposite physical hydrogels based on the synergistic effects of dynamic hydrogen bond and dual coordination bonds. *ACS Appl. Mater. Interfaces* **2017**, *9*, 28305–28318. [CrossRef] [PubMed]
38. Rodríguez-Alloza, A.M.; Gallego, J.; Giuliani, F. Complex shear modulus and phase angle of crumb rubber modified binders containing organic warm mix asphalt additives. *Mater. Struct.* **2017**, *50*, 1–9. [CrossRef]
39. Yang, C.; Wu, S.; Cui, P.; Amirkhanian, S.; Zhao, Z.; Wang, F.; Zhang, L.; Wei, M.; Zhou, X.; Xie, J. Performance characterization and enhancement mechanism of recycled asphalt mixtures involving high RAP content and steel slag. *J. Clean. Prod.* **2022**, *336*, 130484. [CrossRef]
40. McDowell, L.M.; Burzio, L.A.; Waite, J.H.; Schaefer, J. Rotational echo double resonance detection of cross-links formed in mussel byssus under high-flow stress. *J. Biol. Chem.* **1999**, *274*, 20293–20295. [CrossRef] [PubMed]
41. Taylor, S.W.; Chase, D.B.; Emptage, M.H.; Nelson, M.J.; Waite, J.H. Ferric ion complexes of a DOPA-containing adhesive protein from *Mytilus edulis*. *Inorg. Chem.* **1996**, *35*, 7572–7577. [CrossRef]
42. Li, J.; Celiz, A.; Yang, J.; Yang, Q.; Wamala, I.; Whyte, W.; Seo, B.; Vasilyev, N.; Vlassak, J.; Suo, Z. Tough adhesives for diverse wet surfaces. *Science* **2017**, *357*, 378–381. [CrossRef] [PubMed]
43. Silverman, H.G.; Roberto, F.F. Understanding marine mussel adhesion. *Mar. Biotechnol.* **2007**, *9*, 661–681. [CrossRef] [PubMed]
44. Lv, S.; Xia, C.; Yang, Q.; Guo, S.; You, L.; Guo, Y.; Zheng, J. Improvements on high-temperature stability, rheology, and stiffness of asphalt binder modified with waste crayfish shell powder. *J. Clean. Prod.* **2020**, *264*, 121745. [CrossRef]
45. Leng, Z.; Padhan, R.K.; Sreeram, A. Production of a sustainable paving material through chemical recycling of waste PET into crumb rubber modified asphalt. *J. Clean. Prod.* **2018**, *180*, 682–688. [CrossRef]
46. Vatankhah-Varnoosfaderani, M.; Hashmi, S.; GhavamiNejad, A.; Stadler, F.J. Rapid self-healing and triple stimuli responsiveness of a supramolecular polymer gel based on boron–catechol interactions in a novel water-soluble mussel-inspired copolymer. *Polym. Chem.* **2014**, *5*, 512–523. [CrossRef]
47. Xu, Y.J.; Wei, K.; Zhao, P.; Feng, Q.; Choi, C.K.K.; Bian, L. Preserving the adhesion of catechol-conjugated hydrogels by thiourea–quinone coupling. *Biomater. Sci.* **2016**, *4*, 1726–1730. [CrossRef] [PubMed]
48. Wang, J.; Wu, S.; Han, J.; Liu, X. Rheological properties of asphalt modified by supramolecular UV resistant material-LDHs. *J. Wuhan Univ. Technol. Mater. Sci. Ed.* **2012**, *27*, 805–809. [CrossRef]
49. Tao, G.; Xiao, Y.; Yang, L.; Cui, P.; Kong, D.; Xue, Y. Characteristics of steel slag filler and its influence on rheological properties of asphalt mortar. *Constr. Build. Mater.* **2019**, *201*, 439–446. [CrossRef]
50. Cong, P.; Wang, J.; Li, K.; Chen, S. Physical and rheological properties of asphalt binders containing various antiaging agents. *Fuel* **2012**, *97*, 678–684. [CrossRef]

materials

MDPI

Article

Quantitative Assessment of Road Performance of Recycled Asphalt Mixtures Incorporated with Steel Slag

Zipeng Wang [1], Shaopeng Wu [2,*], Chao Yang [2,*], Jun Xie [2], Yongli Xiao [3], Zenggang Zhao [2], Fusong Wang [4] and Lei Zhang [5]

1 Key Laboratory of Road and Traffic Engineering of the Ministry of Education, Tongji University, Shanghai 201804, China; wangzipeng_sjz@126.com
2 State Key Laboratory of Silicate Materials for Architectures, Wuhan University of Technology, Wuhan 430070, China; xiejun3970@whut.edu.cn (J.X.); zhaozenggang@whut.edu.cn (Z.Z.)
3 Baoshan Iron & Steel Co., Ltd., Shanghai 201900, China; xiaoyl@baosteel.com
4 School of Civil Engineering and Mechanics, Huazhong University of Science and Technology, Wuhan 430074, China; wangfs@whut.edu.cn
5 Department of Civil and Environmental Engineering, Norwegian University of Science and Technology, 7491 Trondheim, Norway; lei.zhang@ntnu.no
* Correspondence: wusp@whut.edu.cn (S.W.); hbyangc@whut.edu.cn (C.Y.)

Abstract: Circular utilization of reclaimed asphalt pavement (RAP) has received extensive attention for its economic and environmental benefits. The application of recycled asphalt mixtures (RAM) in the upper layer of asphalt pavement faces the issue of inferior anti-slip performance and durability. This study aims to recycle steel slag as virgin aggregates in RAM and quantitatively evaluate the service performance of RAM with steel slag. Steel slag and basalt RAM were firstly fabricated and the five different RAP contents were involved. Then tests of Marshall stability, indirect tensile strength and Cantabro spatter loss were conducted to investigate the moisture susceptibility of RAM. Moreover, their high temperature stability, crack resistance and skid resistance were characterized. Indirect tensile fatigue test combined with Hamburg wheel tracking test were carried out to discuss the durability of RAM. The comprehensive performance of RAM with steel slag were quantitatively assessed based on an improved radar chart evaluation method. The results show that involving steel slag reveals a remarkable enhancement function on water stability, high and low temperature performance, skid resistance and fatigue resistance of RAM. Steel slag RAM with 50% RAP content demonstrates a rutting depth of 7.60 mm and a creep slope of 2.54×10^{-4}, indicating its superior durability in high temperature and water environment. Compared with the comprehensive evaluation function of 0.5336 for basalt RAM with 30% RAP dosage, steel slag RAM reaches 0.7801, which represents its preferable road performance.

Keywords: steel slag; RAP; recycled asphalt mixtures; road performance; radar chart evaluation

Citation: Wang, Z.; Wu, S.; Yang, C.; Xie, J.; Xiao, Y.; Zhao, Z.; Wang, F.; Zhang, L. Quantitative Assessment of Road Performance of Recycled Asphalt Mixtures Incorporated with Steel Slag. *Materials* **2022**, *15*, 5005. https://doi.org/10.3390/ma15145005

Academic Editor: Francesco Canestrari

Received: 14 June 2022
Accepted: 8 July 2022
Published: 19 July 2022

1. Introduction

With the dual development of economy and society, China's highway construction has also entered a stage of rapid development [1]. Asphalt pavement occupies 95% proportion of high-grade roads due to its series of merits [2]. However, asphalt pavement is prone to the diseases of cracks, rutting and water damage, etc. [3]. By the end of 2020, the highway maintenance mileage was 5.14 million km, accounting for 99.0% of the total highway mileage, which implied that China's highway construction has stepped into the construction and maintenance period [4]. A large amount of reclaimed asphalt pavement (RAP) materials were produced in repair projects, and the recycling of RAP could alleviate resource dependence, energy conservation and emission reduction, which coincided with sustainable development in the transportation industry [5].

The performance of recycled asphalt mixtures was influenced by the material properties and the recycling technology [6,7]. The different sources of RAP had been corroborated to have significant variability in RAM performance [8]. A higher mixing temperature for hot in-place recycling asphalt mixtures revealed superior cracking and moisture resistance [9]. The mixing procedures also showed prominent impact on air voids (AV) and voids in mineral aggregate (VMA) [10]. Chen et al. surveyed the effect of RAP on the performance of recycled asphalt mixtures and found that incorporating RAP improved the resistance to water damage and rutting in contrast to anti-crack performance [11]. Yin et al. endorsed the fact that the moisture stability of recycled asphalt mixtures revealed a descending trend with the addition of RAP. He pointed that hot-mix recycled asphalt mixture (HRAM) exhibited a greater fatigue life than new asphalt mixture (NAM), and its fatigue life increased with the ascending RAP dosage [12]. This is due to generation of a loading buffer interface between aged asphalt and rejuvenator. HRAM had high sensitivity of temperature and strain, resulting in its greater decrement of fatigue life than NAM. These conclusions differed from most of the research results [13,14]. Roja et al. illustrated that the asphalt mixture with 35% RAP content revealed the highest dynamic modulus the least fracture resistance [15]. Zhang et al. found that the recycled asphalt mixture had a lower fatigue life than new asphalt mixture and the gap was enlarged with the ascending RAP content [16]. Some approaches to characterize and classify the RAP were applied to promote its popularization and utilization [17]. Chen et al. investigated the mechanism and performance of mixtures with 100% dosage RAP and bio-rejuvenated additive (BRA) and found that BRA restores the balance of the components of aged bitumen and elevate mechanical performance of asphalt mixtures [18]. Similar studies had also been reported by Zaumanis et al. [19]. Moreover, the controversial moisture susceptibility and fatigue durability of the recycled asphalt mixture were the key elements affecting its implement and application.

Currently, some modifiers are being added to the recycled asphalt and recycled asphalt mixtures to enhance their performance. It was reported that anti-stripping agents was added to recycled asphalt mixtures and demonstrated an outstanding promotion function on rutting resistance and water stability [20]. Zhu et al. explored the fatigue performance of recycled asphalt containing warm mix asphalt (WMA) additive and suggested that the WMA additive could elevate the fatigue potential under stress-loading and strain-loading modes [21]. The fiber was introduced to emulsified asphalt cold recycled mixture for improving its inferior early strength and crack resistance [22]. Kim et al. analyzed the effect of desulfurized gypsum (DG) additive on the mechanical behavior of cold recycled asphalt mixtures (CRAM) [23]. The results stated that DG additive elevated the early strength and stiffness and declined the viscoelasticity of CRAM. Yang et al. considered that cold-recycled mixture with asphalt emulsion (CMAE) with high cement dosage demonstrated larger indirect tensile strength and critical strain energy density [24]. The addition of modifiers brings about high preparation costs and poor storage stability, and it is desirable to improve the crack resistance and durability of recycled asphalt mixtures by involving alternative aggregates with excellent performance.

Steel slag, as the highest amount of solid waste generated and disposed in metallurgical industry, reveals the vantages of anti-slip and abrasion resistance and high adhesion strength [25]. It was concluded that steel slag could be applied as aggregate for road engineering and displayed a remarkable improvement on the performance of asphalt mixture [26]. It was attributed to the fact that steel slag revealed higher polished value and rougher surface texture [27]. Rodríguez-Fernández et al. surveyed the performance parameters of asphalt mixtures with the simultaneous incorporation of RAP and electric arc furnace (EAF) slag [28]. The results proved that recycled asphalt mixtures with EAF slag exhibited a satisfactory resistance to permanent deformation due to excellent polishing resistance and low abrasion coefficient of slag. Pasetto et al. conducted the fatigue analysis of base-binder asphalt mixtures with EAF slag and RAP and found that the stiffness and fatigue resistance of the mixtures was improved due to the addition of EAF slag [29].

Meanwhile, the incorporation of EAF slag also boosted the resilient modulus and dynamic creep property of warm mix asphalt (WMA) mixtures [30]. Yang et al. illustrated that involving steel slag in asphalt mixtures with high RAP content exhibited higher texture depth and BPN than that of basalt [31]. This is closely related to the stronger interlocking structure in steel slag [32].

Overall, incorporating steel slag embodies a commendable performance indexes of recycled asphalt mixtures. Little attention has been paid to the quantitative evaluation of comprehensive performance of recycled asphalt mixtures. The purpose of this research is to explore the service performance of recycled asphalt mixtures incorporated with steel slag, including the volume performance, moisture susceptibility, high temperature stability performance, low temperature performance, skid resistance and durability. Furthermore, an improved radar chart evaluation method was used to quantitatively assess their comprehensive performance. The experimental program is shown in Figure 1. The research results contribute to eliminate the resource dependence and environmental pressure.

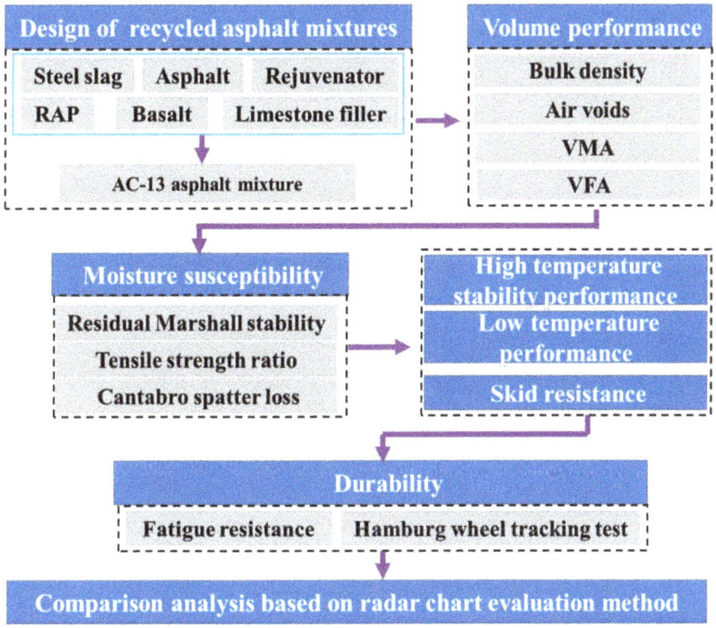

Figure 1. The experimental program of this research.

2. Materials and Methods

2.1. Materials

Styrene–butadiene–styrene modified asphalt (named as SBS asphalt for short) was used in this research, which was acted as virgin asphalt. This was mainly because the recycled asphalt mixtures (RAM) in this paper was mainly used in the upper layer of the asphalt pavement. The upper layer usually had strict requirements on the performance of asphalt mixtures, thus SBS asphalt was commonly adopted [33]. SBS asphalt has a penetration of 68 dmm at 25 °C and a ductility of 486 mm at 5 °C and a softening point of 56 °C. Rejuvenator with a density of 0.943 g/cm^3 and a viscosity of 1.780 Pa·s at 60 °C were included. SBS asphalt and rejuvenator were obtained from Hohhot and Nanjing, China. Steel slag and basalt as virgin aggregates were supplied from Wuhan and Jingshan. Steel slag coupled with SBS modified asphalt were conducted to prepare RAM with excellent performance, so as to create implementation of the maximum additional value of steel slag.

The rejuvenator in this study was mainly used to restore the performance index of aged asphalt in RAP [34]. Their properties were displayed in Table 1.

Table 1. The properties indexes of steel slag and basalt.

Indexes \ Types			Steel Slag	Basalt
Apparent specific gravity	Particle sizes (mm)	9.5–16	3.369	2.983
		4.75–9.5	3.327	2.974
		2.36–4.75	3.138	2.972
Water absorption (%)	Particle sizes (mm)	9.5–16	1.59	0.58
		4.75–9.5	2.38	0.73
		2.36–4.75	2.82	1.04
	Los Angeles abrasion (%)		13.9	20.7
	Crushed value (%)		16.8	17.2
	Polished value (%)		52	46
	Adhesion level		5	4
	Free-CaO content (%)		2.135	-

2.2. Methods

2.2.1. Preparation of RAM

RAP were acquired from surface pavement on Wu-Huang highway of China. RAM with 10%, 20%, 30%, 40% and 50% RAP and 13.2 mm nominal maximum size were prepared based on Marshall design method. The dosage of the rejuvenator in RAM was 6 wt% (mass ratio of rejuvenator and aged asphalt) [35]. Steel slag and basalt were conducted as virgin coarse aggregates and limestone was conducted as virgin fine aggregate to fabricate two types of RAM. Limestone powder as filler was involved. Figure 2 depicted the grading curve of RAM with steel slag and basalt.

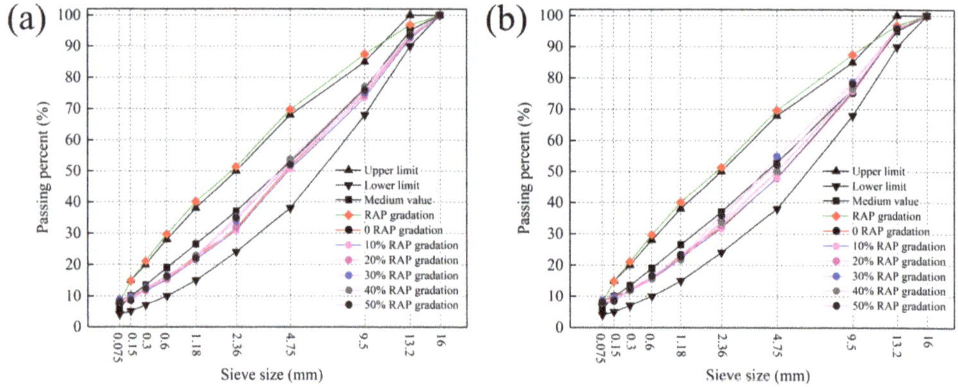

Figure 2. Grading curve of asphalt concrete (AC)-13: (**a**) steel slag; (**b**) basalt.

2.2.2. Performance Evaluation of Recycled Asphalt Mixtures

The bulk density, air voids, VMA and VFA were employed to investigate the volume performance of RAM as the evaluation indexes. Tests of Marshall stability, indirect tensile strength and Cantabro spatter loss were applied to assess the moisture susceptibility of RAM. The standard wheel tracking test with a rolling speed of 42 cycles/min, a testing temperature of 60 °C and a load strength of 0.7 MPa were conducted to discuss the high temperature stability performance. Low temperature performance was characterized through a three-points bending test with a temperature of −10 °C and a loading rate of

50 mm/min. Beam specimens with a size of 250 × 30 × 35 mm were involved. Texture depth (TD) and British Pendulum Number (BPN) were used to evaluate the skid resistance.

A fatigue resistance test combined with a Hamburg wheel tracking (HWT) test were carried out to study the durability of RAM. Indirect tensile fatigue test with a testing temperature of 15 °C, the stress level of 0.35, 0.40, 0.45 and 0.50 MPa and a Poisson's ratio of 0.35 were used. The testing device and schematic diagram of HWT test are described in Figure 3. The creep slope and stripping slope were slope of the tangent line of the creep curve and stripping curve. By calculating the intersection point of the two curves, the stripping inflection point (SIP) could be acquired. The testing temperature and rolling rate were 50 °C and 45 cycles/min [36].

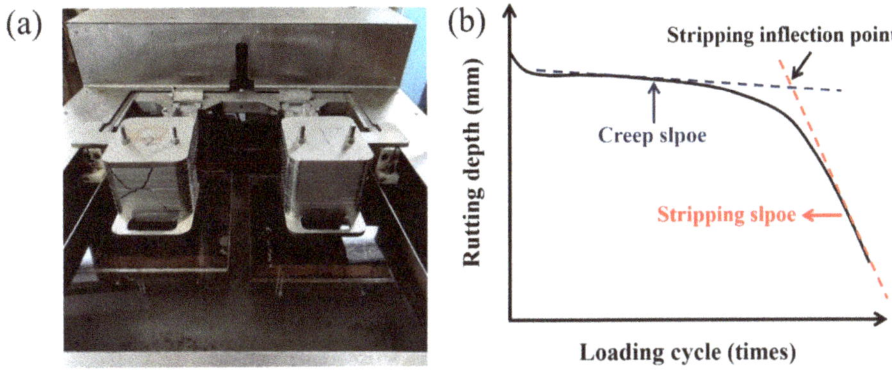

Figure 3. Testing device (**a**) and schematic diagram (**b**) of HWT test.

2.2.3. Radar Chart Evaluation Method

The radar chart method is a common graphical method to display multiple variables, which can map a multidimensional space point to two-dimensional space, indicating the feature of qualitative evaluation of each evaluation object. In the traditional radar map evaluation, the area and perimeter of the graph were extracted as feature vectors, while the feature vector area and perimeter revealed the disadvantage of varying with the ranking of indicators. Therefore, an improved radar chart evaluation method with uniqueness feature was conducted to quantitatively assess the comprehensive performance of RAM.

In the improved radar chart evaluation method, the evaluation vector and evaluation function were constructed by extracting feature vectors to comprehensively reflect the level of RAM and the balanced development degree of each index [37]. Firstly, a matrix A = (a_{ij})n × k for the evaluation indicators was established. Vector X = $\{x_1, x_2, x_3 \ldots x_n\}$ and Y = $\{y_1, y_2, y_3 \ldots y_k\}$ represent a group of objects and a set of indicators for the objects.

Secondly, the data in matrix A were standardized and non-linear transformed through Equations (1) and (2).

$$b_{ij} = \frac{a_{ij} - E(y_j)}{\sigma(y_j)} \tag{1}$$

$$r_{ij} = \frac{2}{\pi}\arctan(b_{ij}) + 1 \tag{2}$$

where b_{ij} and r_{ij} represent each indicator after standardization and non-linear transformation, respectively, and $E(y_j)$ and $\sigma(y_j)$ are the average value and standard deviation indicator j.

Thirdly, the characteristic vectors were calculated according to Equations (3) and (4).

$$u_i = [A_i, L_i] \tag{3}$$

$$\begin{cases} A_i = \sum_{j=1}^{k} \frac{1}{k}\pi r_{ij}^2 \\ L_i = \sum_{j=1}^{k} \frac{2}{k}\pi r_{ij} \end{cases} \tag{4}$$

where A_i and L_i represent the area inside the arcs and sum length of arcs, respectively, and k represents the number of indicators.

Fourthly, the evaluation vector is defined based on the extracted characteristic vector, as shown in Equation (5).

$$v_i = [v_{i1}, v_{i2}] \tag{5}$$

where v_{i1} and v_{i2} are the relative area and perimeter of evaluation object. Calculation method of v_{i1} and v_{i2} are displayed in Equation (6).

$$\begin{cases} v_{i1} = \frac{A_i}{MaxA_i} \\ v_{i2} = \frac{L_i}{2\pi\sqrt{\frac{A_i}{\pi}}} \end{cases} \tag{6}$$

Finally, the comprehensive evaluation function (f) was deduced through the geometric mean of v_{i1} and v_{i2}, as displayed in Equation (7).

$$\mathrm{f}(v_{i1}, v_{i2}) = \sqrt{v_{i1} \times v_{i2}} \tag{7}$$

The road performance indexes of the RAM were obtained through the performance test. Through the above formula, the evaluation indicators of the two RAM with different RAP content were standardized and normalized, and the comprehensive performance of the two materials was evaluated by the obtained evaluation function.

3. Results and Discussion

3.1. Volume Performance

The volume performance parameters of RAM are presented in Table 2. When RAP content is lower than 30%, asphalt aggregate ratio of RAM with steel slag no change with the rising RAP content. As the RAP dosage continues to increase, asphalt aggregate ratio rises. The addition of RAP significantly reduces the bulk density of RAM with steel slag, while fluctuates little on the air voids, VMA and VFA. The asphalt-aggregate ratio of RAM with basalt demonstrates the similar change trend as the RAP content increases. There is a certain degree of increase in bulk density of RAM incorporated with basalt as the ascending RAP dosage. This is due to the diminution of virgin fine aggregate content of limestone with lower density. In general, the volume performance of steel slag and basalt RAM with different RAP content meet the specification requirements [38].

Table 2. Volume performance parameters of RAM.

Mixtures Styles	RAP Content (%)	Asphalt-Aggregate Ratio (%)	Bulk Density (g/cm^3)	Air Voids (%)	VMA (%)	VFA (%)
Steel slag	0	4.9	2.653	4.2	14.1	70.3
	10	4.9	2.645	4.3	14.2	69.6
	20	4.9	2.637	4.5	14.1	68.2
	30	5.0	2.634	3.8	14.1	73.0
	40	5.0	2.624	4.0	14.2	72.0
	50	5.1	2.621	3.9	14.2	72.3
Basalt	0	4.7	2.520	3.9	14.0	72.2
	10	4.7	2.524	4.0	14.3	71.8
	20	4.8	2.528	3.8	14.0	72.6
	30	4.8	2.523	4.1	14.2	70.8
	40	4.8	2.527	4.0	14.3	72.3
	50	4.9	2.531	3.9	14.1	72.5

3.2. Moisture Susceptibility

3.2.1. Residual Marshall Stability

Figure 4 illustrates the results of residual Marshall stability (RMS) of RAM incorporated with steel slag and basalt. Compared with the virgin steel slag asphalt mixtures, the Marshall stability (MS) and immersion Marshall stability (MS_1) of RAM increase with ascending RAP dosage in contrast to RMS. This is attributed to the enhancement of the overall elasticity of RAM with aged asphalt of high modulus. The result is match to the conclusion of Oldham et al. [39]. Meanwhile, aged asphalt reveals inferior adhesion property with aggregates, resulting in the reduction in RMS of RAM as RAP content increases. Steel slag RAM with 50% RAP exhibits an RMS of 90.5% and still remains at a high level. For basalt RAM, its variation rule of MS, MS_1 and RMS are consistent with that of steel slag RAM. While it demonstrates lower performance indexes under the same RAP content. RAM with steel slag and basalt all satisfy the requirements that the RMS of the modified asphalt mixture in the wet areas is not less than 85%.

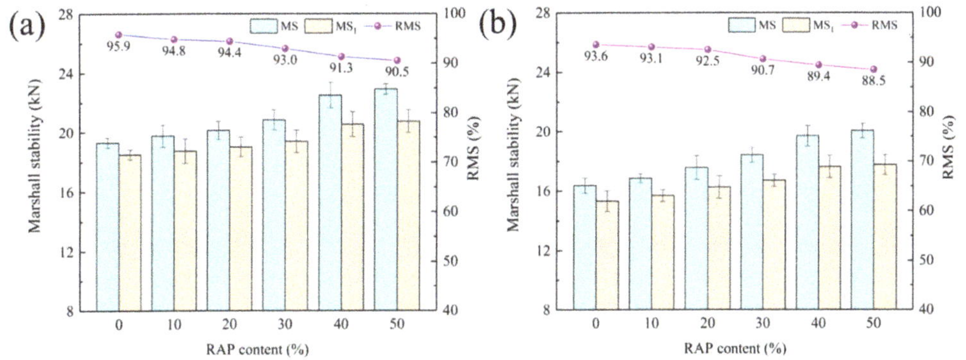

Figure 4. RMS results of RAM: (**a**) steel slag; (**b**) basalt.

3.2.2. Tensile Strength Ratio

Freeze-thaw splitting test can more truly reflect the water damage resistance of asphalt mixtures. Tensile strength ratio (TSR) reveals more stringent requirement than RMS, and TSR of asphalt mixture will not meet the requirements when RMS arrivals design constrain. Figure 5 presents the TSR results of RAM. It is stated that splitting tensile strength and TSR of RAM show a linear decrease with the ascending RAP dosage. This is because that RAP increases the modulus of asphalt in RAM, which diminishes the bonding force between asphalt and aggregate, resulting in the exfoliation of asphalt from the surface layer of the aggregate under water immersion, thus weakening the mechanical properties of the RAM [40]. Steel slag RAM with 50% RAP content exhibits a 6.2% reduction compared to virgin asphalt mixtures and reaches to 89.3%. While for RAM prepared with basalt, the corresponding values are 13.6% and 81.0%. This indicates that incorporating steel slag in RAM can reduce the potential moisture damage risks and elevate the water stability of RAM. This is consistent with the fact that steel slag embodied superior adhesive effect than basalt according to the analysis of molecular simulation [41].

3.2.3. Cantabro Spatter Loss

The Cantabro spatter loss test is commonly applied to assess the adhesion between asphalt and aggregate in open-graded asphalt mixtures. Although a dense gradation was included in this study, the relative high RAP content may reveal a greater impact on the overall bonding of the asphalt mixtures due to inferior adhesion property between aged asphalt and aggregates. Given that AC-13 RAM was adopted as top layer of asphalt pavement, it suffers from surface stripping due to the dual action of rutting and rainwater.

Therefore, Cantabro spatter loss can be conducted to characterize resistance to water damage of RAM.

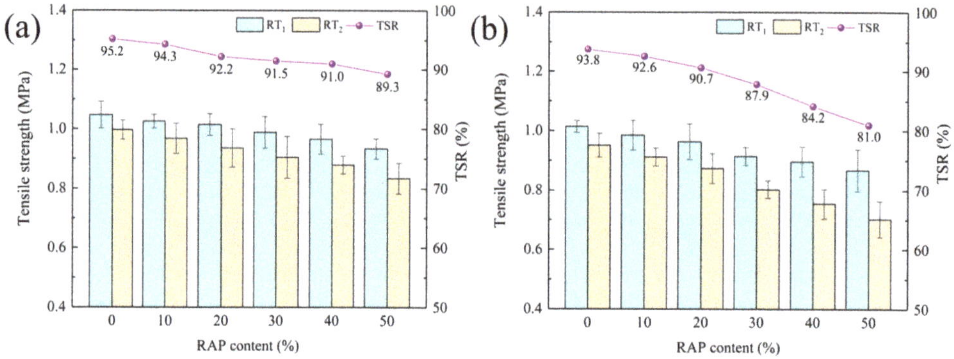

Figure 5. TSR results of RAM: (**a**) steel slag; (**b**) basalt.

Figure 6 depicts the Cantabro spatter loss results of RAM with steel slag and basalt. The spatter loss of steel slag and basalt RAM all boost as the ascending RAP dosage, while steel slag RAM exhibits a lower increment. Steel slag RAM involving 50% RAP content reaches a spatter loss of 5.5% and only increases by 1.9% in comparison with virgin asphalt mixture. This indicates that on the one hand, aged asphalt leads to a reduction in the adhesion of the aggregate to the asphalt in RAM, causing the surface binder to fall off under the action of water. On the other hand, the viscosity of the aged asphalt is restored by the action of the rejuvenator, which leads to an improvement effect of bond performance for overall asphalt in RAM. Comparative analysis of spatter loss results verified that steel slag RAM demonstrates superior moisture susceptibility than basalt RAM. Furthermore, the rising RAP content will not cause structural damage to RAM with steel slag and basalt due to their low spatter loss.

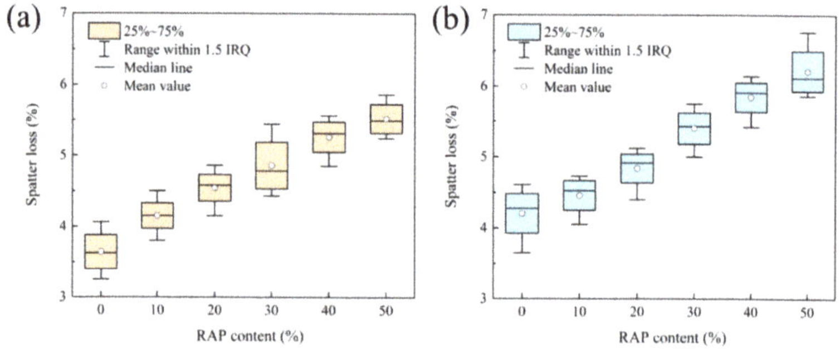

Figure 6. Spatter loss results of RAM: (**a**) steel slag; (**b**) basalt.

3.3. High Temperature Stability Performance

Figure 7 indicates the dynamic stability (DS) results of RAM. The participation of RAP boosts the DS of RAM and elevate its high temperature stability performance. It can be elaborated by the consequence that aged asphalt with high softening point and stiffness can prominently reduce the rutting depth of RAM and enhance its anti-rutting performance. Steel slag RAM reveals larger DS than RAM prepared with basalt, indicating its superior rutting resistance. This is attributed to the outstanding mechanical properties and abundant texture index of steel slag. The DS of steel slag RAM with 50% RAP is 5040 times/mm, which is 1.27 times that of the virgin asphalt mixtures. Steel slag and basalt RAM all exhibit

DS much higher than 2400 times/mm (minimum index in standard) [38], representing their excellent high temperature performance.

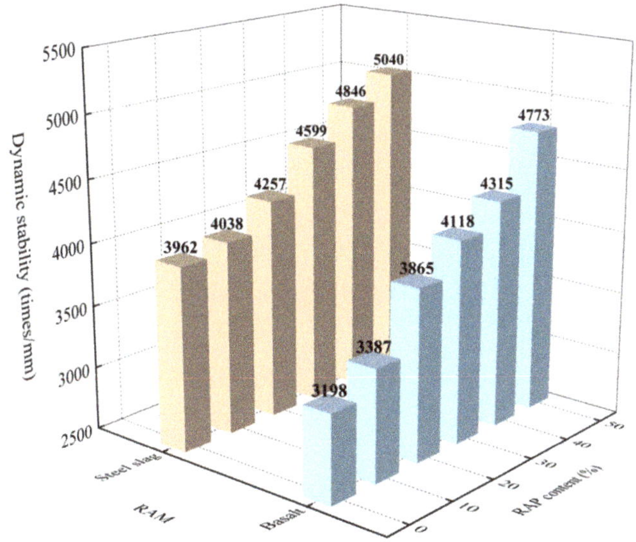

Figure 7. Dynamic stability results of RAM.

3.4. Low Temperature Performance

The flexural tensile strain of asphalt mixture can effectively reflect the possibility of brittle fracture at low temperature and characterize its crack resistance. Asphalt mixture with greater flexural tensile strain means the superior low-temperature crack resistance. The crack resistance indexes from bending test results of RAM are provided in Table 3. As the RAP content rises, the maximum load, tensile strength and tensile strain of RAM ascend in contrast to stiffness modulus, resulting in reducing effect on low temperature performance. It can be explained by the fact that RAM with high RAP content displays lower plasticity and is incline to become hard and brittle, which weakens the resistance to low temperature deformation [42]. Comparison results of steel slag and basalt RAM state that involving steel slag can elevate the flexural tensile strength and tensile strain of RAM. Steel slag RAM with 50% RAP reveals a tensile strain of 2548.4 με. While the corresponding index is 2323.1 με for basalt RAM, which is less than the minimum value of 2500 με in the winter cold area in the specification. This indicates that when RAP content increases to 50%, steel slag RAM can be applied in severe cold areas instead of basalt RAM.

Table 3. The crack resistance indexes of RAM.

Mixtures Styles	RAP Content (%)	Maximum Load (N)	Tensile Strength (MPa)	Tensile Strain (με)	Stiffness Modulus (MPa)
Steel slag	0	1252.6	10.225	3269.2	3127.8
	10	1238.6	10.111	3102.2	3259.3
	20	1211.6	9.891	2955.8	3346.2
	30	1185.4	9.677	2813.5	3439.4
	40	1153.5	9.416	2695.9	3492.9
	50	1122.3	9.162	2548.4	3595.1
Basalt	0	1033.5	8.437	3006.7	2806.0
	10	1010.2	8.247	2892.2	2851.3
	20	975.3	7.962	2745.8	2899.6

Table 3. *Cont.*

Mixtures Styles	RAP Content (%)	Maximum Load (N)	Tensile Strength (MPa)	Tensile Strain (με)	Stiffness Modulus (MPa)
	30	955.7	7.802	2603.5	2996.6
	40	906.5	7.400	2510.0	2948.2
	50	875.1	7.144	2323.1	3075.0

3.5. Skid Resistance

The texture depth and BPN results of RAM are illustrated in Figure 8. Virgin steel slag asphalt mixture demonstrates the highest texture depth and BPN, which reach to 0.95 mm and 77. Incorporating RAP decreases the texture depth and BPN of RAM. This is attributed to the deterioration of polished and weared stone value of RAP aggregate. When the same RAP dosage is involved, steel slag RAM displays larger texture depth and BPN than basalt RAM, indicating its superior skid resistance. This is attributed to the rich texture and impaction structure of steel slag [32]. Steel slag RAM with 50% RAP content exhibits the texture depth of 0.81 mm and BPN of 63, which far exceed the minimum value of 0.55 mm and 45 in requirements of the specification [38]. Texture depth of basalt RAM reduces from 0.86 mm, 0.83 mm, 0.81 mm, 0.78 mm and 0.75 mm to 0.73 mm as RAP content rises from 0 to 50% with an interval of 10%. BPN of basalt RAM with 50% RAP decreases by 20% and only reaches to 52, which is lower than 11 that of steel slag RAM and also arrival design constrain.

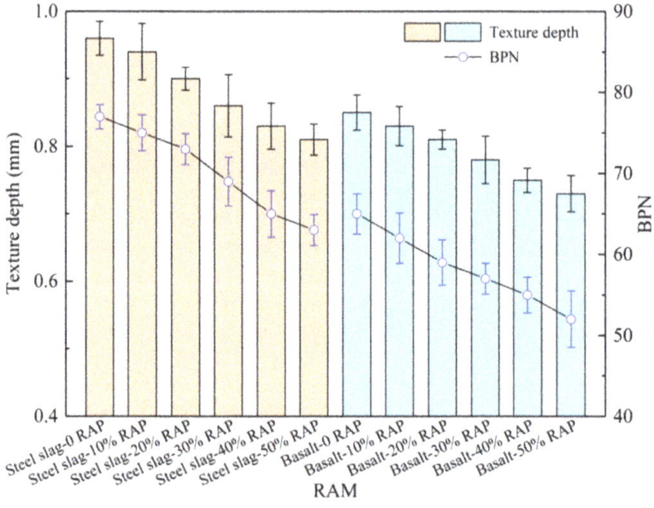

Figure 8. Texture depth and British Pendulum Number (BPN) results of RAM.

3.6. Durability

3.6.1. Fatigue Resistance

Fatigue life, as the critical assessment indicator, can reflect the number of stress cycles of sample experienced failure. The fatigue life curves of RAM are depicted in Figure 9. Both of steel slag RAM and basalt RAM reveal a downward trend of fatigue life as RAP dosage increases. The conclusion is inconsistent with Yin et al.'s results [12]. This is due to the increase in the stiffness of blended asphalt and reduction in the response rate to stress for RAM [43]. Under the same RAP content, steel slag RAM embodies the higher fatigue life than basalt RAM, representing its superior fatigue performance.

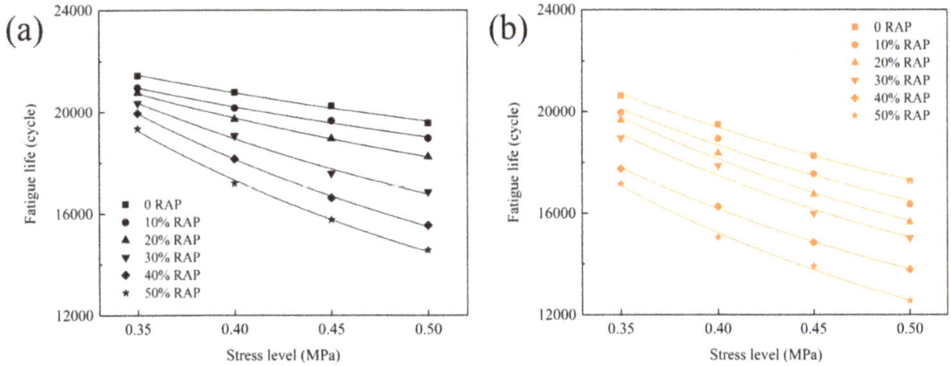

Figure 9. Fatigue life curve of RAM: (**a**) steel slag; (**b**) basalt.

Fitting coefficient of fatigue equation for RAM are presented in Table 4. The satisfactory correlation is found in all RAM samples for their high correlation coefficient. RAM with larger K and smaller n means preferable fatigue resistance. Steel slag virgin asphalt mixture possesses the highest K value, indicating its excellent fatigue performance. The addition of RAP decreases the K value and boosts n value of steel slag RAM. Compared with basalt RAM, incorporating steel slag in RAM can elevate its K value and fatigue resistance. This conclusion is match to the analysis results of fatigue life. This is because steel slag emerges better interfacial adhesion for its high angularity and texture.

Table 4. Fitting coefficient of fatigue equation for RAM.

Mixtures Styles	RAP Content (%)	Fitting Formula ($N_f = K(\sigma)^{-n}$)				R^2
		K		n		
		Value	Standard Deviation	Value	Standard Deviation	
Steel Slag	0	16,506.4	234.4	−0.251	0.016	0.9877
	10	15,716.9	198.5	−0.274	0.014	0.9919
	20	14,190.3	105.4	−0.362	0.008	0.9984
	30	11,486.9	344.0	−0.546	0.034	0.9888
	40	9495.0	111.5	−0.707	0.013	0.9990
	50	8346.9	204.7	−0.797	0.027	0.9965
Basalt	0	12,229.8	218.0	−0.502	0.020	0.9953
	10	11,233.9	500.1	−0.555	0.049	0.9764
	20	10,039.6	358.2	−0.646	0.039	0.9888
	30	9472.0	595.9	−0.669	0.070	0.9680
	40	8427.0	125.2	−0.712	0.017	0.9984
	50	6940.3	275.3	−0.858	0.044	0.9922

3.6.2. Hamburg Wheel Tracking Test

The Hamburg wheel tacking (HWT) test can comprehensively evaluate the rutting resistance, moisture susceptibility and stripping resistance under the coupling action of temperature and load. The deformation development of RAM with the ascending loading cycle are presented in Figure 10. The compaction process and creep process are discovered in steel slag RAM, while the stripping process only appears in basalt RAM with 40% and 50% RAP. The creep slope and stripping slop represent the deformation rate of RAM in rutting deformation and moisture damage, and larger values imply higher possibility of rutting destruction and moisture damage. The stripping inflection point (SIP), as a shift point from creep process to stripping process, its presence means the beginning of asphalt film peeling off from the surface of the aggregate [16]. As shown in Figure 10, all steel slag RAM do not reach the SIP. While SIP is observed in basalt RAM with 40% and 50% RAP, indicating their inferior durability.

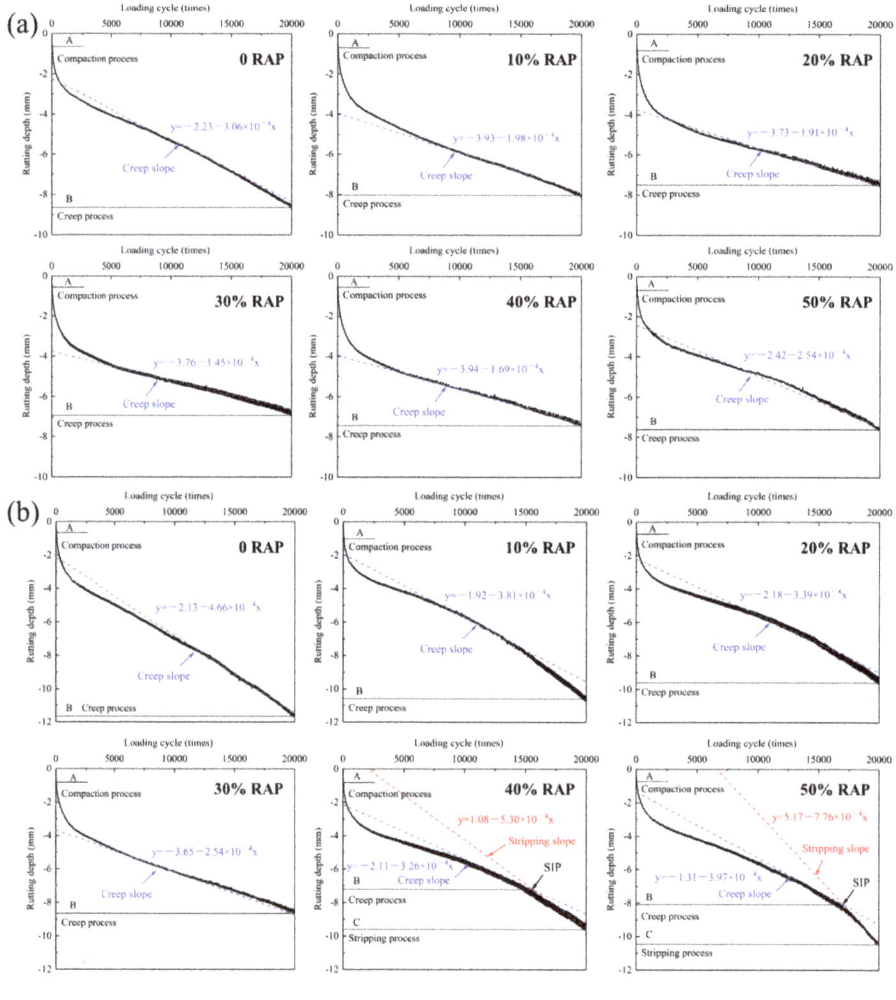

Figure 10. Deformation development of RAM with the increasing loading cycle: (**a**) steel slag; (**b**) basalt.

Table 5 summarizes the HWT test results of RAM. Basalt virgin asphalt mixture reveals the largest rutting depth of 11.69 mm after 20,000 cycles loading. Under the same RAP dosage, steel slag RAM exhibits the lower rutting depth and creep slope than basalt RAM. When RAP content is lower than 40%, the rutting depth and creep slope of RAM reduce as the ascending RAP content, and then start to elevate after the RAP content reaches 40%. This is attributed to the greater attenuation function of RAP on water stability than its enhancement effect on rutting resistance for RAM with over 40% RAP content. Steel slag RAM with 50% RAP demonstrates the rutting depth of 7.60 mm and creep slope of 2.54×10^{-4}, which are lower than corresponding values of 10.46 mm and creep slope of 3.97×10^{-4} for basalt RAM. It is worth mentioning that basalt RAM with 40% RAP content undergoes the stripping process under the cycles loading of 15,589 times and reaches the SIP. As the RAP dosage rises to 50%, the SIP moves back and appears at 17,065 cycles loading. The reason is that the ascending RAP content retards the cycles loading of SIP occurrences for RAM, but also elevates its stripping slop, resulting in the dramatic increase in the rutting depth at a later stage. Comparative analysis of HWT test results of RAM with steel slag and basalt state that involving steel slag endows the superior durability than basalt in high temperature and water environment.

Table 5. The characterization parameters of RAM in HWT test.

Mixtures Styles	RAP Content (%)	Rutting Depth (mm)	Creep Slope ($\times 10^{-4}$)	Stripping Slope ($\times 10^{-4}$)	Stripping Inflection Point (SIP)
Steel slag	0	−8.61	−3.06	/	/
	10	−7.98	−1.98	/	/
	20	−7.51	−1.91	/	/
	30	−6.87	−1.45	/	/
	40	−7.40	−1.69	/	/
	50	−7.60	−2.54	/	/
Basalt	0	−11.69	−4.66	/	/
	10	−10.65	−3.81	/	/
	20	−9.58	−3.39	/	/
	30	−8.49	−2.54	/	/
	40	−9.15	−3.26	−5.30	X = 15,589, Y = −7.19
	50	−10.46	−3.97	−7.76	X = 17,065, Y = −8.08

3.7. Comparison Analysis Based on Radar Chart Evaluation Method

The radar chart evaluation method is conducted to quantitatively compare the effect of steel slag and basalt on the performance of RAM. The nine evaluation indicators, including residual Marshall stability, tensile strength ratio, spatter loss, dynamic stability, tensile strain, texture depth, British Pendulum Number, intercept K of fatigue equation, absolute value of creep slope in HWT test, were used in radar chart. They were denoted as RMS, TSR, SL, DS, TS, TD, BPN, Intercept K, CS, respectively. Table 6 provides the nine indicators of twelve groups in matrix A. Then the indicators are standardized and non-linear transformed referring to Equations (1) and (2), as shown in Table 7.

Table 6. Evaluation indicators of RAM with different RAP content.

Mixtures Styles	RAP Content (%)	Water Stability			Rutting Resistance	Crack Resistance	Skid Resistance		Durability	
		RMS (%)	TSR (%)	SL (%)	DS (Times/mm)	TS (με)	TD (mm)	BPN	Intercept K	CS ($\times 10^{-4}$)
Steel slag	0	95.9	95.2	3.6	3962	3269.2	0.95	77	16,506.4	3.06
	10	94.8	94.3	4.2	4038	3102.2	0.94	75	15,716.9	1.98
	20	94.4	92.2	4.5	4257	2955.8	0.90	73	14,190.3	1.91
	30	93.0	91.5	4.9	4599	2813.5	0.86	69	11,486.9	1.45
	40	91.3	91.0	5.3	4846	2695.9	0.83	65	9,495.0	1.69
	50	90.5	89.3	5.5	5040	2548.4	0.81	63	8,346.9	2.54
Basalt	0	93.6	93.8	4.2	3198	3006.7	0.86	65	12,229.8	4.66
	10	93.1	92.6	4.5	3387	2892.2	0.83	62	11,233.9	3.81
	20	92.5	90.7	4.8	3865	2745.8	0.81	59	10,039.6	3.39
	30	90.7	87.9	5.4	4118	2603.5	0.78	57	9472.0	2.54
	40	89.4	84.2	5.8	4315	2510.0	0.75	55	8427.0	3.26
	50	88.5	81.0	6.2	4773	2323.1	0.73	52	6940.3	3.97

Table 7. Evaluation indicators after standardization and non-linear transformation.

Mixtures Styles	RAP Content (%)	Water Stability			Rutting Resistance	Crack Resistance	Skid Resistance		Durability	
		RMS	TSR	SL	DS	TS	TD	BPN	Intercept K	CS
Steel slag	0	1.6550	1.5622	0.3213	0.7357	1.6839	1.6624	1.6544	1.6843	1.1337
	10	1.5449	1.4979	0.5064	0.8145	1.5587	1.6359	1.6043	1.6399	0.5300
	20	1.4894	1.2801	0.6726	1.0672	1.3631	1.4828	1.5400	1.5142	0.5055
	30	1.1971	1.1836	0.9927	1.4055	1.0601	1.2093	1.3490	1.0688	0.3821
	40	0.7223	1.1085	1.3161	1.5571	0.7816	0.9279	1.0554	0.6645	0.4393
	50	0.5567	0.8435	1.4367	1.6366	0.5252	0.7486	0.8900	0.5064	0.7985
Basalt	0	1.3427	1.4555	0.5064	0.3145	1.4437	1.2093	1.0554	1.2234	1.6884
	10	1.2234	1.3300	0.6726	0.3731	1.2406	0.9279	0.8113	1.0133	1.4978
	20	1.0563	1.0619	0.9053	0.6464	0.8956	0.7486	0.6121	0.7615	1.3232
	30	0.5929	0.6562	1.3803	0.9042	0.6060	0.5436	0.5131	0.6607	0.7985
	40	0.4071	0.3704	1.5663	1.1339	0.4781	0.4113	0.4367	0.5156	1.2538
	50	0.3288	0.2594	1.6752	1.5192	0.3244	0.3504	0.3531	0.3808	1.5470

Figure 11 summaries the radar charts for RAM with different RAP content. The discrepancies in the enhancement effect of steel slag are presented clearly. Incorporating steel slag reveals a significant improvement on RMS, TSR, DS, TS, TD, BPN and Intercept K in contrast to SL and CS. The appropriate RAP content in steel slag RAM can be identified from the charts considering the improvement of a certain performance indicator.

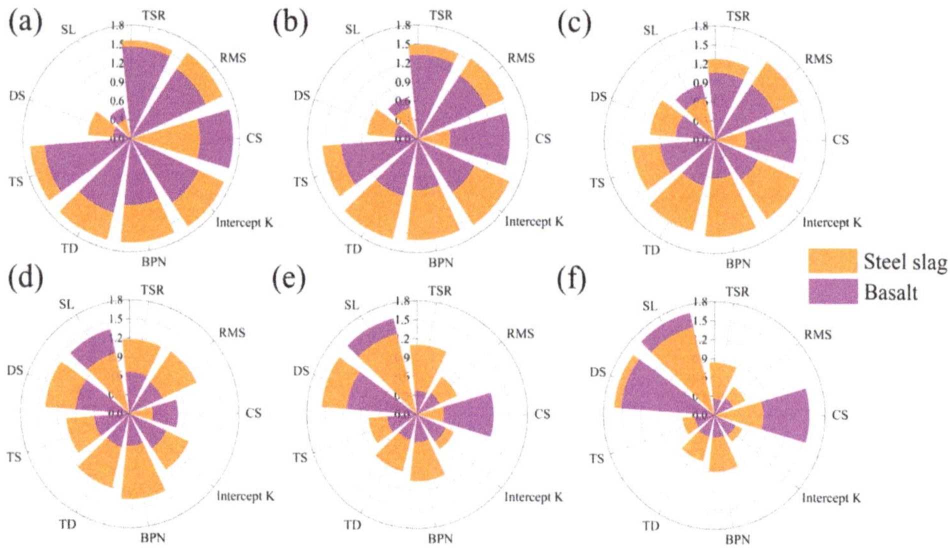

Figure 11. Radar charts for RAM with different RAP content: (**a**) 0 RAP; (**b**) 10% RAP; (**c**) 20% RAP; (**d**) 30% RAP; (**e**) 40% RAP; (**f**) 50% RAP.

According to Equations (3)–(7), the characteristic vectors (u_i) and evaluation vectors (v_i) in matrices can be calculated, as listed in Table 8. Calculation results of comprehensive evaluation function (f) for twelve groups are elaborated in Figure 12. It is indicated that steel slag virgin asphalt mixture reveals the largest f value of 0.9710. The addition of RAP decreases f value of RAM with steel slag. When RAP dosage elevates from 0 to 50% with an interval of 10%, the f value of steel slag RAM are 0.9119, 0.8694, 0.7801, 0.6872 and 0.6456 with a decrement of 6.1%, 10.5%, 19.7%, 29.2% and 33.5%. Distinct from steel slag RAM, basalt RAM exhibits an upward and then downward trend as the increasing RAP dosage. This is due to the greater enhancement effect of RAP content on DS of basalt RAM than steel slag RAM. The f value of basalt RAM with 50% RAP content reduces by 28.1% and reaches to 0.5928. While RAM incorporated with steel slag possesses the larger f value than basalt under the same RAP content.

Table 8. The characteristic vectors and evaluation vectors of matrices.

Mixtures Styles	RAP Content (%)	$u_i = [A_i, L_i]$	$v_i = [v_{i1}, v_{i2}]$
Steel slag	0	[6.3817, 8.4425]	[1.0000, 0.9427]
	10	[5.6549, 7.9116]	[0.8861, 0.9385]
	20	[5.0353, 7.6201]	[0.7890, 0.9579]
	30	[4.0104, 6.8752]	[0.6284, 0.9685]
	40	[3.1861, 5.9849]	[0.4993, 0.9458]
	50	[2.8924, 5.5446]	[0.4532, 0.9197]
Basalt	0	[4.6372, 7.1484]	[0.7266, 0.9364]
	10	[3.5555, 6.3461]	[0.5571, 0.9494]
	20	[2.6350, 5.5926]	[0.4129, 0.9719]
	30	[1.9216, 4.6464]	[0.3011, 0.9455]
	40	[2.2578, 4.5889]	[0.3538, 0.8615]
	50	[2.8556, 4.7042]	[0.4475, 0.7853]

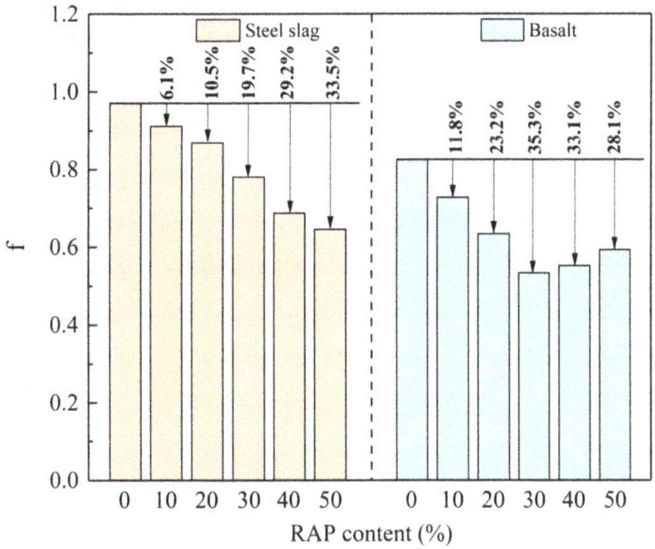

Figure 12. Comprehensive evaluation index of steel slag and basalt RAM.

4. Conclusions

In this study, the pavement performances of recycled asphalt mixtures (RAM) with steel slag and basalt were examined using volume performance, moisture susceptibility, high temperature stability performance, low temperature performance, skid resistance and durability. Then, an improved radar chart evaluation method was applied to quantitatively assess their comprehensive performance. The following conclusions can be summarized.

- The incorporation of steel slag elevates the residual Marshall stability (RMS), tensile strength ratio (TSR) and diminishes the Cantabro spatter loss of RAM, endowing its superior moisture susceptibility than basalt RAM. Steel slag RAM involving 50% RAP dosage demonstrates a RMS of 90.5%, TSR of 89.3% and spatter loss of 5.5%, which remains at a high level;

- Steel slag RAM reveals larger dynamic stability and tensile strain in comparison to basalt RAM, representing its better rutting resistance and low temperature cracking resistance. The texture depth and BPN of steel slag RAM are 0.81 mm and 63, which are higher than those of basalt RAM by 0.08 mm and 11, respectively;
- RAM with steel slag embodies the larger K value and higher fatigue life, representing its superior fatigue performance compared with basalt RAM. All steel slag RAM do not reach the stripping process and stripping inflection point, which exhibit the lower rutting depth and creep slope than basalt RAM, resulting in preferable durability under high temperature and water condition;
- An improved radar chart evaluation method is capable of quantitatively assessing the discrepancies in the improvement effect of steel slag for RAM. Steel slag virgin asphalt mixture possesses the largest comprehensive evaluation function (f) of 0.9710, and the ascending RAP content diminishes the f value of steel slag RAM. Steel slag RAM displays larger f value compared with basalt RAM under the same RAP dosage, indicating its more desirable comprehensive road performance.

The above conclusions have verified that incorporating steel slag reinforces comprehensive road performance of recycled asphalt mixtures, which is conducive to promoting environmental sustainability. Further study is essential to investigate the engineering implementation and energy consumption of recycled asphalt mixtures involving steel slag.

Author Contributions: Conceptualization, S.W. and C.Y.; Data curation, Z.W. and C.Y.; Funding acquisition, S.W. and J.X.; Investigation, Z.W., C.Y. and Z.Z; Methodology, Z.W., F.W. and C.Y.; Project administration, S.W. and Y.X.; Supervision, S.W.; Validation, Z.Z., Y.X. and L.Z.; Writing—original draft, Z.W.; Writing—review & editing, J.X., F.W. and L.Z. All authors have read and agreed to the published version of the manuscript.

Funding: This research was sponsored by National Key R&D Program of China (No. 2018YFB1600200), Key R&D Program of Guangxi Province (No. 2021AB26023), Key R&D Program of Hubei Province (No. 2020BCB064) and Hebei Provincial Communication Department project (No. YC-201926).

Institutional Review Board Statement: Not applicable.

Informed Consent Statement: Not applicable.

Data Availability Statement: Not applicable.

Conflicts of Interest: The authors declare no conflict of interest.

References

1. Wang, F.; Xie, J.; Wu, S.; Li, J.; Barbieri, D.M.; Zhang, L. Life cycle energy consumption by roads and associated interpretative analysis of sustainable policies. *Renew. Sustain. Energ. Rev.* **2021**, *141*, 110823. [CrossRef]
2. Li, N.; Jiang, Q.; Wang, F.; Cui, P.; Xie, J.; Li, J.; Wu, S.; Barbieri, D.M. Comparative Assessment of Asphalt Volatile Organic Compounds Emission from field to laboratory. *J. Clean. Prod.* **2021**, *278*, 123479. [CrossRef]
3. Wan, P.; Liu, Q.; Wu, S.; Zhao, Z.; Chen, S.; Zou, Y.; Rao, W.; Yu, X. A novel microwave induced oil release pattern of calcium alginate/ nano-Fe3O4 composite capsules for asphalt self-healing. *J. Clean. Prod.* **2021**, *297*, 126721. [CrossRef]
4. Ministry of Transport of the People's Republic of China. *Statistical Bulletin of Transport Industry Development in 2020*; Ministry of Transport of the People's Republic of China: Beijing, China, 2021.
5. Wei, M.; Wu, S.; Zhu, L.; Li, N.; Yang, C. Environmental Impact on VOCs Emission of a Recycled Asphalt Mixture with a High Percentage of RAP. *Materials* **2021**, *14*, 947. [CrossRef]
6. Moins, B.; Hernando, D.; Buyle, M.; France, C.; Van den bergh, W.; Audenaert, A. On the road again! An economic and environmental break-even and hotspot analysis of reclaimed asphalt pavement and rejuvenators. *Resour. Conserv. Recy* **2022**, *177*, 106014. [CrossRef]
7. Salehi, S.; Arashpour, M.; Kodikara, J.; Guppy, R. Sustainable pavement construction: A systematic literature review of environmental and economic analysis of recycled materials. *J. Clean. Prod.* **2021**, *313*, 127936. [CrossRef]
8. Montañez, J.; Caro, S.; Carrizosa, D.; Calvo, A.; Sánchez, X. Variability of the mechanical properties of Reclaimed Asphalt Pavement (RAP) obtained from different sources. *Constr. Build. Mater.* **2020**, *230*, 116968. [CrossRef]
9. Ma, Y.; Polaczyk, P.; Park, H.; Jiang, X.; Hu, W.; Huang, B. Performance evaluation of temperature effect on hot in-place recycling asphalt mixtures. *J. Clean. Prod.* **2020**, *277*, 124093. [CrossRef]
10. Xie, Z.; Rizvi, H.; Purdy, C.; Ali, A.; Mehta, Y. Effect of rejuvenator types and mixing procedures on volumetric properties of asphalt mixtures with 50% RAP. *Constr. Build. Mater.* **2019**, *218*, 457–464. [CrossRef]

11. Chen, Y.; Chen, Z.; Xiang, Q.; Qin, W.; Yi, J. Research on the influence of RAP and aged asphalt on the performance of plant-mixed hot recycled asphalt mixture and blended asphalt. *Case Stud. Constr. Mater.* **2021**, *15*, e00722. [CrossRef]
12. Yin, P.; Pan, B. Effect of RAP content on fatigue performance of hot-mixed recycled asphalt mixture. *Constr. Build. Mater.* **2022**, *328*, 127077. [CrossRef]
13. Babagoli, R.; Norouzi, N.; Ameli, A. Laboratory investigation of the influence of aging and compaction effort on low temperature performance of asphalt mixture containing different percentage of RAP. *Constr. Build. Mater.* **2021**, *298*, 123899. [CrossRef]
14. Antunes, V.; Freire, A.C.; Neves, J. Investigating aged binder mobilization and performance of RAP mixtures for surface courses. *Constr. Build. Mater.* **2021**, *271*, 121511. [CrossRef]
15. Roja, K.L.; Masad, E.; Mogawer, W. Performance and blending evaluation of asphalt mixtures containing reclaimed asphalt pavement. *Road Mater. Pavement* **2022**, *22*, 2441–2457. [CrossRef]
16. Zhang, J.; Guo, C.; Chen, T.; Zhang, W.; Yao, K.; Fan, C.; Liang, M.; Guo, C.; Yao, Z. Evaluation on the mechanical performance of recycled asphalt mixtures incorporated with high percentage of RAP and self-developed rejuvenators. *Constr. Build. Mater.* **2021**, *269*, 121337. [CrossRef]
17. Guduru, G.; Kuna, K.K. Classification of Reclaimed Asphalt Pavement (RAP) material using simple indicative tests. *Constr. Build. Mater.* **2022**, *328*, 127075. [CrossRef]
18. Chen, A.; Qiu, Y.; Wang, X.; Li, Y.; Wu, S.; Liu, Q.; Wu, F.; Feng, J.; Lin, Z. Mechanism and Performance of Bituminous Mixture Using 100% Content RAP with Bio-Rejuvenated Additive (BRA). *Materials* **2022**, *15*, 723. [CrossRef]
19. Zaumanis, M.; Arraigada, M.; Poulikakos, L.D. 100% recycled high-modulus asphalt concrete mixture design and validation using vehicle simulator. *Constr. Build. Mater.* **2020**, *260*, 119891. [CrossRef]
20. Li, J.; Xiao, F.; Amirkhanian, S.N.; Xu, O. Dynamic and rutting characteristics of recycled asphalt mixtures containing natural sand and anti-stripping agents. *J. Clean. Prod.* **2021**, *280*, 124365. [CrossRef]
21. Zhu, X.; Sun, Y.; Du, C.; Wang, W.; Liu, J.; Chen, J. Rutting and fatigue performance evaluation of warm mix asphalt mastic containing high percentage of artificial RAP binder. *Constr. Build. Mater.* **2020**, *240*, 117860. [CrossRef]
22. Kong, L.; Lu, Z.; He, Z.; Shen, Z.; Xu, H.; Yang, K.; Yu, L. Characterization of crack resistance mechanism of fiber modified emulsified asphalt cold recycling mixture based on acoustic emission parameters. *Constr. Build. Mater.* **2022**, *327*, 126939. [CrossRef]
23. Kim, Y.; Lee, S. Experiments and numerical analysis of cold-recycled asphalt mixture modified with desulfurization gypsum additive. *Constr. Build. Mater.* **2022**, *326*, 126803. [CrossRef]
24. Yang, W.; Ouyang, J.; Meng, Y.; Han, B.; Sha, Y. Effect of curing and compaction on volumetric and mechanical properties of cold-recycled mixture with asphalt emulsion under different cement contents. *Constr. Build. Mater.* **2021**, *297*, 123699. [CrossRef]
25. Li, J.; Yu, J.; Wu, S.; Xie, J. The Mechanical Resistance of Asphalt Mixture with Steel Slag to Deformation and Skid Degradation Based on Laboratory Accelerated Heavy Loading Test. *Materials* **2022**, *15*, 911. [CrossRef] [PubMed]
26. Cui, P.; Wu, S.; Xiao, Y.; Hu, R.; Yang, T. Environmental performance and functional analysis of chip seals with recycled basic oxygen furnace slag as aggregate. *J. Hazard. Mater.* **2021**, *405*, 124441. [CrossRef]
27. Song, Q.; Guo, M.-Z.; Wang, L.; Ling, T.-C. Use of steel slag as sustainable construction materials: A review of accelerated carbonation treatment. *Resour. Conserv. Recy* **2021**, *173*, 105740. [CrossRef]
28. Rodríguez-Fernández, I.; Lastra-González, P.; Indacoechea-Vega, I.; Castro-Fresno, D. Recyclability potential of asphalt mixes containing reclaimed asphalt pavement and industrial by-products. *Constr. Build. Mater.* **2019**, *195*, 148–155. [CrossRef]
29. Pasetto, M.; Baldo, N. Dissipated energy analysis of four-point bending test on asphalt concretes made with steel slag and RAP. *Int. J. Pavement Res. Technol.* **2017**, *10*, 446–453. [CrossRef]
30. Fakhri, M.; Ahmadi, A. Recycling of RAP and steel slag aggregates into the warm mix asphalt: A performance evaluation. *Constr. Build. Mater.* **2017**, *147*, 630–638. [CrossRef]
31. Yang, C.; Wu, S.; Cui, P.; Amirkhanian, S.; Zhao, Z.; Wang, F.; Zhang, L.; Wei, M.; Zhou, X.; Xie, J. Performance characterization and enhancement mechanism of recycled asphalt mixtures involving high RAP content and steel slag. *J. Clean. Prod.* **2022**, *336*, 130484. [CrossRef]
32. Cui, P.; Wu, S.; Xiao, Y.; Yang, C.; Wang, F. Enhancement mechanism of skid resistance in preventive maintenance of asphalt pavement by steel slag based on micro-surfacing. *Constr. Build. Mater.* **2020**, *239*, 117870. [CrossRef]
33. Cao, Z.; Huang, X.; Yu, J.; Han, X.; Wang, R.; Li, Y. Study on all-components regeneration of ultraviolet aged SBS modified asphalt for high-performance recycling. *J. Clean. Prod.* **2020**, *276*, 123376. [CrossRef]
34. Yang, C.; Xie, J.; Wu, S.; Amirkhanian, S.; Wang, Z.; Song, J.; Zhou, X.; Wang, F.; Zhang, L. Enhancement mechanism of induction heating on blending efficiency of RAP—Virgin asphalt in steel slag recycled asphalt mixtures. *Constr. Build. Mater.* **2021**, *269*, 121318. [CrossRef]
35. Yang, C.; Zhang, J.; Yang, F.; Cheng, M.; Wang, Y.; Amirkhanian, S.; Wu, S.; Wei, M.; Xie, J. Multi-scale performance evaluation and correlation analysis of blended asphalt and recycled asphalt mixtures incorporating high RAP content. *J. Clean. Prod.* **2021**, *317*, 128278. [CrossRef]
36. Song, J.; Xie, J.; Wu, S.; Yang, C.; Wang, Z.; Chen, H.; Shi, Y. Study on properties and improving mechanism of OGFC-13 asphalt mixtures modified by novel rubber pellets. *Constr. Build. Mater.* **2022**, *325*, 126799. [CrossRef]
37. Dong, S.; Wang, D.; Hao, P.; Zhang, Q.; Bi, J.; Chen, W. Quantitative assessment and mechanism analysis of modification approaches for cold recycled mixtures with asphalt emulsion. *J. Clean. Prod.* **2021**, *323*, 129163. [CrossRef]

38. Ministry of Transport of the People's Republic of China. *Technical Specifications for Construction of Highway Asphalt Pavements*; JTG F40–2004; China Communications Press: Beijing, China, 2004.
39. Oldham, D.; Rajib, A.; Dandamudi, K.P.R.; Liu, Y.; Deng, S.; Fini, E.H. Transesterification of Waste Cooking Oil to Produce A Sustainable Rejuvenator for Aged Asphalt. *Resour. Conserv. Recycl.* **2021**, *168*, 105297. [CrossRef]
40. Ma, Y.; Polaczyk, P.; Hu, W.; Zhang, M.; Huang, B. Quantifying the effective mobilized RAP content during hot in-place recycling techniques. *J. Clean. Prod.* **2021**, *314*, 127953. [CrossRef]
41. Zhou, X.; Zhao, G.; Tighe, S.; Chen, M.; Wu, S.; Adhikari, S.; Gao, Y. Quantitative comparison of surface and interface adhesive properties of fine aggregate asphalt mixtures composed of basalt, steel slag, and andesite. *Constr. Build. Mater.* **2020**, *246*, 118507. [CrossRef]
42. Erdem, S.; Blankson, M.A. Environmental performance and mechanical analysis of concrete containing recycled asphalt pavement (RAP) and waste precast concrete as aggregate. *J. Hazard. Mater.* **2014**, *264*, 403–410. [CrossRef]
43. Ai, X.; Cao, J.; Feng, D.; Gao, L.; Hu, W.; Yi, J. Performance evaluation of recycled asphalt mixtures with various percentages of RAP from the rotary decomposition process. *Constr. Build. Mater.* **2022**, *321*, 126406. [CrossRef]

materials

MDPI

Article

Research on Design and Performance of Self-Compacting Cement Emulsified Bitumen Mixture (CEBM)

Jinming Yi [1,2], Jianlin Feng [1,*], Yuanyuan Li [1,*], Tao Bai [1], Anqi Chen [3], Yangming Gao [4], Fan Wu [1], Shaopeng Wu [3], Quantao Liu [3] and Chuangmin Li [5]

[1] School of Civil Engineering and Architecture, Wuhan Institute of Technology, Wuhan 430205, China; jinmingyi_2005@163.com (J.Y.); baigs08@wit.edu.cn (T.B.); 22004010123@stu.wit.edu.cn (F.W.)
[2] Poly Changda Engineering Co., Ltd., Guangzhou 510062, China
[3] State Key Laboratory of Silicate Materials for Architectures, Wuhan University of Technology, Wuhan 430070, China; angelchen@seu.edu.cn (A.C.); wusp@whut.edu.cn (S.W.); liuqt@whut.edu.cn (Q.L.)
[4] Faculty of Civil Engineering & Geosciences, Delft University of Technology, Stevinweg 1, 2628 CN Delft, The Netherlands; y.gao-3@tudelft.nl
[5] School of Traffic and Transportation Engineering, Changsha University of Science and Technology, Changsha 410114, China; lichuangmin@csust.edu.cn
* Correspondence: 22004010110@stu.wit.edu.cn (J.F.); liyy@wit.edu.cn (Y.L.)

Citation: Yi, J.; Feng, J.; Li, Y.; Bai, T.; Chen, A.; Gao, Y.; Wu, F.; Wu, S.; Liu, Q.; Li, C. Research on Design and Performance of Self-Compacting Cement Emulsified Bitumen Mixture (CEBM). *Materials* **2022**, *15*, 4840. https://doi.org/10.3390/ma15144840

Academic Editor: Francesco Canestrari

Received: 9 June 2022
Accepted: 7 July 2022
Published: 12 July 2022

Abstract: To meet the needs of the road industry for maintenance operations, a new cement emulsified bitumen mixture (CEBM) with early-strength, self-compacting, and room-temperature construction characteristics was designed. The strength formation mechanism of CEBM was revealed with a scanning electron microscope (SEM) and the surface free energy (SFE) theory. The mechanical properties and road performance of the CEBM were investigated extensively. The results show that before the demulsification of emulsified bitumen, the SFE of the bitumen–aggregate–water three-phase system was reduced due to the replacement of the bitumen–aggregate interface with water. The adhesion work between the emulsified bitumen and the aggregate is negative, which means the adhesion between the emulsified bitumen and the aggregate will not occur spontaneously due to the existence of water. The liquid emulsified bitumen improves the workability of the mixture and ensures that the mixture can be evenly mixed and self-compacted. After demulsification, the work of adhesion between the residual bitumen and the aggregate is positive, which means residual bitumen and aggregate can bond spontaneously. In addition, the hydration products of cement and aggregate form a skeleton, and the emulsified bitumen film wraps and bonds the cement and aggregate together, creating strength. The emulsified bitumen, cement content, and curing conditions have significant effects on the stability of CEBM. The recommended dosage of emulsified bitumen and cement is 8% and 8–10%, respectively. This material integrates the hardening effect of cement and the viscoelastic performance of bitumen and has good workability, mechanical properties, and road performance. Therefore, the CEBM is technically feasible for application to bitumen pavement.

Keywords: bitumen/cement composite mixture; strength formation mechanism; early-strength; self-compacted; mixture performance

1. Introduction

Due to the increase in the traffic volume and heavy load of the service process of asphalt pavement [1], a series of distresses such as high-temperature rutting, low-temperature cracking, fatigue cracking, and water damage gradually appear in the asphalt pavement, which significantly reduces the service level of asphalt pavement [2,3]. In addition, the mileage of the roads needing maintenance also increases rapidly year by year. The pothole and grooves of asphalt pavement are the main distressed types of asphalt pavement. Water damage, looseness, spalling, cracks, and other distresses of asphalt pavement may evolve into potholes and grooves of asphalt pavement as well [4,5]. These damages will significantly reduce the service level of asphalt pavement and affect the driving comfort and safety of asphalt pavement [6]. If they are not repaired in time, these distresses will develop rapidly under the comprehensive actions of traffic load and water, resulting in an increase in maintenance costs and seriously endangering driving safety [7,8]. Therefore, to meet the needs of the road industry for maintenance operations, it is of great significance to develop fast-setting and environment-friendly road maintenance materials.

The traditional repair materials for the potholes in asphalt pavement include hot-mix asphalt mixture (HMA) and cold-mix asphalt mixture (CMA) [9,10]. HMA has a great road performance and long life. However, the aggregate and bitumen need to be heated during the construction of asphalt pavement in processes such as hot mixing, hot paving, and hot compaction. After paving, asphalt pavement needs to be rolled by compaction machines, which is not only extremely inconvenient but also entails high labor costs and a slow construction speed. Therefore, the construction and production of HMA requires a lot of fuel and results in the emission of a amounts of greenhouse gases [11,12]. The fuel cost accounts for 15% of the total cost of HMA [13,14]. In addition, the high temperature will accelerate the release rate of volatile organic compounds in the asphalt mixture [15], which causes a great deal of harm to the environment and construction workers. On the other hand, the CMA [16,17] needs only simple mixing techniques. It requires the heating of the aggregate and bitumen during construction. However, it still needs to be compacted after paving, and the strength formation rate of CMA is slow, so the road performance of CMA is not as good as that of HMA. Based on the above analysis, the HMA has performance advantages, and the CMA has construction technology advantages, but they also have disadvantages. The cold-mixed cement/emulsified asphalt mixture can repair the potholes of asphalt pavement quickly and conveniently, which has both the advantages of the rigidity of cement concrete and the flexibility of asphalt mixture [18,19]. Compared with ordinary asphalt mixture, the cement/emulsified asphalt mixture can save a lot of energy, which is good for energy conservation and emission reduction [20]. However, the cement/emulsified asphalt mixture requires a certain amount of time to demulsify and hydrate. It also needs time for curing, which will prolong the construction period and affect traffic [21]; moreover, the fatigue and water damage resistance of cement/emulsified asphalt mixture is not good [22,23], which also limits the wide application of cement/emulsified asphalt mixture [24]. Furthermore, the above mixtures are not self-compacting, and all of them are required to be compacted after the paving [25], which virtually makes the operation and construction harder.

Many scholars have studied the strength formation mechanism of cement emulsified asphalt mixture. Anmar [26–28] studied the influence of emulsified asphalt and cement content on binary blended cement filler (BBCF), and the results showed that when the emulsified asphalt content was 8% and the cement content increased from 0 to 4%, the indirect tensile strength first increased and then decreased. When the cement content is constant at 3% and the emulsified asphalt content increases from 6% to 9%, the indirect tensile strength, compressive strength, and elastic modulus first increase and then decrease. The depth of the BBCF rutting test is about 1–2 mm. Mechanism research shows that the strength of cement emulsified asphalt mainly comes from two parts: hydration products formed by the hydration of cement and the demulsification of emulsified asphalt [23,26]. With the increase in the cement dosage, the number of hydration products increases correspondingly, especially in the immersion Marshall test, where the samples are tested in a water bath environment to ensure that the cement particles have enough water for hydration. Therefore, the Marshall stability value of the repair material increases with the increase of the cement dosage [28]. When the amount of emulsified asphalt is constant, with the increase in the cement content, the compressive strength of the repair material sample increases [29]. There are three main reasons: the increase in cement dosage is equivalent to the decrease in the water-binder ratio, so it has a positive effect on the strength of repair the materials [30]; the quantity of cement hydration products increases with the increase of cement dosage, so the strength of the repair materials increases accordingly [31]; the process of increasing the cement consumption consumes more free water and increases the demulsification process of emulsified asphalt, which is more conducive to the formation of the spatial network structure of emulsified asphalt-hydration products, thus providing strength for the repair materials [32]. However, the early strength development in the cement emulsified asphalt mixture presently studied is relatively slow, and it is not suitable for the roads that need to be repaired quickly, on the spot, and opened to traffic in a short time, such as earthquake-resistant roads, military roads, etc. Particularly, the repaired pavement is damaged again under the coupling action of rain and vehicle loads, which seriously affects the road capacity [33]. In addition, there is a lack of systematic research on cement emulsified asphalt mixture.

To compensate for the disadvantages of the above conventional materials in terms of construction technology, economy, environmental protection, and road performance, the use of a new bitumen/cement composite stabilized mixture with early-strength, self-compacting, and room-temperature construction characteristics (CEBM) for road maintenance is proposed, which may have adequate applications for repairing potholes in the asphalt pavement quickly and conveniently. Using sulphoaluminate fast-hardening cement and emulsified bitumen as the composite binders for CEBM can not only achieve the rigidity [34] enabled by cement but also meet the load-bearing capacity and improve high-temperature anti-rutting. The bitumen also has the flexibility [35] to meet the requirements of low-temperature cracking resistance and medium-temperature fatigue resistance [36]. It can be mixed under room temperature conditions to fill and repair potholes in asphalt pavement. The aggregate gradation is determined according to Gussasphalt concrete, commonly used for steel bridge deck pavement, which has a large ratio of bituminous material [37]. This mixture is self-compacting [38], so it is no need to compact during the construction [39]. At the same time, sulfoaluminate fast-hardening cement can quickly develop strength [40], Which is beneficial to quickly repair the road.

The surface micromorphology and interface energy of CEBM were studied through scanning electron microscopy (SEM) and surface energy theory, which can be used to reveal the strength formation mechanism of CEBM. The influences of the emulsified bitumen dosage, cement dosage, curing time, and other factors on the strength of CEBM were investigated to determine the design and preparation parameters of CEBM. In addition, the road performance properties, such as high-temperature rutting, low-temperature cracking, and water damage resistances of CEBM are also studied.

2. Materials and Experimental Methods

2.1. Materials

2.1.1. Emulsified Bitumen

The emulsified bitumen that was used was cationic modified emulsified bitumen (BCR). BCR is prepared by emulsifying SBS modified bitumen. To prevent the segregation of the BCR, it should be stirred evenly before using and testing. The indicator test results are shown in Table 1.

Table 1. Technical properties of emulsified bitumen.

Test Item		Unit	Results	Requirements	Experimental Method
Demulsification speed		–	Slow crack	Slow crack	JTG E20-2011 T0658
Particle charge		–	+	+	JTG E20-2011 T0653
Evaporation residual degree		%	58.6	≥55	JTG E20-2011 T0651
Penetration of evaporated residue bitumen (25 °C)		0.1 mm	56	45–150	JTG E20-2011 T0604
Ductility of evaporated residue bitumen (15 °C)		cm	45.5	≥40	JTG E20-2011 T0606
Softening point of evaporated residue bitumen		°C	49.4	–	JTG E20-2011 T0605
Normal temperature storage stability	1 d	%	0.47	≤1	JTG E20-2011 T0655
	5 d		2.17	≤5	

2.1.2. Cement

Cement that was used was sulphoaluminate fast-hardening and high-strength cement (SFHC). For comparison, the specific indicators of the SFHC and commonly used Portland cement ($PO_{42.5}$) silicate cement are shown in Table 2. The technical performance of the SFHC met the technical requirements of Specification for Design of Highway Cement Concrete Pavement (JTG D40-2011).

Table 2. Technical properties of SFHC and $PO_{42.5}$ silicate cement.

Cement	Specific Surface Area (m^2/kg)	Setting Time (min)		Compressive Strength (MPa)			Flexural Strength (MPa)		
		Initial Setting	Final Setting	1 Day	3 Days	28 Days	1 Day	3 Days	28 Days
SFHC	431	15	31	30	41.2	52.1	3.2	4.1	7.3
$PO_{42.5}$ silicate cement	396	175	235	8	27.5	49	1.3	5.5	8.0

2.1.3. Aggregates and Fillers

The coarse and fine aggregates that were used comprised crushed limestone. The particle size of the coarse aggregate is ≥2.36 mm, while that of the fine aggregate is 0.075–2.36 mm. The filler that was used was the mineral powder ground from the limestone; the results of its technical properties are shown in Tables 3–5.

Table 3. Technical properties of coarse aggregate.

Parameters	Unit	Results	Requirements	Experimental Method
Stone crushing value	%	23.2	≤28	JTG E42-2005 T0316
Needle flake content	%	8.2	≤15	JTG E42-2005 T0312
Los Angeles wear value	%	15	≤28	JTG E42-2005 T0317
Water absorption	%	0.9	≤2.0	JTG E42-2005 T0308
Apparent specific gravity	–	2.667	≥2.6	JTG E42-2005 T0605

Table 4. Technical properties of fine aggregate.

Parameters	Unit	Results	Requirements	Experimental Method
Sediment percentage	%	2	≤3	JTG E42-2005 T0335
Sand equivalent	%	63	≥60	JTG E42-2005 T0334
Angularity (flow time method)	s	45	≥30	JTG E42-2005 T0345
Apparent specific gravity	–	2.650	≥2.5	JTG E42-2005 T0328

Table 5. Technical properties of filler.

Parameters		Unit	Results	Requirements	Experimental Method
Apparent specific gravity		–	2.612	≥2.5	JTG E42-2005 T0352
Particle size range (%)	<0.6	mm	100	100	JTG E42-2005 T0351
	<0.15	mm	92.4	90–100	
	<0.075	mm	86.3	75–100	
Plasticity coefficient		–	3.5	<4	JTG E42-2005 T0354
Hydrophilic coefficient		–	0.82	<1	JTG E42-2005 T0353

2.2. Mixture Gradation

Gussasphalt concrete is a kind of asphalt mixture with high asphalt content, high mineral powder content, and a void ratio of less than 1%, which is mixed at high temperature (220~260 °C) and paved by the fluidity of the mixture itself without rolling. The CEBM has great flowability with a self-compaction property after paving, it has high bitumen and high mineral powder content, and the gradation of CEBM is similar to that of Gussasphalt concrete. The gradation range of Gussasphalt concretes varies among different countries. The gradation of CEBM was selected from the range intersection of Gussasphalt concrete in China, Germany, and the EU [41]. The aggregate composite gradation is shown in Figure 1. Table 6 is the proportion of each mineral aggregate. The cement is used to replace the part of the mineral powder in equal volume, and the sum of cement and mineral powder accounts for 24% the weight of the mineral aggregates. The emulsified bitumen content is measured in the weight of aggregates, and its proportion is calculated with the residue bitumen after demulsification.

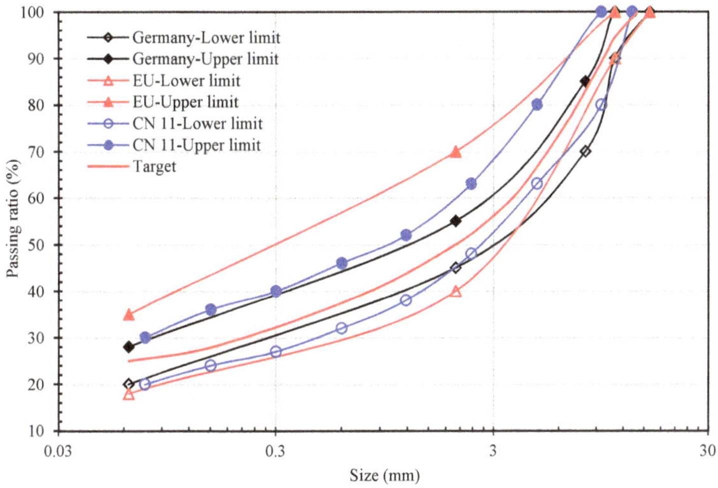

Figure 1. Gradation curve.

Table 6. Grading and proportion of mineral materials.

Aggregate	10–15 mm	5–10 mm	3–5 mm	0–3 mm	Filler
Proportion	4%	12%	32%	28%	24%

2.3. Mechanism Characterization Test

2.3.1. Surface Micro-Morphology Testing (SEM)

A Scanning Electron Microscope (SEM, JSW-5510LV) was applied to investigate the surface micro-morphology of CEBM, and the test voltage was 15 mv. The sample was collected from the middle of the CEBM specimen, cured for 24 h, and dried in an oven at 45 °C. Then, a gold spraying treatment was conducted on the surface of the sample, to ensure the sample conduction. The sample was placed on the sample table of SEM for detection.

2.3.2. Surface Free Energy Measurements

The surface free energy (SFE) of the residual bitumen from the emulsified bitumen, 70# base bitumen, styrene-butadiene-styrene (SBS) modified bitumen, limestone aggregate, and cement concrete were tested. For the preparation of the bitumen sample, the bitumen was dropped on a glass slide and placed horizontally in an oven at 150 $\pm$ 5 °C for 3 min to level the bitumen surface. Then, it was naturally cooled in a dust-free and dry environment for 6 h. The heated and liquified bitumen were spread onto a dried heat-resistant microscope slide to form a homogeneous film. The bitumen specimen is shown in Figure 2a. For the aggregate and cement specimen, limestone rock and 7 d cured cement block were cut into 50 mm $\times$ 50 mm $\times$ 5 mm and a polishing treatment of their surfaces was performed to obtain a smooth specimen. The specimen is shown in Figure 2b.

Figure 2. Contact angle test of bitumen and cement mortar. ((**a**) The bitumen specimen; (**b**) limestone rock specimen).

The bitumen (film), limestone, and cement concrete were tested using three liquids with known SFE parameters, namely distilled water, ethylene glycol, and glycerol, respectively. The distilled water, ethylene glycol, and glycerol were used to measure the contact angle; the surface free energy parameters of these three kinds of liquid are shown in Table 7. The pendant-drop method was applied and the contact angle was measured with a contact angle tester, as shown in Figure 3, the test temperature was 25 °C. The experiment was carried out using an SDC-100 contact angle device that was sourced from Dongguan Dingsheng Precision Instrument Co., Ltd.

Table 7. SFE parameters of test liquid (25 °C, mJ/m^2).

Reagent	SFE (γ_L)	Dispersion Component (γ_L^d)	Polarity Component (γ_L^p)	SFE Acidity Pa-rameter (γ_L^+)	SFE Alkalinity Parameter (γ_L^-)
Distilled water	72.8	21.8	51.0	25.5	25.5
Ethylene glycol	48.3	29.3	19.0	3.0	30.1
Glycerol	64.0	34.0	30.0	3.92	57.4

2.4. *Manufacturing of CEBM*

CEBM adopts a Gussasphalt concrete gradation, based on a C/EB formula system with high asphalt content, high mineral powder (including cement) content, and a small amount of water, which grants it great flowability. Using its great flowability "pouring, leveling, and compacting", it can form a uniform pavement with high density and low voids without rolling.

Firstly, Gussasphalt concrete gradation was adopted to ensure that the air voids of the molding mixture sample were less than 1%. The emulsified bitumen, cement, water, aggregate, and mineral powder were measured according to the proportion ratio of CEBM.

In addition, the water-cement ratio was determined to be 0.5:1 [18]. The Additional water quantity of CEBM was then calculated according to Equation (1).

$$C_W = R_{w/c}C_C - (1 - C_{Br})C_B \tag{1}$$

where C_W is the additional water content in CEBM, $R_{w/c}$ is the water-cement ratio, C_C is the cement content in CEBM, C_B is the emulsified bitumen content in CEBM, and C_{Br} is the evaporation residue of the emulsified bitumen.

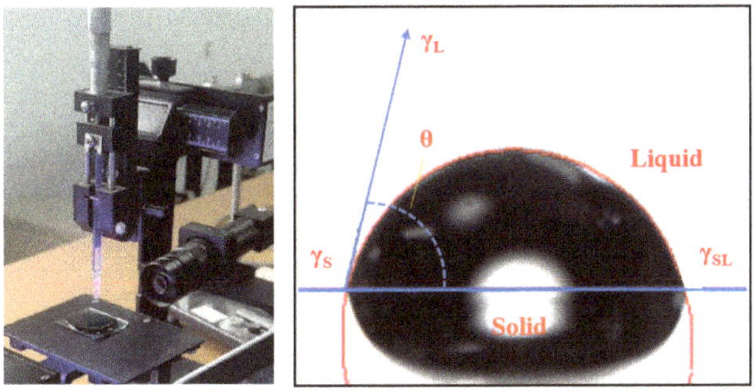

Figure 3. Diagram of contact angle test.

Then, the aggregates were mixed in dry conditions for an even mixture; the designed amount of water was sprayed to the uniformly mixed aggregate, it was stirred quickly and evenly, thus granting the mixture the desired flowability, and flowability experiment's outflow time was less than 20 s (JTG/T3364-02—2019). The mixture was conditioned at room temperature (25 °C) for curing for more than 24 h. The molding process and specimens of CEBM are shown in Figure 4. The Volume parameters of CEBM are shown in Table 8.

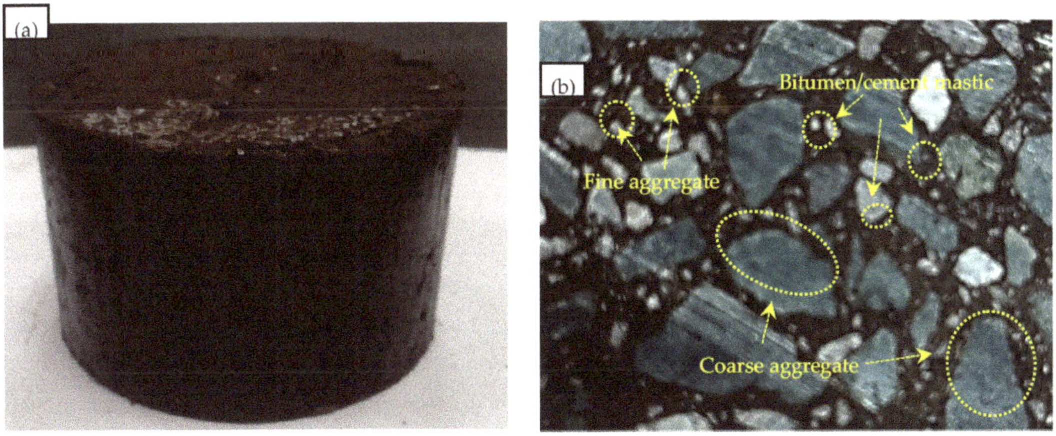

Figure 4. Manufacturing of CEBM ((**a**) Marshall specimen appearance; (**b**) apparent section of Marshall specimen).

Table 8. Volume parameter of CEBM.

Volume Parameter	Emulsified Bitumen Contents (%)		
	8	10	12
Bulk specific gravity	2.594	2.589	2.581
Theoretical maximum density	2.610	2.605	2.599
Air voids (%)	0.61	0.61	0.69

2.5. Experimental Methods

2.5.1. Mechanical Property Test of CEBM

The Marshall stability of the Marshall specimen with different cement and emulsified bitumen contents and the curing time were detected to investigate the influential parameters of the mechanical property test of CEBM. Each group had 8 specimens, 4 of which were conditional and 4 were non-conditional. After curing them at 25 °C for 6 h, 12 h, and 24 h, respectively, the Marshall stability of CEBM was tested under the drying and bath water conditions of 25 °C and 60 °C, respectively.

2.5.2. Wheel Track Test of CEBM

The size of the rutting specimen of CEBM was 300 mm × 300 mm × 50 mm, and the specimens were cured at 25 °C for 24 h and 72 h (Complete curing). Before the test, the samples were placed in the rutting instrument for more than 5 h to maintain a constant temperature, and the test temperature, wheel loading, and rate were 60 °C, 0.7 MPa, and 42 times/min, respectively.

2.5.3. Low-Temperature Bending Test of CEBM

The Universal testing machine (UTM-100) was used for testing the low-temperature crack resistance of CEBM. The specimen size was 250 mm × 30 mm × 35 mm. Each group had 3 specimens. The test temperature was −10 °C, the span of the beam was 200 mm, and the loading rate was 50 mm/min. The schematic diagram is shown in Figure 5.

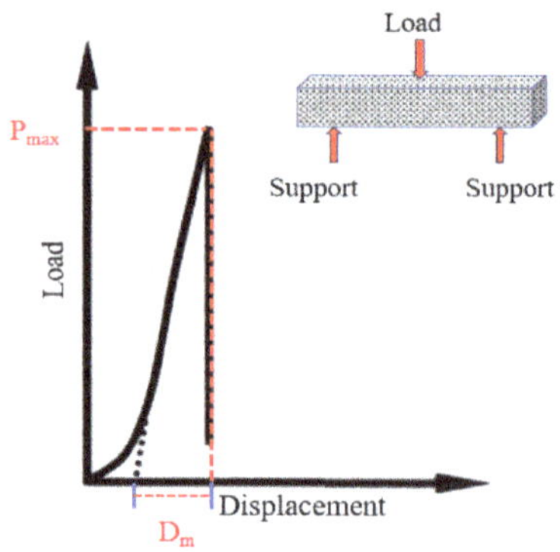

Figure 5. Three-point bending test [42].

2.5.4. Water Sensitivity Test of CEBM

The immersion Marshall stability test and the freeze-thaw splitting test were conducted according to the criteria of the "Standard Test Method of Bitumen and Bituminous Mixtures for Highway Engineering" JTG E20-2011. Each group had 8 specimens, 4 of which were conditional and 4 were non-conditional, which were used to test the water stability of CEBM with different cement and emulsified bitumen contents.

2.6. Test Flow Chart

The test flow chart is shown in Figure 6.

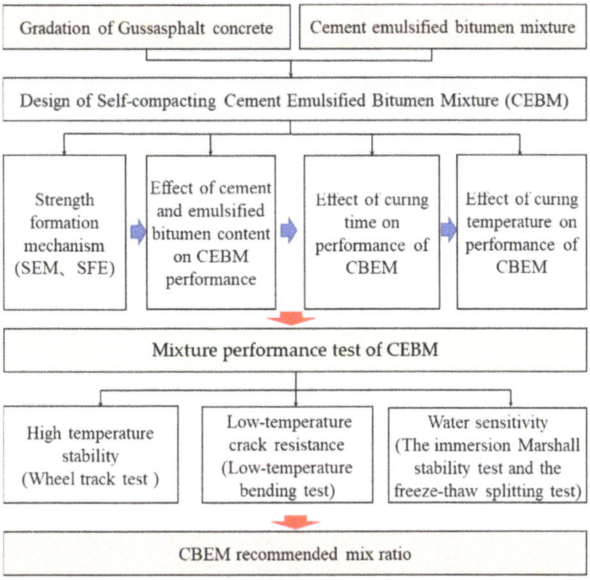

Figure 6. Test flow chart.

3. Experimental Results and Discussion

3.1. Analysis of Surface Morphology and Surface Energy

3.1.1. Surface Micro-Morphology of CEBM

Cement and emulsified bitumen were used as the composite binders in the CEBM. A SEM was used to observe the micro-morphology of CEBM cured for 24 h. The 1500×, 3000×, 5000×, and 8000× magnified figures of the sample are shown in Figure 6. The hydration products of cement mainly include acicular and reticulated hydrated calcium silicate (C-S-H), flaky calcium hydroxide (C-H), and columnar ettringite (C-A-S-H) [43,44]. In Figure 7a,b, there are many micropores in the CEBM, which are the air voids where water evaporates from the demulsification of emulsified bitumen. In Figure 7c, the flake calcium hydroxide is well wrapped by the demulsified bitumen. From Figure 7c,d, the hydration products of cement that can be observed are uniformly distributed in the demulsified bitumen. And the circular hydrated calcium silicate, columnar ettringite, and flaky calcium hydroxide can be clearly observed in the CEBM [18,21]. In Figure 7d, the emulsified bitumen forms a film after demulsification and wraps the surface of the aggregate, mineral powder, and cement concrete, and the hydration products of cement are also evenly distributed in the CEBM. The hydration products can pierce the bitumen film and smooth areas, and form bonds with other hydration products or the aggregate surface. The bitumen and cement composite bind the aggregate and mineral power together [45].

On the one hand, water and liquid emulsified bitumen can significantly increase the flowability of the CEBM, so it can mix at room temperature and has the advantage of self-compaction. On the other hand, the cementitious phase in the CEBM was dispersed within the emulsified bitumen. The cement hydration consumed a portion of the water that occupies the micro air void spaces between the emulsified bitumen and aggregate, which had a stiffening effect on emulsified bitumen [43,46]. The hydration products of the aggregate and cement, the skeleton function of the cement, and the encapsulation and adhesion function of the emulsified bitumen complement each other, and together with the hydration products of cement interweave with demulsified bitumen to enhance the overall stability and form the strength of CEBM.

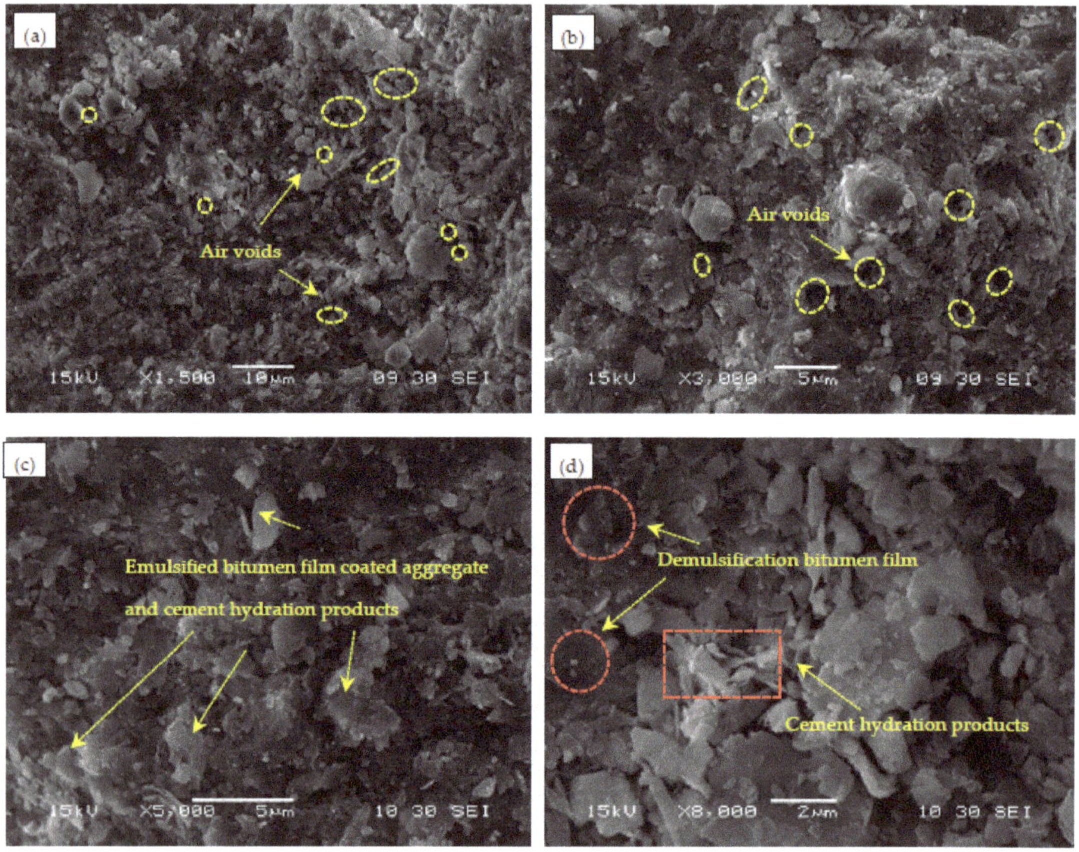

Figure 7. Microscopic morphology figures of CEBM ((**a**) 1500×; (**b**) 3000×; (**c**) 5000×; (**d**) 8000×).

3.1.2. Surface Free Energy Analysis of CEBM

In the SFE theory, the surface tension of matter is composed of the Lifshitz-Van Der Waals interaction (γ^d) and the Lewis acid-base interaction ($\gamma^{+/-}$) [47,48]. The contact angles of distilled water, glycol, and glycerol titration on the surfaces of bitumen, aggregate, and cement blocks were measured to calculate their surface energy. The surface energy parameters of the bitumen and aggregate were calculated by Young's Equations (2)–(4) [47,49]. Via substitution into Young's equation, the three unknowns ($\gamma^d, \gamma^+, \gamma^-$) were solved simultaneously. The results are shown in Table 9.

$$\gamma_L \frac{1+\cos\theta}{2} = \sqrt{\gamma_S^d \gamma_L^d} + \sqrt{\gamma_S^+ \gamma_L^-} + \sqrt{\gamma_S^- \gamma_L^+} \tag{2}$$

$$\gamma_L = \gamma^d + \gamma_L^p \tag{3}$$

$$\gamma_L^p = 2\sqrt{\gamma_L^+ \gamma_L^-} \tag{4}$$

where $\gamma_S^d, \gamma_S^+, \gamma_S^-$ represent the dispersion component, Lewis acid number, and Lewis base number of the tested solid, respectively; $\gamma_L, \gamma_L^d, \gamma_L^+, \gamma_L^-$ express the surface free energy, dispersion component, Lewis acid number, and Lewis base number of the test liquor, respectively.

Table 9. SFE parameters of bitumen and aggregate (25 °C, mJ/m^2).

Reagent	SFE (γ_L)	Dispersion Component (γ_L^d)	Polarity Component (γ_L^p)	SFE Acidity Parameter (γ_L^+)	SFE Alkalinity Parameter (γ_L^-)
demulsification bitumen	21.43	18.65	2.78	0.114	17.013
70# bitumen	20.02	19.88	0.14	0.001	6.557
SBS modified bitumen	17.71	13.17	4.54	1.859	2.770
Limestone aggregate	71.52	37.33	34.18	4.036	72.369
Cement block	205.11	125.95	79.16	74.786	20.947

Adhesion work refers to the work performed when separating the two phases that contact (adhere to) each other on two new surfaces. The adhesion work >0 indicates that the adhesion process can proceed spontaneously. The adhesion process between the residual bitumen of the emulsified bitumen and aggregate can be explained by SFE theory, and SFE can calculate the adhesion work between the bitumen and aggregate. The cohesive work of the same (single-phase) substance was computed using Equation (5); the interface between the two substances (two-phase) can be calculated according to Equation (6) [48,50]. However, The volatilization of water and hydration of cement are the main reasons for the rapid strength formation of CEBM. Therefore, it is necessary to study the adhesion work between the emulsified bitumen, the cement block, and the aggregate before and after the demulsification of the emulsified bitumen. According to the surface energy theory, before bitumen demulsification, water, bitumen, and aggregate coexist, and the adhesion work W_{bsw} of the three-phase system of water, bitumen, and aggregate is calculated by the Equation (7) [51].

$$\gamma_{ii} = 2\gamma_i = 2\left(\gamma_i^d + \gamma_i^p\right) \tag{5}$$

$$\gamma_{ij} = 2(\sqrt{\gamma_i^d \gamma_j^d} + \sqrt{\gamma_i^+ \gamma_j^-} + \sqrt{\gamma_i^- \gamma_j^+}) \tag{6}$$

$$W_{bsw} = \gamma_{bs} + \gamma_{ww} - \gamma_{bw} - \gamma_{sw} \tag{7}$$

where γ_{ii} is the cohesive work of the same substance; γ_{ij} is the interface energy of two substances; W_{bsw} is the adhesion work of the water, bitumen, and aggregate three-phase system before demulsification; γ_{bs}, γ_{bw}, γ_{sw} represent the adhesion work between the bitumen and aggregate, bitumen and water, and aggregate and water, respectively; γ_{ww} is the cohesive force between water and water; and $\gamma_i^d, \gamma_i^+, \gamma_i^-$ represent the dispersion component, Lewis acid, and base number (*i* and *j* are the *b*, *s*, and *w*; the *b*, *s*, and *w* represent bitumen, aggregate, and water, respectively).

The adhesion work of bitumen towards the aggregate and cement mortar block was calculated by substituting the parameters in Table 9. The adhesion work >0 indicates that the adhesion process can proceed spontaneously. The results are shown in Figure 8. Before demulsification, the SFE of the bitumen–aggregate–water three-phase system was reduced due to the existence of water in the bitumen–aggregate interface. The adhesion work between the emulsified bitumen and aggregate is negative, and the adhesion between emulsified bitumen and aggregate may not happen spontaneously due to the existence of water. Therefore, the liquid emulsified bitumen can improve the workability of the mixture and ensures that the mixture can be evenly mixed and self-compacted. The adhesion work of the emulsified bitumen with the limestone aggregate and cement mortar before demulsification is −129.97 mJ/m^2 and −108.80 mJ/m^2, that is, the surface free energy changes are 129.97 mJ/m^2 and 108.80 mJ/m^2, which shows that the adhesion of the emulsified bitumen with limestone aggregate and cement mortar cannot occur without other physicochemical effects.

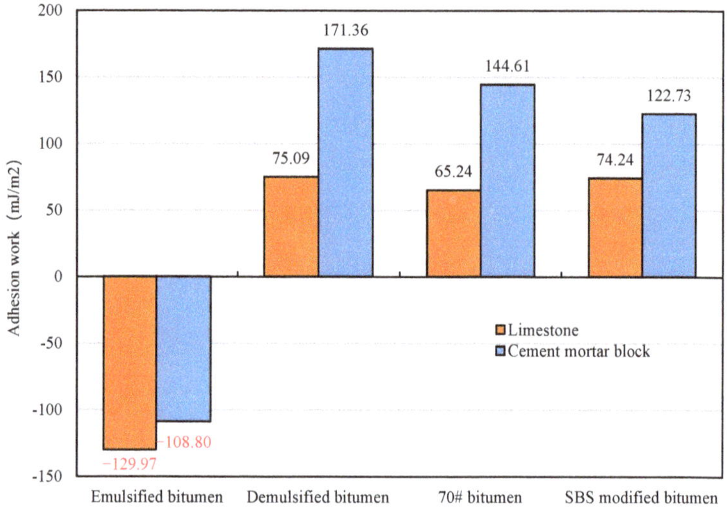

Figure 8. Bitumen-aggregate interface adhesion work.

After demulsification, the adhesion work between the residual bitumen and aggregate is positive, and the residual bitumen and aggregate can bond spontaneously. The demulsification film of bitumen further wraps and adheres to the aggregate, and establishes a spatial network structure in the mixture, thus forming strength. After the demulsification of emulsified bitumen, the adhesion work between 70# bitumen and SBS modified bitumen with the aggregate is positive, which means the change of surface free energy is negative, indicating that their adhesion is spontaneous and can be bonded without external work. In addition, after the demulsification of the emulsified bitumen, the adhesion work of emulsified bitumen towards the limestone and cement mortar block is 75.09 mJ/m^2 and 171.36 mJ/m^2, respectively; the adhesion work of 70# bitumen towards the limestone and cement mortar block is 65.24 mJ/m^2 and 144.61 mJ/m^2, respectively; and the adhesion work of SBS modified bitumen towards the limestone and cement mortar block is 74.24 mJ/m^2 and 122.73 mJ/m^2, respectively. The adhesion work of emulsified bitumen after demulsification towards limestone is 15.1% and 1.1% higher than that of 70# bitumen and SBS modified bitumen, and the adhesion work of emulsified bitumen after demulsification towards cement mortar is 18.5% and 39.6% higher than that of 70# bitumen and SBS modified bitumen. The results show that the adhesion work of emulsified bitumen after demulsification with limestone and cement mortar is higher than that of 70# bitumen and SBS modified bitumen, which can ensure that the CEBM possesses good water damage resistance.

3.2. *Influence Factors of Mechanical Performance of CEBM*

3.2.1. Influence of Cement and Emulsified Bitumen Content on Mechanical Performance of CEBM

The Marshall specimens are prepared with 8% and 10% emulsified bitumen and 8%, 10%, and 12% sulphoaluminate cement. The flowability of CEBM with different cement and emulsified bitumen contents is shown in Table 10, and the test results of its Marshall stability are shown in Figure 9. In Figure 9, the Marshall stability of CEBM increases with the increase of the emulsified bitumen (8–12%) and cement (8–12%) contents; when the cement content is 12%, no large difference is observed for the Marshall stabilities of CEBM with 8% and 10% emulsified bitumen content. The CEBM with 12% cement and 8% emulsified bitumen has the largest Marshall stability of 10.88 kN, while with 8% cement and 8% emulsified bitumen, it has the lowest stability (8.26 kN). However, it still satisfies the requirement of the Marshall stability of HMA (≥8 kN).

Table 10. Flowability of CEBM with different cement and emulsified bitumen contents.

Emulsified Bitumen	Flowability of Different Cement Content (s)		
	8%	10%	12%
8% BCR	18.2	19.6	22.1
10% BCR	16.4	18.6	20.2

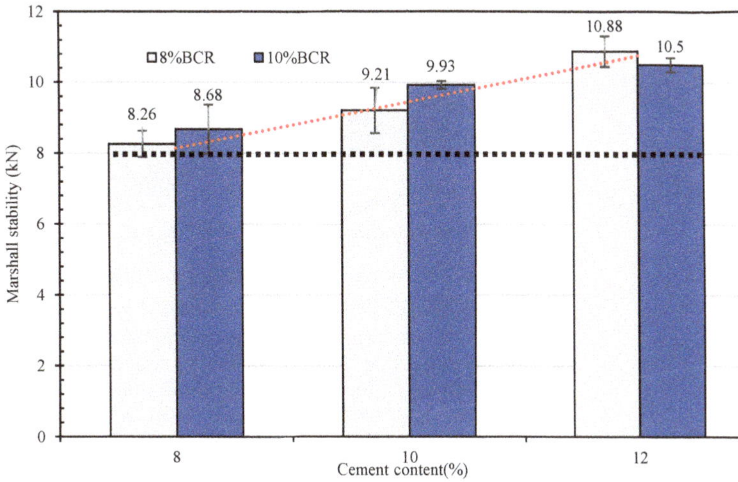

Figure 9. Marshall stability of CEBM with different cement and emulsified bitumen contents.

The hydration of cement makes use of the free water produced by the demulsification of emulsified bitumen, which accelerates the demulsification of bitumen [16,44]. Emulsified bitumen residues and cement hydration products are the composite binders of the CEBM, which explains why the strength of the mixture increases gradually with the increase of the cement content. For emulsified bitumen, the free water produced by demulsification not only promotes the hydration reaction of cement but also enables the mixture to have great flowability and self-compacting characteristics [45,47]. With the increase of the emulsified bitumen content, the flowability of the mixture is enhanced. However, with the further increase of the emulsified bitumen content, both the free bitumen and free water content will increase [52,53] such that the 12% cement content may increase the free water content and free bitumen content, and the free water will produce the air void in the CEBM after its evaporation; therefore, the free bitumen act as the lubricant between the aggregates and cement concrete, thereby reducing the mechanical performance of CEBM.

3.2.2. Curing Time Effect on the Mechanical Performance of CEBM

The strengths of the CEBM with 8% emulsified bitumen and 8% and 10% cement are tested after 6 h, 12 h, and 24 h curing, respectively, and the test results are shown in Figure 10. Figure 10 shows that the Marshall stability of CEBM increases continuously with the increase of the curing time (0–24 h). In detail, the Marshall stability of CEBM with 8% BCR and 10% cement after 24 h curing is 68.1% and 167.7% higher than that of CEBM after 12 h and 6 h curing. When the curing time is 6 h, the Marshall stability of the CEBM with 8% emulsified bitumen and 10% cement is 3.44 kN, which satisfies the requirement of the Marshall stability CMA (≥3 kN). After 24 h curing, the Marshall stability of the CEBM meets the requirements of HMA (≥8 kN); therefore, it has a good early-strength property.

the maximum failure loading of CEBM is 0.46 kN, and the corresponding mid-span deflection is 0.52 mm. From Table 11, the bending tensile strength of the trabeculae is 4.28 MPa. Compared with ordinary cement emulsified bitumen mixture, the maximum bending tensile strain of CEBM increases by 19.19%., and the bending stiffness modulus decreases by 4.98%. The Low-temperature performance has been improved to some extent. Although there is no technical requirement for the failure strain in the low-temperature bending test of CMA, the maximum flexural tensile strain of CEBM still meets the requirement of China's criterion (JTG D50-2017) that the flexural strain of the HMA should be greater than 2000με. This performance is improvement due to the existence of the emulsified bitumen, which can provide flexibility for CEBM and improve the Low-temperature crack resistance of CEBM.

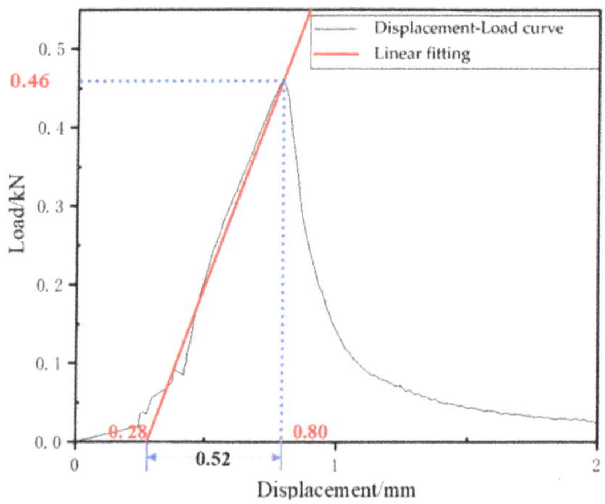

Figure 14. Loading-deformation curve of low-temperature bending test.

Table 11. Low temperature bending test results.

Mixture	Mid-Span Deflection (mm)	Bending Tensile Strength (MPa)	Bending Tensile Strain (με)	Bending Stiffness Modulus (MPa)
CEBM	0.52	4.28	2503.08	1710.21

3.3.3. Water sensitivity of CEBM

(1) Immersion Marshall test

The results of the water immersion Marshall test of CEBM at 25 °C and 60 °C are shown in Table 12. According to Table 12, the Marshall stabilities of CEBM both before and after water immersion are higher than 8 kN (the HMA requirement), and the residual Marshall stability of CEBM after water immersion increases at first and then decreases with the increase of the cement content. The water-immersed residual stability (IRS) of ordinary cement emulsified bitumen mixture is 85.3% [28]. The IRS of CEBM is shown in Figure 15. The 25 °C water immersion-conditioned CEBM with 8% BCR and 8% cement has the smallest IRS value of 92.4%, and the 60 °C IRS of CEBM with 8% BCR and 10% cement has the smallest IRS of 130.3%. The 25 °C IRS of CEBM increases with the increase of cement content, while the 60 °C IRS of CEBM initially increases and then decreases with the increase of the cement content, and reaches the maximum value at 10% cement content. All of the IRS values of CEBM are greater than ordinary cement emulsified bitumen mixture, which is obviously better than the requirements of higher than 80%.

Table 12. Marshall stabilities of CEBM with and without water immersion.

Emulsified Bitumen and Cement Content	Marshall Stability (25 °C)		Marshall Stability (60 °C)	
	Unconditioned Group (kN)	Conditioned Group (kN)	Unconditioned Group (kN)	Conditioned Group (kN)
8% BCR + 8% cement	10.12	9.35	8.26	8.35
8% BCR + 10% cement	14.01	15.70	8.71	11.35
8% BCR + 12% cement	13.85	16.52	10.88	12.30
10% BCR + 8% cement	9.56	9.34	8.68	8.54
10% BCR + 10% cement	12.26	14.68	9.93	11.50
10% BCR + 12% cement	13.64	16.62	10.50	11.89

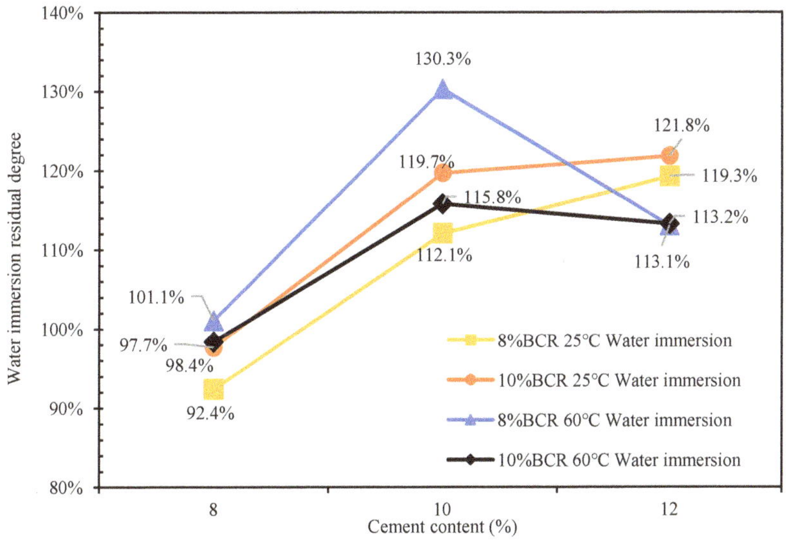

Figure 15. Water-immersed residual stability (IRS) of CEBM.

(2) Freeze-thaw splitting test

The results of the freeze-thaw indirect tensile strength test are shown in Figure 16. According to Figure 16, the freeze-thaw indirect tensile strength ratio (TSR) of the CEBM with 10% BCR and 8% cement content is the smallest value, which is 83.3%, and the TSR value of the CEBM with 8% BCR and 12% cement content is the highest, which is 105.3%. The TSR of the mixture increases with the increase of the cement content. Since the splitting strength of the conditional group in the freeze-thaw splitting test is about 0.4 MPa, the TSR increases with the decrease of the splitting strength of the unconditional group. This explains why with the increase in the cement content, although the TSR increases, its splitting strength also decreases. In addition, all of the TSR values of the above CEBMs meet the requirements of higher than 75%. Therefore, the CEBM has good water stability.

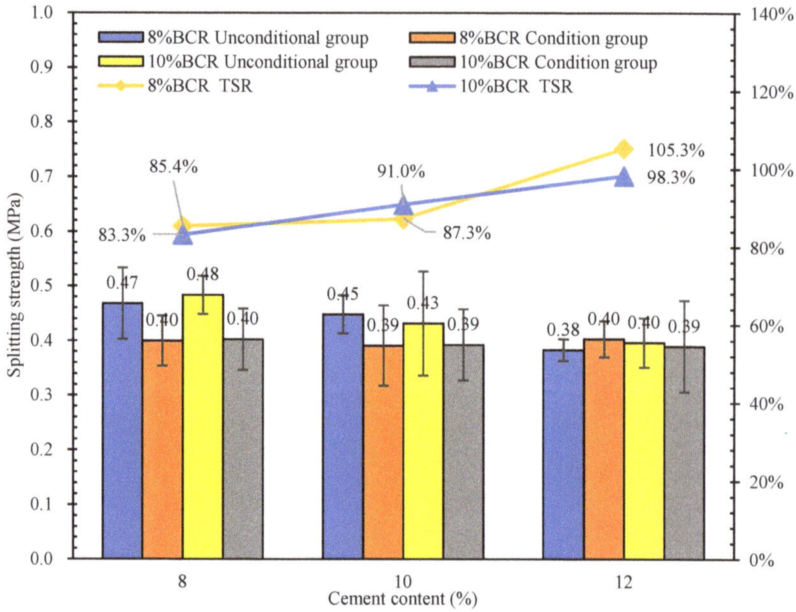

Figure 16. Splitting strength and freeze-thaw splitting ratio of CEBM.

4. Conclusions

The SEM and surface energy theory of the CEBM were studied using modern testing technology, and the strength formation mechanism of CEBM was revealed. The influence of the emulsified bitumen, the cement dosage, and the curing time on the strength of CEBM was studied. In addition, the road performance of CEBM is evaluated by road performance tests. The following conclusions were obtained.

(1) Before demulsification, the SFE of the bitumen–aggregate–water three-phase system was reduced due to the replacement of the bitumen–aggregate interface with water. The adhesion work between the emulsified bitumen and aggregate is negative, and the adhesion between the emulsified bitumen and aggregate may not happen spontaneously due to the existence of water. Meanwhile, water exists in bitumen and aggregate, which improves the workability of CEBM and ensures its uniform mixing and self-compacting. After demulsification, the adhesion work between the residual bitumen and aggregate is positive, and the residual bitumen and aggregate can bond spontaneously. The free water produced by the demulsification of bitumen reacts with the cement, the hydration products of cement form a skeleton in aggregate, and the demulsified bitumen further encapsulates the aggregate and cement and bonds them together. The skeleton of the cement and the adhesion of bitumen complement each other, and establish a spatial network structure in the CEBM, thus forming high strength.

(2) The emulsified bitumen content, cement content, and curing conditions have significant effects on the mechanical stability of CEBM. When the cement content is 12% and the emulsified bitumen content is 8%, the CEBM has the maximum Marshall stability of 10.88 kN; when the cement content is 8% and the emulsified bitumen content is 8%, the CEBM has the maximum Marshall stability of 8.26 kN. All of these values are even higher than the requirement for the hot mix bitumen mixture ($\geq$8 kN). In addition, when the curing time is 6 h, all the Marshall stabilities of CEBM can reach the stability requirement of CMA ($\geq$3 kN).

(3) Due to the hardening effect of cement, the CEBM has an excellent rutting resistance at high temperatures, and the dynamic stability is 18,333 times/mm cured for 24 h. On the other hand, due to the viscoelasticity of bitumen, the maximum flexural-tensile strain at

low temperature is 2503 με, which even meets the requirement of the flexural-tensile strain of hot mix bitumen mixture (≥2000 με). The water immersion residue is higher than 110% with 10% cement, and the TSR is higher than 85%, indicating the CEBM has good water stability.

(4) CEBM has good working and mechanical properties; therefore, it is technically feasible to use CEBM as a new material for road construction and maintenance. For comprehensive economic considerations, the recommended dosage of CEBM emulsified bitumen is 8%, and that of cement is 8–10%. The cement dosage can be determined according to the relevant engineering requirements.

Author Contributions: Conceptualization, T.B., A.C. and Y.G.; Data curation, J.Y., J.F., Y.L., T.B., A.C., Y.G., F.W. and Q.L.; Funding acquisition, J.Y. and T.B.; Investigation, J.Y. and Y.L.; Methodology, J.Y., Y.L., A.C., S.W. and C.L.; Project administration, T.B.; Resources, F.W.; Software, Y.L., F.W. and Q.L.; Supervision, J.F.; Validation, Y.L. and F.W.; Writing—original draft, J.F. and A.C.; Writing—review & editing, J.Y., Y.L., Y.G., F.W., S.W., Q.L. and C.L. All authors have read and agreed to the published version of the manuscript.

Funding: The authors acknowledge the financial support from the National Natural Science Foundation of China (No. 52108415), the National Key Research and Development Program of China (No. 2018YFB1600200), the Natural Science Foundation of China (No. 51778515), the key technical innovation projects of Hubei Province (No. 2019AEE023), the plan for outstanding young and middle-aged scientific and technological innovation team in universities of Hubei Province (No. T2020010), the scientific research fund of Hunan Provincial Education Department (No. 18A117), the Key R&D Program of Hubei Province (No. 2020BCB064), and the support for the project from the science and technology projects fund of Changsha city (No. kq2004065).

Data Availability Statement: The data presented in this study are available on request from the corresponding author.

Conflicts of Interest: The authors declare no competing financial interest.

References

1. Guo, R.; Nian, T.; Zhou, F. Analysis of factors that influence anti-rutting performance of asphalt pavement. *Constr. Build. Mater.* **2020**, *254*, 119237. [CrossRef]
2. Yao, L.; Dong, Q.; Jiang, J.; Ni, F. Establishment of Prediction Models of Asphalt Pavement Performance based on a Novel Data Calibration Method and Neural Network. *Transp. Res. Rec.* **2019**, *2673*, 66–82. [CrossRef]
3. Yang, L.; Hu, Y.; Zhang, H. Comparative study on asphalt pavement rut based on analytical models and test data. *Int. J. Pavement Eng.* **2020**, *21*, 781–795. [CrossRef]
4. Nhat, T.T.; Takahashi, O. Investigation on indices of workability and rutting resistance for wearing course mixtures. *Balt. J. Road Bridge Eng.* **2017**, *12*, 30–37. [CrossRef]
5. Li, S.; Gu, X.; Xu, X.; Xu, D.; Zhang, T.; Liu, Z.; Dong, Q. Detection of concealed cracks from ground penetrating radar images based on deep learning algorithm. *Constr. Build. Mater.* **2021**, *273*, 121949. [CrossRef]
6. Lyu, Z.; Shen, A.; Qin, X.; Yang, X.; Li, Y. Grey target optimization and the mechanism of cold recycled asphalt mixture with comprehensive performance. *Constr. Build. Mater.* **2019**, *198*, 269–277. [CrossRef]
7. Gong, M.; Zhang, H.; Liu, Z.; Fu, X. Study on PQI standard for comprehensive maintenance of asphalt pavement based on full-cycle. *Int. J. Pavement Eng.* **2021**. [CrossRef]
8. Wang, F.; Xie, J.; Wu, S.; Li, J.; Barbieri, D.M.; Zhang, L. Life cycle energy consumption by roads and associated interpretative analysis of sustainable policies. *Renew. Sust. Energy Rev.* **2021**, *141*, 110823. [CrossRef]
9. Tan, Y.; Meng, G.; Xu, H.; Zhang, R. Comparative Study on Laboratory Performance of Hot- and Warm-mix Asphalt Mixtures. *J. Test. Eval.* **2012**, *40*, 20120056. [CrossRef]
10. Xiao, F.; Yao, S.; Wang, J.; Li, X.; Amirkhanian, S. A literature review on cold recycling technology of asphalt pavement. *Constr. Build. Mater.* **2018**, *180*, 579–604. [CrossRef]
11. Li, N.; Jiang, Q.; Wang, F.; Cui, P.; Xie, J.; Li, J.; Wu, S.; Barbieri, D.M. Comparative assessment of asphalt volatile organic compounds emission from field to laboratory. *J. Clean. Prod.* **2021**, *278*, 123479. [CrossRef]
12. Thives, L.P.; Ghisi, E. Asphalt mixtures emission and energy consumption: A review. *Renew. Sustain. Energy Rev.* **2017**, *72*, 473–484. [CrossRef]
13. Cui, P.; Wu, S.; Xiao, Y.; Hu, R.; Yang, T. Environmental performance and functional analysis of chip seals with recycled basic oxygen furnace slag as aggregate. *J. Hazard. Mater.* **2021**, *405*, 124441. [CrossRef] [PubMed]

14. Wang, T.; Xiao, F.; Zhu, X.; Huang, B.; Wang, J.; Amirkhanian, S. Energy consumption and environmental impact of rubberized asphalt pavement. *J. Clean. Prod.* **2018**, *180*, 139–158. [CrossRef]
15. Cui, P.; Wu, S.; Xiao, Y.; Wan, M.; Cui, P. Inhibiting effect of Layered Double Hydroxides on the emissions of volatile organic compounds from bituminous materials. *J. Clean. Prod.* **2015**, *108*, 987–991. [CrossRef]
16. Almeida-Costa, A.; Benta, A. Economic and environmental impact study of warm mix asphalt compared to hot mix asphalt. *J. Clean. Prod.* **2016**, *112*, 2308–2317. [CrossRef]
17. Ling, C.; Bahia, H.U. Development of a Volumetric Mix Design Protocol for Dense-Graded Cold Mix Asphalt. *J. Transp. Eng. Part B Pavements* **2018**, *144*, 04018039. [CrossRef]
18. Zhang, H.; Liang, S.; Ma, Y.; Fu, X. Study on the mechanical performance and application of the composite cement–asphalt mixture. *Int. J. Pavement Eng.* **2019**, *20*, 44–52. [CrossRef]
19. Xu, O.; Wang, Z.; Wang, R. Effects of aggregate gradations and binder contents on engineering properties of cement emulsified asphalt mixtures. *Constr. Build. Mater.* **2017**, *135*, 632–640. [CrossRef]
20. Dong, W.; Xie, L.; Pan, P. A comparative study on early-stage strength development and mechanical properties of cement emulsified asphalt mixture using brake pad waste. *Constr. Build. Mater.* **2018**, *184*, 515–523. [CrossRef]
21. Rutherford, T.; Wang, Z.; Shu, X.; Huang, B.; Clarke, D. Laboratory investigation into mechanical properties of cement emulsified asphalt mortar. *Constr. Build. Mater.* **2014**, *65*, 76–83. [CrossRef]
22. Pi, Y.; Huang, Z.; Pi, Y.; Li, G.; Li, Y. Composition Design and Performance Evaluation of Emulsified Asphalt Cold Recycled Mixtures. *Materials* **2019**, *12*, 2682. [CrossRef] [PubMed]
23. Wang, L.; Shen, A.; Yao, J. Effect of different coarse aggregate surface morphologies on cement emulsified asphalt adhesion. *Constr. Build. Mater.* **2020**, *262*, 120030. [CrossRef]
24. Du, S. Influence of chemical additives on mixing procedures and performance properties of asphalt emulsion recycled mixture with reclaimed cement-stabilized macadam. *Constr. Build. Mater.* **2016**, *118*, 146–154. [CrossRef]
25. Zhang, C.; Wang, H.; You, Z.; Yang, X. Compaction characteristics of asphalt mixture with different gradation type through Superpave Gyratory Compaction and X-ray CT Scanning. *Constr. Build. Mater.* **2016**, *129*, 243–255. [CrossRef]
26. Dulaimi, A.; Al Nageim, H.; Ruddock, F.; Seton, L. New developments with cold asphalt concrete binder course mixtures containing binary blended cementitious filler (BBCF). *Constr. Build. Mater.* **2016**, *124*, 414–423. [CrossRef]
27. Dulaimi, A.; Al Nageim, H.; Ruddock, F.; Seton, L. High performance cold asphalt concrete mixture for binder course using alkali-activated binary blended cementitious filler. *Constr. Build. Mater.* **2017**, *141*, 160–170. [CrossRef]
28. Dulaimi, A.; Shanbara, H.K.; Al-Rifaie, A. The mechanical evaluation of cold asphalt emulsion mixtures using a new cementitious material comprising ground-granulated blast-furnace slag and a calcium carbide residue. *Constr. Build. Mater.* **2020**, *250*, 118808. [CrossRef]
29. Ferrotti, G.; Pasquini, E.; Canestrari, F. Experimental characterization of high-performance fiber-reinforced cold mix asphalt mixtures. *Constr. Build. Mater.* **2014**, *57*, 117–125. [CrossRef]
30. Ge, Z.; Li, H.; Han, Z.; Zhang, Q. Properties of cold mix asphalt mixtures with reclaimed granular aggregate from crushed PCC pavement. *Constr. Build. Mater.* **2015**, *77*, 404–408. [CrossRef]
31. Fu, Q.; Xie, Y.; Long, G.; Niu, D.; Song, H. Dynamic mechanical thermo-analysis of cement and asphalt mortar. *Powder Technol.* **2017**, *313*, 36–43. [CrossRef]
32. Lin, J.; Hong, J.; Xiao, Y. Dynamic characteristics of 100% cold recycled asphalt mixture using asphalt emulsion and cement. *J. Clean. Prod.* **2017**, *156*, 337–344. [CrossRef]
33. Wan, P.; Liu, Q.; Wu, S.; Zhao, Z.; Chen, S.; Zou, Y.; Rao, W.; Yu, X. A novel microwave induced oil release pattern of calcium alginate/nano-Fe3O4 composite capsules for asphalt self-healing. *J. Clean. Prod.* **2021**, *297*, 126721. [CrossRef]
34. Ma, Y.W.; Zhao, H.Y.; Li, G.; Wang, Z.J.; Tang, H.; Wang, A.Q. Property Improvement of Cement Emulsified Asphalt Paste Modified by Graphene Oxide. *Adv. Mater. Sci. Eng.* **2020**, *2020*, 3462342. [CrossRef]
35. Sun, H.; Ding, Y.; Jiang, P.; Wang, B.; Zhang, A.; Wang, D. Study on the interaction mechanism in the hardening process of cement-asphalt mortar. *Constr. Build. Mater.* **2019**, *227*, 116663. [CrossRef]
36. Zhuang, C.; Zhao, X.; Ye, Y. Low Temperature Crack Resistance and Fatigue Properties of Asphalt Mixture with Pre-buried Heat Conduction Tube. *IOP Conf. Ser. Earth Environ. Sci.* **2020**, *455*, 012093. [CrossRef]
37. Zou, G.; Zhang, X.; Wu, C. Evaluation of Steel Bridge Deck Ma Mixture Properties during Construction. *J. Mar. Sci. Technol.* **2015**, *23*, 293–301.
38. Fettahoglu, A.; Genes, M.C.; Kunt, M.M. Creep Compliance and Relaxation Moduli of PmB 25 A and PmB 45 A Gu ss asphalts at Different Temperatures. *J. Mater. Civ. Eng.* **2018**, *30*, 04018032. [CrossRef]
39. Zou, G.; Xu, X.; Li, J.; Yu, H.; Wang, C.; Sun, J. The Effects of Bituminous Binder on the Performance of Gussasphalt Concrete for Bridge Deck Pavement. *Materials* **2020**, *13*, 364. [CrossRef]
40. Tchekwagep, J.J.K.; Chen, D.; Mukhopadhyay, A.K.; Wang, S.; Huang, S.; Cheng, X. Quantitative Rietveld analysis of the decomposition of hardened rapid sulphoaluminate cement after exposure to elevated temperatures. *Arch. Civ. Mech. Eng.* **2021**, *21*, 119. [CrossRef]
41. López-Carreño, R.D.; Pujadas, P.; Cavalaro, S.H.; Aguado, A. Cavalaro, Antonio Aguado. Bond strength of white toppings and bonded overlays constructed with self-compacting high-performance concrete. *Constr. Build. Mater.* **2017**, *153*, 835–845. [CrossRef]

42. Li, Y.; Feng, J.; Chen, A.; Wu, F.; Wu, S.; Liu, Q.; Gong, R. Effects of Low-Temperature Construction Additives (LCAs) on the Performance of Asphalt Mixtures. *Materials* **2022**, *15*, 677. [CrossRef] [PubMed]
43. Zhang, J.; Fan, Z.; Wang, H.; Sun, W.; Pei, J.; Wang, D. Prediction of dynamic modulus of asphalt mixture using micromechanical method with radial distribution functions. *Mater. Struct.* **2019**, *52*, 49. [CrossRef]
44. Qu, F.; Lv, S.; Gao, J.; Liu, C. Performance and Mechanism of Asphalt Modified by Buton-Rock Asphalt and Different Types of Styrene-Butadiene-Rubber. *Appl. Sci.* **2020**, *10*, 3077. [CrossRef]
45. Umar, H.A.; Zeng, X.; Lan, X.; Zhu, H.; Li, Y.; Zhao, H.; Liu, H. A Review on Cement Asphalt Emulsion Mortar Composites, Structural Development, and Performance. *Materials* **2021**, *14*, 3422. [CrossRef]
46. Wu, X.; Fan, X.L.; Wang, J.F. Temperature Sensitivity of Mechanical Properties of Cement Asphalt Mortar with Nanoparticles. *Adv. Civ. Eng.* **2020**, *2020*, 3109612. [CrossRef]
47. Tan, Y.; Guo, M. Using surface free energy method to study the cohesion and adhesion of asphalt mastic. *Constr. Build. Mater.* **2013**, *47*, 254–260. [CrossRef]
48. Hamedi, G.H.; Saedi, D.; Ghahremani, H. Effect of Short-Term Aging on Low-Temperature Cracking in Asphalt Mixtures Using Mechanical and Thermodynamic Methods. *J. Mater. Civ. Eng.* **2020**, *32*, 04020288. [CrossRef]
49. Shaevich, R.B. Measurement of the Specific Free Surface free energy of Solids. *Meas. Tech.* **2007**, *50*, 1121–1123. [CrossRef]
50. Azarhoosh, A.; Hamedi, G.H.; Azarhoosh, M.J. The influence of cohesion and adhesion parameters on the moisture sensitivity of modified Asphalt mixtures with polymer additive. *J. Adhes.* **2020**, *97*, 1486–1508. [CrossRef]
51. Ouyang, J.; Han, B.; Cao, Y.; Zhou, W.; Li, W.; Shah, S.P. The role and interaction of superplasticizer and emulsifier in fresh cement asphalt emulsion paste through rheology study. *Constr. Build. Mater.* **2016**, *125*, 643–653. [CrossRef]
52. Ouyang, J.; Zhao, J.; Tan, Y. Modeling Mechanical Properties of Cement Asphalt Emulsion Mortar with Different Asphalt to Cement Ratios and Temperatures. *J. Mater. Civ. Eng.* **2018**, *30*, 04018263. [CrossRef]
53. Liu, B.; Liang, D. Effect of mass ratio of asphalt to cement on the properties of cement modified asphalt emulsion mortar. *Constr. Build. Mater.* **2017**, *134*, 39–43. [CrossRef]

materials

MDPI

Article

Research on the Road Performance of Asphalt Mixtures Based on Infrared Thermography

Wei Chen [1], Kesen Wei [2], Jincheng Wei [1,3,*], Wenyang Han [3], Xiaomeng Zhang [3], Guiling Hu [1], Shuaishuai Wei [3], Lei Niu [3], Kai Chen [3], Zhi Fu [3], Xizhong Xu [3], Baogui Xu [1] and Ting Cui [3]

1 School of Transportation Engineering, Shandong Jianzhu University, Jinan 250101, China; chenwei13572468@163.com (W.C.); huguilingtech@foxmail.com (G.H.); xubaogui0805@163.com (B.X.)

2 Zibo Transportation Management Center, Shandong Hi-Speed Company Limited, Zibo 255000, China; m17865438990@163.com

3 Science and Technology Innovation Center, Shandong Transportation Institute, Jinan 250102, China; hanwenyang@sdjtky.cn (W.H.); zhangxiaomeng@sdjtky.cn (X.Z.); shuaishuai_wei@163.com (S.W.); nlago@foxmail.com (L.N.); cck133@163.com (K.C.); fufranklin@163.com (Z.F.); xxz137152@163.com (X.X.); tingcui1016@126.com (T.C.)

* Correspondence: c2523860697@163.com

Abstract: Temperature segregation during the paving of asphalt pavements is one of the causes of asphalt pavement distress. Therefore, controlling the paving temperature is crucial in the construction of asphalt pavements. To quickly evaluate the road performance of asphalt mixtures during paving, in this work, we used unmanned aerial vehicle infrared thermal imaging technology to monitor the construction work. By analyzing the temperature distribution at the paving site, and conducting laboratory tests, the relationship between the melt temperature, high-temperature stability, and water stability of the asphalt mix was assessed. The results showed that the optimal temperature measurement height for an unmanned aerial vehicle (UAV) with an infrared thermal imager was 7–8 m. By coring the representative temperature points on the construction site and then conducting a Hamburg wheel tracking (HWT) test, the test results were verified through the laboratory test results in order to establish a prediction model for the melt temperature and high-temperature stability of $y = 10.73e^{0.03x} + 1415.78$, where the predictive model for the melt temperature and water was $y = -19.18e^{-0.02x} + 98.03$. The results showed that using laboratory tests combined with UAV infrared thermography could quickly and accurately predict the road performance of asphalt mixtures during paving. We hope that more extensive evaluations of the roadworthiness of asphalt mixtures using paving temperatures will provide reference recommendations in the future.

Keywords: infrared thermal imaging technology of an unmanned aerial vehicle (UAV); optimal temperature measurement height; melt temperature threshold; molding temperature-road performance prediction model

Citation: Chen, W.; Wei, K.; Wei, J.; Han, W.; Zhang, X.; Hu, G.; Wei, S.; Niu, L.; Chen, K.; Fu, Z.; et al. Research on the Road Performance of Asphalt Mixtures Based on Infrared Thermography. *Materials* **2022**, *15*, 4309. https://doi.org/10.3390/ma15124309

Academic Editor: Rui Vasco Silva

Received: 14 May 2022
Accepted: 13 June 2022
Published: 17 June 2022

1. Introduction

With China's rapid economic development and growing public demand for quality travel, the country's total road length reached 5,198,100 km by the end of 2020 [1]. Due to the rapid development of China's roads, the volume of traffic that must be carried on these roads has also increased, leading to a significant increase in the occurrence of damage to the road surface, resulting in travel inconveniences. Therefore, construction quality control has become a significant issue for the durability of asphalt pavements [2]. One reason for the various issues with asphalt pavements is the difficulty in controlling the temperature of the asphalt mixture during the paving process [3], as the temperature of the asphalt mixture must be strictly controlled during transportation and paving to prevent temperature segregation in the asphalt pavement. Otherwise, the road performance of the asphalt pavement will be seriously affected [4]. Thus, an increasing number of researchers have assessed the effects of paving temperatures on asphalt pavements.

In 1929, Czerny et al. [5] developed the first generation of thermal imaging cameras, which introduced the idea of active infrared thermal imaging for non-destructive testing, whereas Mahoney et al. [6,7] investigated the effects of various factors, such as the temperature difference, the density of the compacted material, transport distance, paving equipment, weather, and other variables, on the results of infrared thermography. In addition, Han, Dongdong et al. [8] used infrared thermography to monitor the heating process at the construction site of hot in-place recycling asphalt pavement to obtain the optimal heating method and temperature threshold. Hiasa et al. [9] proposed a method of data threshold analysis processing using infrared thermography, which was capable of visually measuring the delaminated areas of concrete slabs. Ellenberg et al. [10] developed a new image processing algorithm to obtain spectral images using small infrared reflections to help identify delamination, enabling the rapid identification of bridge deck delamination locations and the detection of bridge deck distress. In their study of two asphalt pavements with different aging conditions, Nishar et al. [11] used infrared spectral region data to assess defects in asphalt pavements to distinguish newly paved asphalt pavements from old asphalt pavements. Using infrared thermography on paving sites, researchers such as Nevalainen [12] found that the long transport of asphalt mixes was the leading cause of temperature segregation. Rahman et al. [13] used thermal imaging to record the temperature changes at a construction site and suggested that a temperature difference of 14 °C in the mix would result in an abnormal void fraction in asphalt pavements. Javilla et al. [14] used wheel tracking tests on AC-13 and AC-20 asphalt mixtures and concluded that the A/C index could provide a suitable indirect assessment of rutting index performance. Furthermore, Polaczyk et al. [15] tested the locking point concept by using applied performance tests and evaluating the impact of aggregate interlocking on rutting and fatigue cracking to obtain the best-defined locking point.

In this work, we used an unmanned aerial vehicle (UAV) infrared thermal imaging technique to inspect an asphalt pavement mixture paving site, to determine the optimal inspection height and to predict the performance of the asphalt pavement after grinding. In addition, to establish a new evaluation system for asphalt pavements based on infrared thermography, we provide a theoretical basis for future applications in practical engineering.

2. Test Methods and Principles

2.1. Infrared Thermal Imaging Technology of the UAV and Principles

In this study we used an M300 RTK UAV from DJI (Shenzhen, China), which was paired with a Zenmuse H20T infrared thermal imaging camera, as shown in Figure 1. Using the M300 RTK's flight control system, the UAV could view the temperature distribution of the photographic interface on any trajectory and at any angle to obtain the required infrared images. Then, the infrared image data processing software was used to visualize the temperature distribution of the measured area. The principle of operation of infrared thermography involves the use of infrared detection elements. Thus, the infrared radiation emitted by the object was received during the test, the power signal radiated by the object was converted into an electrical signal and amplified, and then, after digital processing and image processing, a thermal image with better temperature values was finally presented on a display device [16], as shown in Figure 2.

2.2. Rutting Test

According to the Chinese standard JTG E20-2011 (T0719) [17], a rutting test can be used to evaluate the rutting resistance of asphalt mixtures, and its dynamic stability can more accurately reflect the ability of asphalt mixture pavements to reduce rut formation under high-temperature conditions [18]. The asphalt mixes at different melt temperatures were placed in standard test molds of 300 mm in length, 300 mm in width, and 50 mm in height, and were set up with three parallel test pieces for each compaction temperature.

During the test, the specimen, along with the test mold, was placed in a wheel rutting test machine at a temperature of 60 °C ± 1 °C for no less than 5 h and no more than 12 h. The test wheel was pressed against the rutting plate specimen at 0.7 MPa, and the test wheel was rolled back and forth at a rate of 42 times ± 1 times per min for 1 h or stopped when maximum deformation reached 25 mm. Deformations d1 and d2 were recorded at 45 and 60 min.

Figure 1. M300 RTK UAV.

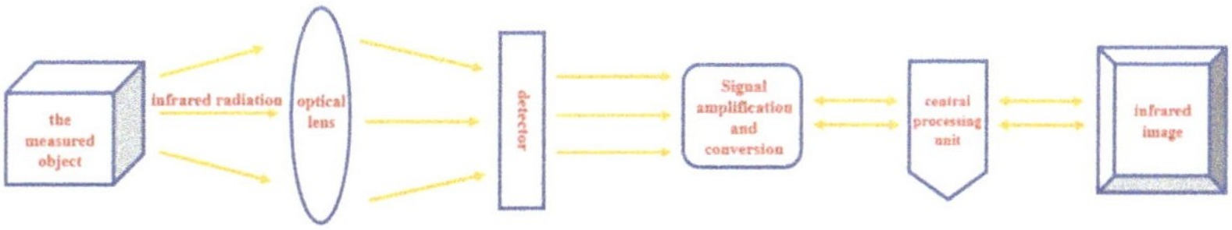

Figure 2. Principal infrared thermal imaging instrument temperature measurement diagram.

2.3. Freeze-Thaw Splitting Test

Freeze-thaw splitting tests were used to analyze the water stability of asphalt mixtures at different melt temperatures, according to the Chinese Standard JTG E20-2011 (T0719) [17], by measuring the splitting tensile strength ratio of asphalt mixture specimens subjected to water damage before and after different melt temperatures and then evaluating the effect of melt temperature on the water stability of the asphalt mixtures [19]. For each asphalt mixture with different melt temperatures, eight standard Marshall test pieces were created, which were divided into two random groups of four pieces. The first group was maintained at room temperature as the control test pieces, and another group of test pieces was vacuumed for 15 min, restored to normal pressure, and placed in water for 0.5 h. Then, the test pieces were removed and placed in a plastic bag with 10 mL of water and placed in a refrigerator at a constant temperature of −18 °C ± 2 °C for 16 h ± 1 h of freezing. After the freezing was complete, the test pieces were immediately placed into a water bath at 60 °C ± 0.5 °C for 24 h after removing the plastic bag. Then, the two groups of test pieces were placed together into a constant temperature bath at 25 °C ± 0.5 °C for 2 h, for the final freeze-thaw splitting test. The splitting tensile strength ratio was calculated as follows:

$$R_{T1} = \frac{2p_{T1}}{\pi D_1 h_1} \quad (1)$$

$$R_{T2} = \frac{2p_{T2}}{\pi D_2 h_2} \quad (2)$$

$$TSR = \frac{\overline{R}_{T_2}}{\overline{R}_{T_1}} \tag{3}$$

where R_{T1} indicates group 1 splitting tensile strength specimens (MPa), R_{T2} denotes group 2 splitting tensile strength specimens (MPa), P_T is the value of the test load for a single specimen (N), D is the Marshall specimen diameter (mm), h is the Marshall specimen height (mm), TSR is the splitting tensile strength ratio (%), $\overline{R}_{T1}$ is the average splitting tensile strength of group 1 specimens (MPa), and $\overline{R}_{T2}$ is the average splitting tensile strength of group 2 specimens (MPa).

2.4. Hamburg Wheel Tracking Test

The Hamburg wheel tracking (HWT) was used to evaluate the water sensitivity and resistance to rutting of the asphalt mixtures [20] according to the Chinese standard JTG E20-2011 (T0719) [17]. The asphalt pavement was core-cut on site, and the core samples were 150 mm in diameter and 38 mm thick. Then, standard specimens were cut and placed in the Hamburg test mold. When the water tank temperature reached 50 °C, the steel wheel with a wheel load of 705 N ± 4.5 N was reciprocally crushed at a speed of 52 ± 2 times per minute. The maximum speed of the wheel through the midpoint of the specimen was 0.305 m/s. When the loading time reached 20,000, the test steel wheel was automatically lifted, at which point the rutting depth was recorded.

3. Field Measurement Basis with UAV Infrared Thermography

3.1. Project Overview

This work was based on the reconstruction and expansion project of a section of a highway in Shandong Province, China, which was expanded from four to eight lanes in both directions, with a design speed of 120 km/h. This study combined the actual asphalt pavement top layer paving site conditions with UAV infrared thermography to measure the temperature of the asphalt pavement during paving, to predict the actual road performance of the asphalt pavement after grinding was complete.

3.2. Identifying the Optimal Temperature Measurement Height

The Zenmuse H20T imaging system was affected by the ambient temperature, humidity, object emissivity, and the measurement height. Therefore, to investigate the accuracy of temperature measurement results at different heights, it was necessary to carry out temperature measurement studies at different heights on the asphalt pavements in the same area, to determine the optimal height for temperature measurements.

3.2.1. Parameter Calibration

Before testing, it was necessary to set the instrument parameters according to the site environment, including the ambient temperature, ambient humidity, emissivity, and measurement height [21]. The temperature at the site was 17 °C, the relative humidity was 58%, and the emissivity of each object is shown in Table 1. The parameters were imported into DJI Thermal Analysis Tool software, where the images showed the temperature data for the desired location.

3.2.2. Temperature Measurement Results and Analysis of Different Measuring Heights

To determine the optimal temperature measurement height for the UAV with the infrared thermal imager, we randomly selected an area where grinding was complete and which retained the residual temperature. We placed four objects with a lower temperature in this area to form a 1 m × 1 m detection area and obtained an infrared thermal image of the area using the UAV thermal imager, as shown in Figure 3. Starting at a distance of 3 m from the pavement, the measurement height was continuously adjusted upwards at 1 m intervals to investigate the pattern of temperature change with the shooting height, to determine the optimal height for temperature measurements. To prevent heat loss caused by external environmental factors during the temperature measurement process, the temperature acquisition time was limited to 20 s to ensure detection accuracy as much as possible.

Table 1. Emissivity of various substances.

Materials	Emissivity
Asphalt	0.97
Water	0.95
Iron	0.94
Cement	0.96
Basalt	0.72
Concrete	0.94

Figure 3. On-site temperature measurements.

According to the initial on-site measurements, when the temperature measurement height was more than 12 m, the temperature measured by the UAV infrared thermal imager was more dissimilar than the temperature measured by the handheld temperature measurement gun, which deviated from the actual temperature of the asphalt pavement; thus, a temperature measurement height range of 3–11 m was selected. By entering the actual height data for the temperature measurements in the DJI Thermal Analysis Tool software, the actual temperature of the measured area could be displayed directly, as shown in Figure 4. We chose six temperature points from each measurement area and obtained the average value as a representative of the temperature at this height.

For each height, the maximum, minimum, and average measured temperatures were analyzed to investigate their variation patterns and to determine the optimal height for temperature measurements, as shown in Figures 5–7.

As shown in Figures 5–7, the distribution curves of the maximum and average temperatures followed the same pattern of low change. Although fluctuations in temperature occurred between the different heights, the variations in temperature between the adjacent heights were minor and did not exceed 1 °C. The maximum temperature fluctuated around 35 °C, whereas the average temperature was around 34 °C. The variations in the minimum temperature curve were different; however, the temperature fluctuated around 30 °C. With increasing measurement height, the measurement area increased and was more subject to various uncertainties. The combined temperature data showed that the measured temperature values were all single-point, and the maximum and minimum temperatures were not the same between the different heights in the same area. Hence, the average temperature was more representative. When the measurement height was between 7 and 8 m, the three curves were close to each other, and the variations in each temperature were the smallest. Combined with the measurement results of the point-type temperature-measuring gun, the best temperature measurement height was determined to be 7–8 m.

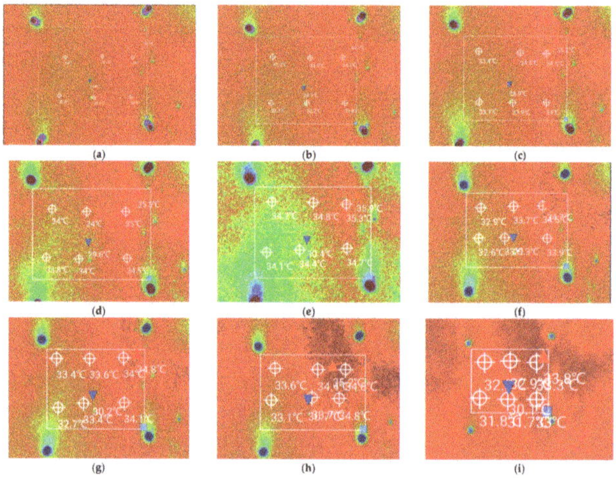

Figure 4. Infrared images at different measurement heights: (**a**) h = 3 m, (**b**) h = 4 m, (**c**) h = 5 m, (**d**) h = 6 m, (**e**) h = 7 m, (**f**) h = 8 m, (**g**) h = 9 m, (**h**) h = 10 m, and (**i**) h = 11 m.

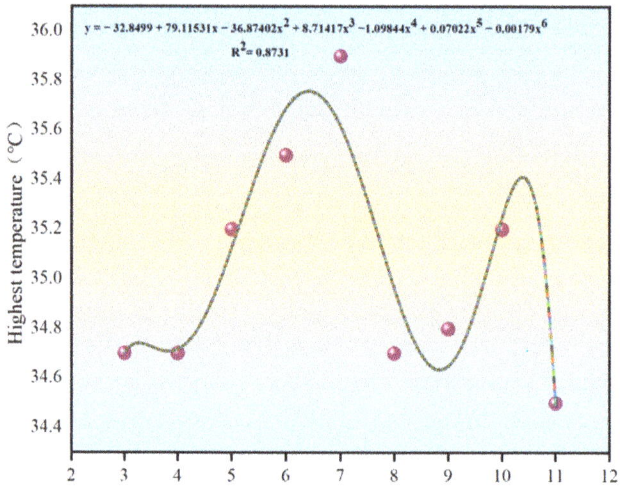

Figure 5. Distribution curve of the maximum temperature.

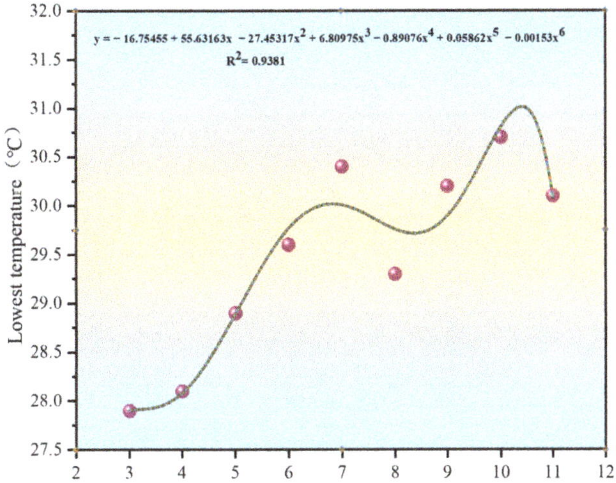

Figure 6. Distribution curve of the minimum temperature.

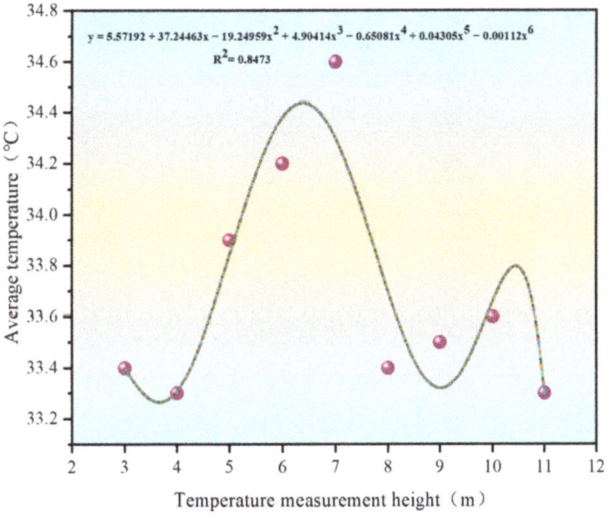

Figure 7. Distribution curve of the average temperature.

3.3. On-Site Measurements

During the construction of the top layer of the asphalt pavement, the site temperature was 20 °C and the distance from the mixing plant to the construction site was 3 km. The tipper truck delivered the asphalt mix to the paving site and then poured it into the spreading machine's receiving hopper, where it was transported via a conveyor to the spreading machine for secondary mixing and finally paving. During this process, multiple UAVs with infrared thermal imagers measured the temperatures of the dump trucks and the spreading machine throughout the process, as shown in Figure 8. Infrared thermography was used to measure the temperature distribution of the asphalt mixture during the paving process, focusing on the pavement that had just been paved but which was not yet ground, as shown in Figure 9.

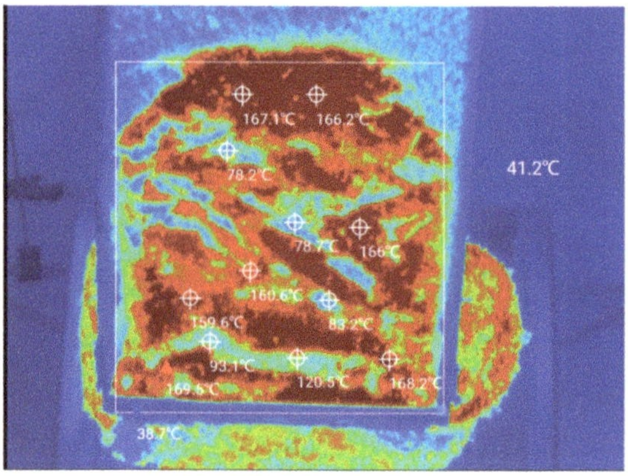

Figure 8. Infrared thermal image of the asphalt mix on the tipper truck.

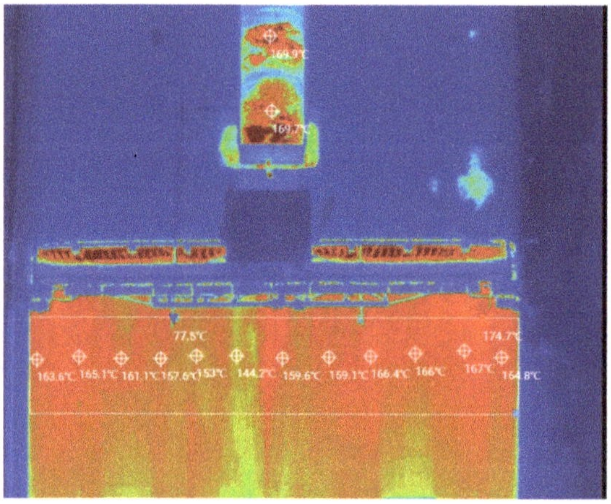

Figure 9. Infrared thermal image of asphalt pavement top layer construction.

Due to the slow travel speed of the spreading machines, it was not possible to measure the temperature data over long distances at once. Therefore, the UAV thermal imager had to remain in the air for continuous temperature measurements and then the captured data images were combined, which reduced the temperature losses during the measurement process. After measurement, a randomly selected area of 50 m in length and 7.5 m in width was studied and analyzed. The research area was divided into 0.5 m × 0.5 m squares [2] with 1500 squares, with the average of the temperatures measured in each square taken as the measured value for that square, as shown in Figure 10.

As shown in Figure 10, the temperature distribution of the SMA-13 asphalt mix pavement after paving was very uneven, with a large temperature span. The Chinese standard JTG F40-2004 [22] specifies that the paving temperature has to be greater than or equal to 160 °C, and the temperature at the start of the initial grind must be greater than 150 °C. In the research area, the maximum temperature was 168 °C, and the average temperature was 155 °C, with the lowest temperatures (142 °C) located in the marginal parts. Therefore, we inferred that the proximity of this area to the shoulder resulted in edge temperatures that did not meet the specification requirements [22], and temperature segregation could occur, affecting the road performance.

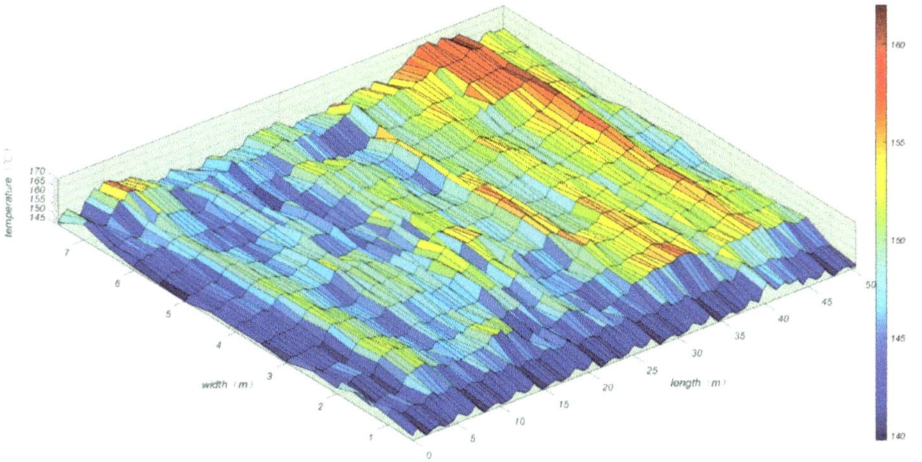

Figure 10. Study area temperature distribution map.

According to Figures 9 and 10, the reasons for the uneven paving temperatures were twofold. First, the asphalt mixture after a certain distance of transport and the surface temperature of the asphalt mixture in the delivery truck dropped considerably, whereas the internal temperature loss of the mixture was small; thus, the temperature difference of the asphalt mixture was significant. Second, the asphalt mixture dumped from the delivery truck entered the spreading machine receiving hopper. The spiral distributor evenly paved it, and this process is exposed to air; thus, temperature dissipation inevitably occurred.

To reduce the significant temperature differences that occurred in the asphalt mixes during transport, reducing the heat dissipation from the asphalt mixes was necessary. During transport, temperature segregation was more severe at the top of the truck and on both sides of the carriage; thus, attention had to be paid to strengthening the insulation of the internal structure of the material carrier. To avoid temperature segregation, when loading the material carriage, the following method was used: the front part was loaded first, then the rear part, and finally the middle part. The bottom of the carriage had to be coated with a lubricant to achieve unloading of the entire asphalt mixture. During the paving process, spacing between the spiral blades on both sides could be reduced to an unequal distance so that the asphalt mixtures on both sides could be mixed evenly through the spiral blades, reducing the temperature difference. In addition, it was possible to adjust the size and tilt angle of the spiral blades to avoid heat exchange between the spiral blades and the atmosphere, as well as to reduce the temperature dissipation of the asphalt mix. It was also possible to adjust the height of the spiral distributor to speed up the flow of the asphalt mixture so that low-temperature aggregates could be quickly remixed with high-temperature aggregates, thus reducing the occurrence of uneven temperature distributions in the asphalt mix pavements [23].

3.4. Evaluation of Road Performance of Asphalt Mixtures Based on the HWT Test

To investigate the effects of different paving temperatures on the road performance of the SMA-13 asphalt pavements, the asphalt pavements were marked at different paving temperatures, and after grinding, the marked locations were cored according to the HWT test to analyze their road performance. According to Figure 10, the research area was divided into five temperature gradients: 130–140 °C, 140–150 °C, 150–160 °C, 160–170 °C, and 170–180 °C. Values of 135 °C, 145 °C, 155 °C, 165 °C, and 175 °C were selected as representative values for each temperature gradient, and then the cores were taken, cut, and subjected to HWT testing. Because the area selected was close to the road border, the overall temperature was slightly lower, with few areas where the paving temperature reached 175 °C. Therefore, only the core samples from areas at 135 °C, 145 °C, 155 °C, and 165 °C were selected for the Hamburg wheel mill test, and the test results are shown in Figure 11.

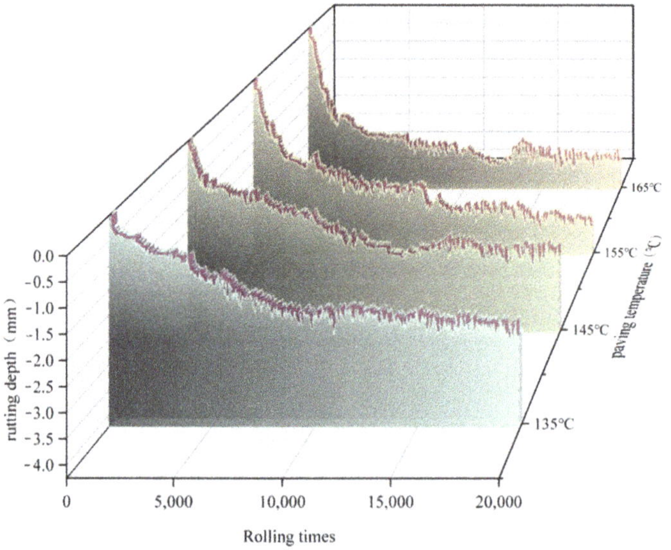

Figure 11. Results of the HWT test.

As shown in Figure 11, the rutting depths at 10,000 and 20,000 wheel rolls met the requirements of the Chinese specifications of ≤4 mm and ≤10 mm [17], and the paving temperatures of 135 °C, 145 °C, 155 °C, and 165 °C at a wheeling depth of 20,000 were 3.576, 3.472, 2.639, and 2.401 mm, respectively. The rutting depth at a paving temperature of 135 °C increased by 48.9% compared to that at 165 °C. The rutting depth at the 145 °C positions increased by 35.5% compared to that at the 165 °C positions, whereas the rutting depth at the 155 °C positions increased by only 3.0% compared to that at the 165 °C positions. This indicated that when the paving temperature of the asphalt pavement was below 155 °C, temperature segregation may have occurred, and its resistance to rutting was significantly reduced after grinding, affecting the performance of the road.

The analysis showed that the void ratio was large at the location where temperature segregation occurred. A good nested structure could not form between the aggregates, resulting in a greater-than-standard rutting depth under the action of wheel rolling, which in the long run would form ruts and affect the service life of the pavement [18]. Therefore, for the temperature segregation produced in the construction of the asphalt pavement, although the final quality acceptance of the pavement performance indicators could meet the requirements, to increase the service life of the pavement, the construction process could be used to monitor the paving temperature of the asphalt mixture using UAV infrared thermal imaging technology. To extend the service life of the road, for areas below 155 °C, there should be a timely adjustment of the paver parameters to reduce the occurrence of the temperature segregation phenomenon.

4. Effect of Melt Temperature on the Road Performance of Asphalt Mixtures

The range of the temperature distribution during asphalt paving was obtained via infrared thermographic inspection of the SMA-13 asphalt mixes at the construction site. Due to the construction site conditions, it was impossible to cut many rutting plate specimens; thus, it was difficult to accurately evaluate the rutting resistance, water stability, and other road properties of the asphalt mixtures at different paving temperatures. In this study, the paving temperature of the asphalt mixture in the field was combined with the road performance measured in the indoor tests, with the paving temperature at the construction site set as the mixing and melt temperature for the indoor asphalt mixture tests. The prediction model of asphalt mixture road performance and melt temperature was established through indoor tests to improve the accuracy and efficiency of road performance testing.

4.1. Indoor Experimental Design

In this study, the asphalt mixes were mixed at 135 °C, 145 °C, 155 °C, 165 °C, and 175 °C to simulate the paving temperatures in the field and analyze the high-temperature stability and water stability of the mixes at these temperatures. The results of the tests were compared with those of the HWT test, to verify the effect of SMA-13 asphalt mix paving temperature on the on-road performance.

4.1.1. Raw Material

The use of aggregates and asphalt brought back from the construction site for the indoor tests prevented errors from arising due to differences in the test materials. The performance specifications of the materials used in this work are presented below.

(1) *Asphalt*

The SBS modified asphalt was selected to meet the requirements of the Chinese Code [15], and the test results are shown in Table 2.

Table 2. Technical indices of the SBS modified asphalt.

Item	Specification	Test Values
Penetration (25 °C, 0.1 mm)	40–60	53
Softening point (°C)	≥60	77.0
Ductility (5 °C, cm)	≥20	30
Dynamic viscosity at 135 °C (Pa·s)	≤3	1.3

(2) *Coarse aggregate*

The coarse aggregate consisted of lava rock, with 5–10 mm and 10–15 mm sizes. The physical properties of the coarse aggregate were determined in accordance with the Chinese specification JTG E42-2005 [24], and the technical indicators satisfied the specification, as shown in Table 3.

Table 3. Technical indices of the coarse aggregate.

Items		Specification	Test Values
Crush value (%)		≤26	9.7
Apparent relative density (g/cm^3)	10–15 mm	≥2.6	2.926
	5–10 mm		2.967
Water absorption (%)	10–15 mm	≤2.0	1.6
	5–10 mm		1.2
Abrasion value (%)		≤28	9.5

(3) *Fine aggregates*

The fine aggregates in the asphalt mix consisted of 0–3 mm limestone, as shown in Table 4.

Table 4. Technical indices of the fine aggregates.

Items	Specification	Test Values
Apparent relative density (g/cm^3)	≥2.5	2.955
Water absorption (%)	-	1.7
Sand equivalent (%)	≥60	62.7
Angularity (s)	≥30	39

4.1.2. Gradation Composition

According to the technical indicators of the production mix ratio of the asphalt mix for the top layer of the expansion project, it was known that the optimal asphalt content was 5.9% and the amount of fiber was 0.35%. The composition of the grade is shown in Table 5.

Table 5. SMA-13 gradation composition.

Seive size(mm)	16	13.2	9.5	4.75	2.36	1.18	0.6	0.3	0.15	0.075
Passing rate(%)	99.8	91.6	61.7	25.6	20.6	17.9	15.9	14.5	12.1	9.8

4.2. Effect of Melt Temperature on the High-Temperature Stability of Asphalt Mixtures

As shown in Figure 12, the wheel tracking test was used to evaluate the high-temperature stability of the asphalt mixture specimens at different melt temperatures. By measuring the dynamic stability and rutting depth of the asphalt mixture specimens at different melt temperatures, the effect of melt temperature on the rutting resistance of the asphalt mixture could be evaluated. The test results are shown in Figure 13.

The dynamic stability of the asphalt mixture specimens at different melt temperatures in the wheel tracking test satisfied the requirements of JTG E20-2011 (T0719), as shown in Figure 13 [17]. The rutting depth decreased as the melt temperature of the specimen increased, with the rutting depth decreasing by 0.15 mm as the melt temperature increased from 135 °C to 155 °C. The rutting depth was reduced by 0.73 mm when the melt temperature increased from 155 °C to 175 °C, and as the melt temperature gradually increased, the dynamic stability showed a steady increase, reaching 4221 cycles/mm when the melt temperature was 175 °C.

Figure 12. Wheel tracking test.

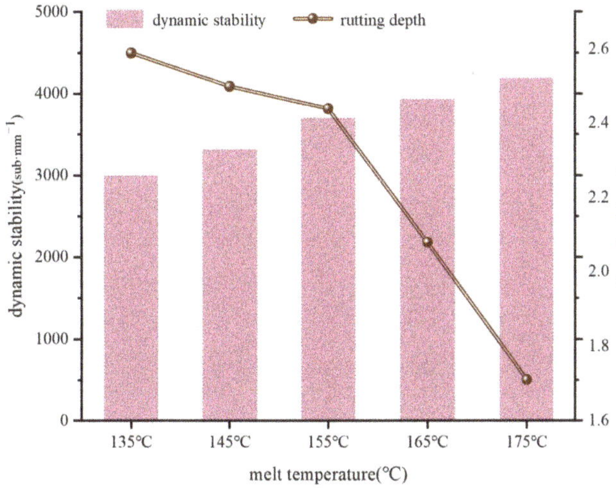

Figure 13. Results of the wheel tracking tests on SMA-13 asphalt mixtures with different melt temperatures.

In this case, for the asphalt used in the SBS modified asphalt, the modifier inside the asphalt was cured and did not flow as well at lower temperatures, which caused the inter-aggregates not to be filled with bitumen and to exhibit a large internal void ratio, resulting in poor high-temperature stability. When the melt temperature increased, the asphalt fully wrapped the aggregate, bonding between the aggregates was tight, and compaction gradually increased after compaction, which resulted in an asphalt mixture with increased rutting resistance. Therefore, the more significant the difference in the melt temperature, the larger the difference in the internal voids of the mix, and the more likely it was that pressure-tight rutting would occur with traffic.

4.3. Effect of Melt Temperature on the Water Stability of Asphalt Mixtures

The indoor tests utilized freeze-thaw splitting tests to analyze the water stability of the asphalt mixes at different melt temperatures. This was carried out by calculating the freeze-thaw splitting strength ratios before and after water damage to the five groups of asphalt mix specimens with different melt temperatures and then evaluating the effect of the melt temperature on the water stability of the asphalt mixes. The specific test results are shown in Figure 14.

As shown in Figure 14, when the melt temperature was 135 °C, the freeze-thaw splitting strength ratio was 68.8%, which did not meet the specification requirements. However, when the melt temperature was above 145 °C, the freeze-thaw splitting strength ratios of all five groups of specimens met the specification requirements. With an increase in the melt temperature, the freeze-thaw splitting strength ratio of the asphalt mix increased slightly; however, the increase was minimal. When the melt temperature was 175 °C, the freeze-thaw splitting strength ratio of the asphalt mix had a maximum value of 88.9%. This showed that the effect of melt temperature on the freeze-thaw splitting strength ratio was small, and the asphalt mixture specimens at different melt temperatures can maintain good water stability under freeze-thaw cycles, making the pavement less susceptible to water damage. This was because the melt temperature was more significant than 145 °C, the aggregate was covered by the asphalt, and the adhesion strength of the asphalt to the aggregate was greater, which improved the water stability of the asphalt mix.

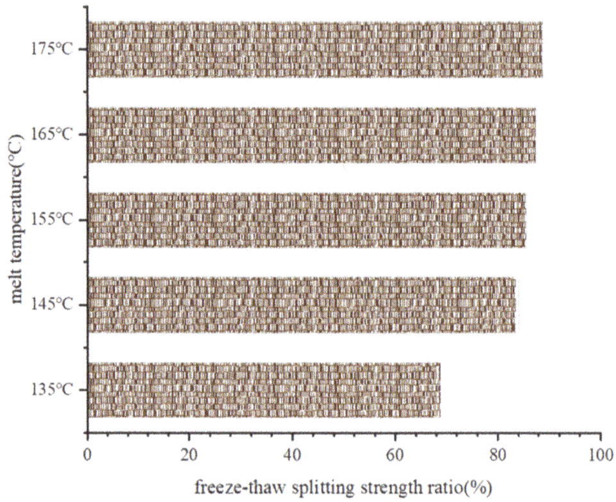

Figure 14. Results of the freeze-thaw splitting tests on the SMA-13 asphalt mixes with different melt temperatures.

4.4. Research on the Predictive Modeling of Melt Temperature and Road Performance

According to the indoor test results of the asphalt mixtures with different melt temperatures, the asphalt mixture road performance during the paving of the SMA-13 asphalt pavement was simulated, and the relationship between the melt temperature and the dynamic stability and freeze-thaw splitting strength ratio was established. In the construction of asphalt pavements, UAV infrared thermography was used for testing, thus predicting the road performance of the asphalt mix.

4.4.1. Building the Model for Melt Temperature/High-Temperature Stability Prediction

Using UAV infrared thermal imaging technology to detect the melt temperature at the construction site, and combining the results with the dynamic stability results obtained from the indoor tests, we established a prediction model between the asphalt pavement melt temperature and dynamic stability. This enabled the dynamic stability distribution of the asphalt pavements to be visually observed in real-time during the construction process, thus allowing us to evaluate the high-temperature stability of the asphalt pavements. To establish the relationship between the melt temperature and dynamic stability, each melting temperature was divided into five intervals: 130–140 °C, 140–150 °C, 150–160 °C, 160–170 °C, and 170–180 °C. Based on the temperature distributions during the paving process, the temperature intervals were equally distributed in combination with the dynamic stability measured in the room to establish a prediction model for melt temperature/dynamic stability, as shown in Figure 15.

As shown in Figure 15, the predictive model for melting temperature/dynamic stability followed $y = 10.73e^{0.03x} + 1415.78$, $R^2 = 0.98$; therefore, in the process of asphalt pavement construction, the melt temperature could be used to further predict the dynamic stability distribution of the asphalt pavement through temperature acquisition by means of UAV thermal imaging technology. This would make it possible to quickly determine whether the high-temperature stability of asphalt pavements would meet the specification requirements during construction, significantly improving the efficiency of testing the quality of asphalt pavement construction.

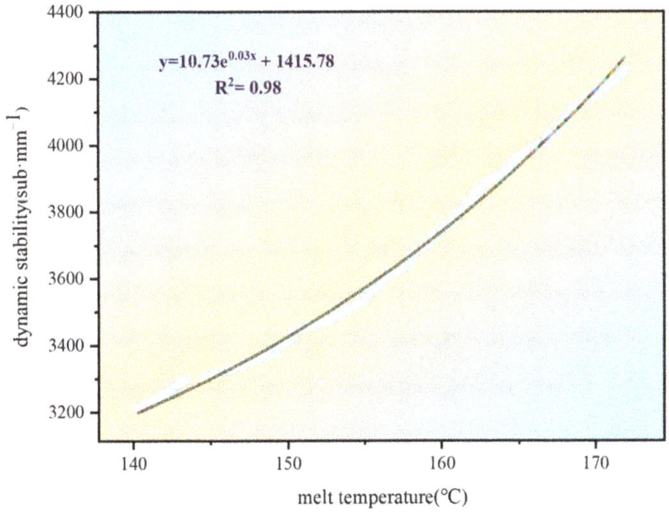

Figure 15. Predictive model for melt temperature/dynamic stability.

4.4.2. Building the Model for Forming Temperature/Water Stability Predictions

According to the experimental design, each paving temperature was divided into five temperature intervals, and the five paving temperature intervals were equated to the melt temperature of the freeze-thaw splitting test to obtain a predictive model of the melt temperature to the freeze-thaw splitting strength ratio during construction, as shown in Figure 16. This allowed the water stability of asphalt pavements to be further evaluated based on the distribution of paving temperatures during construction.

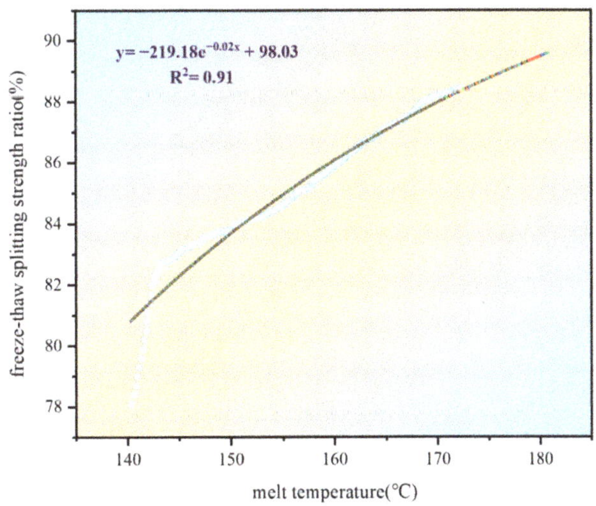

Figure 16. Predictive model for the melt temperature/freeze-thaw splitting strength ratio.

As shown in Figure 16, the predictive model for the resulting melting temperature/freeze-thaw splitting strength ratio was $y = -219.18e^{-0.02x} + 98.03$. During construction, the freeze-thaw splitting strength ratio of the asphalt mixture could be quickly predicted based on the paving temperature to evaluate whether the water stability of the asphalt pavement could meet the requirements of the standard JTG E20-2011 (T0719) [17]. For locations that would not satisfy the requirements, adjustments could be made when appropriate.

4.5. Evaluation System of Asphalt Mixture Road Performance Based on Infrared Thermography

Using UAV infrared thermography on the paving site of asphalt pavement, the temperature distribution of freshly paved, but not yet rolled, asphalt pavement was obtained. Then, for the representative temperature points, we cored, cut, and sent the samples indoors for HWT testing to obtain the water and high-temperature stability of the asphalt mixture at each representative melt temperature point. The materials used at the construction site were used in the indoor tests to obtain the high-temperature stability and water stability of the asphalt mix at representative melt temperature values. By comparing these two datasets, we found that the high-temperature stability and water stability of the asphalt pavement met the requirements of the JTG E20-2011 (T0719) specification [17] after grinding, and during grinding, the paving temperature was more significant than at 155 °C. However, paving temperatures below 145 °C required timely adjustments to the paving process on-site to reduce the potential for later pavement distress.

The above tests were successfully carried out. Before on-site paving operations, the aggregates on-site were subjected to indoor tests in advance to establish a predictive model of the melt temperature/road performance applicable to the section. According to the minimum roadworthiness specified by the design unit, the threshold value for the melt temperature could be calculated. The melt temperature thresholds obtained from the indoor tests were converted into paving temperature thresholds during the paving process. Then, real-time temperature monitoring of the paving site was carried out using UAV infrared thermal imaging technology, with the UAV's temperature measurement height controlled at 7–8 m. Manual intervention was carried out for areas where temperatures were lower than the threshold to reduce post-maintenance costs. For areas where temperatures were close to the threshold, stakes and locations were recorded, and future essential pavement maintenance could be carried out on the area to prevent early damage and increase the lifespan of the road. Based on this method, the paving process could be quickly tested and the road performance of the asphalt pavement could be evaluated in real-time.

5. Conclusions

In this work, we combined UAV infrared thermography with laboratory tests to investigate how to more efficiently and accurately analyze the high-temperature stability and water stability of asphalt pavements during the paving process, which resulted in the following conclusions.

(1) Using UAV infrared thermal imaging technology to continuously measure the temperatures at different heights in the same area, and by studying the highest temperature, lowest temperature, and average temperature of the temperature measurement area, we obtained the best temperature measurement height of 7–8 m for UAV infrared thermal imaging technology.

(2) For the road section in this study, when the melt temperature increased from 135 °C to 155 °C, the rutting depth decreased by 0.15 mm and the freeze-thaw splitting strength ratio increased by 24.3%. When the melt temperature increased from 155 °C to 175 °C, the rutting depth decreased by 0.73 mm and the freeze-thaw splitting strength ratio increased by 4.1%. Therefore, infrared thermography could be used to monitor the paving temperature of asphalt pavements during paving, and quick predictions of road performance could be made.

(3) The prediction model for melt temperature/dynamic stability was obtained as $y = 10.73e^{0.03x} + 1415.78$, where the prediction model for the melt temperature-freeze-thaw splitting strength ratio was $y = -219.18e^{-0.02x} + 98.03$, which could be used to predict the field paving temperature threshold.

Author Contributions: Writing—original draft, W.C. and K.W.; data curation, W.C. and K.W.; software, Z.F., X.X. and B.X.; methodology, W.H., X.Z., S.W., X.X. and G.H.; validation, S.W., L.N., K.C., Z.F. and T.C.; formal analysis, G.H., L.N., K.C. and T.C.; project administration, W.H. and B.X.; investigation, J.W. and X.Z.; supervision, J.W. All authors have read and agreed to the published version of the manuscript.

Funding: This research was funded by the National Key R&D Program of China (grant number 2018YFB1600100), the Shandong Natural Science Foundation Committee (ZR2020QE271, ZR2020KE024), the Shandong Provincial Natural Science Foundation (grant number ZR2020QE272), the Shandong Jianzhu University Doctoral research foundation (grant number X18073Z).

Institutional Review Board Statement: Not applicable.

Informed Consent Statement: Not applicable.

Data Availability Statement: Data sharing is not applicable for this article.

Acknowledgments: The authors would like to thank the staff at Shandong Transportation Research Institute for their support.

Conflicts of Interest: The authors declare no conflict of interest.

References

1. Ahmedzade, P.; Sengoz, B. Evaluation of steel slag coarse aggregate in hot mix asphalt concrete. *J. Hazard. Mater.* **2009**, *165*, 300–305. [CrossRef] [PubMed]
2. Chen, W.; Hu, G.L.; Han, W.Y.; Zhang, X.M.; Wei, J.C.; Xu, X.Z.; Yan, X.P. Research on the Quality of Asphalt Pavement Construction Based on Nondestructive Testing Technology. *Coatings* **2022**, *12*, 379. [CrossRef]
3. Li, X.L.; Zhou, Z.G.; Lv, X.C.; Xiong, K.Y.; Wang, X.J.; You, Z.P. Temperature segregation of warm mix asphalt pavement: Laboratory and field evaluations. *Constr. Build. Mater.* **2017**, *136*, 436–445. [CrossRef]
4. Tan, Y.Q.; Zhang, L.; Xu, H.N. Evaluation of low-temperature performance of asphalt paving mixtures. *Cold Reg. Sci. Technol.* **2012**, *70*, 107–112. [CrossRef]
5. Merchant, C.J.; Borgne, P.L.; Marsouin, A.; Roquet, H. Optimal estimation of sea surface temperature from split-window observations. *Remote Sens. Environ.* **2008**, *112*, 2469–2484. [CrossRef]
6. Janků, M.; Stryk, J. Application of infrared camera to bituminous concrete pavements: Measuring vehicle. *IOP Conf. Ser. Mater. Sci. Eng.* **2017**, *236*, 012104. [CrossRef]
7. Mohammad, L.N.; Hassan, M.M.; Kim, M. Effects of Paver Stoppage on Temperature Segregation in Asphalt Pavements. *J. Mater. Civ. Eng.* **2016**, *29*, 04016200. [CrossRef]
8. Han, D.; Zhao, Y.; Pan, Y.; Liu, G.; Yang, T. Heating process monitoring and evaluation of hot in-place recycling of asphalt pavement using infrared thermal imaging. *Autom. Constr.* **2020**, *111*, 103055. [CrossRef]
9. Hiasa, S.; Birgul, R.; Catbas, F.N. A data processing methodology for infrared thermography images of concrete bridges. *Comput. Struct.* **2017**, *190*, 205–218. [CrossRef]
10. Ellenberg, A.; Kontsos, A.; Moon, F.; Bartoli, I. Bridge deck delamination identification from unmanned aerial vehicle infrared imagery. *Autom. Constr.* **2016**, *72*, 155–165. [CrossRef]
11. Nishar, A.; Richards, S.; Breen, B.; Breen, D.; Robertson, J. Thermal infrared imaging of geothermal environments and by an unmanned aerial vehicle (UAV): A case study of the Wairakei-Tauhara geothermal field, Taupo, New Zealand. *Renew. Energy* **2016**, *86*, 1256–1264. [CrossRef]
12. Nevalainen, N.; Pellinen, T. The use of a thermal camera for quality assurance of asphalt pavement construction. *Int. J. Pavement Eng.* **2016**, *17*, 626–636. [CrossRef]
13. Rahman, M.M.; Grenfell, J.R.A.; Arulanandam, S.J.; Ianakiev, A. Influence of Thermal Segregation on Asphalt Pavement Compaction. *Transp. Res. Rec. J. Transp. Res. Board* **2013**, *2347*, 71–78. [CrossRef]
14. Javilla, B.; Fang, H.; Mo, L.T.; Shu, B.N.; Wu, S.P. Test evaluation of rutting performance indicators of asphalt mixtures. *Constr. Build. Mater.* **2017**, *155*, 1215–1223. [CrossRef]
15. Polaczyk, P.; Ma, Y.; Xiao, R.; Hu, W.; Jiang, X.; Huang, B. Characterization of aggregate interlocking in hot mix asphalt by mechanistic performance tests. *Road Mater. Pavement Des.* **2021**, *22*, S498–S513. [CrossRef]
16. Qu, Z.; Jiang, P.; Zhang, W.X. Development and Application of Infrared Thermography Non-Destructive Testing Techniques. *Sensors* **2020**, *20*, 3851. [CrossRef] [PubMed]
17. *JTG E20-2011;* Standard Test Methods of Bitumen and Bituminous Mixtures for Highway Engineering. Ministry of Transport: Beijing, China, 2011.
18. Pasetto, M.; Baldo, N. Mix design and performance analysis of asphalt concretes with electric arc furnace slag. *Constr. Build. Mater.* **2011**, *25*, 3458–3468. [CrossRef]
19. Chen, W.; Wei, J.; Xu, X.; Zhang, X.; Han, W.; Yan, X.; Hu, G.; Lu, Z. Study on the Optimum Steel Slag Content of SMA-13 Asphalt Mixes Based on Road Performance. *Coatings* **2021**, *11*, 1436. [CrossRef]

20. Chaturabong, P.; Bahia, H.U. Mechanisms of asphalt mixture rutting in the dry Hamburg Wheel Tracking test and the potential to be alternative test in measuring rutting resistance. *Constr. Build. Mater.* **2017**, *146*, 175–182. [CrossRef]
21. Gogolla, T. Imaging Measuring System and Measuring Method for Measuring Thermal Output to a Target Object. CN102809433A, 14 September 2016.
22. *JTG F40-2004*; Technical Specifications for Construction of Highway Asphalt Pavements. Ministry of Transport: Beijing, China, 2004.
23. Mcleod, N.W. Using Paving Asphalt Rheology to Impair or Improve Asphalt Pavement Design and Performance. In *Asphalt Rheology: Relationship to Mixture*; Briscoe, O.E., Ed.; American Society for Testing and Materials: Ann Arbor, MI, USA, 1987.
24. *JTG E42-2005*; Standard Test Methods of Aggregate for Highway Engineering. Ministry of Transport: Beijing, China, 2005.

materials MDPI

Article

Physical Properties and Storage Stability of Buton Rock Asphalt Modified Asphalt

Yue Su [1], Xiaodi Hu [1], Jiuming Wan [1,*], Shaopeng Wu [2], Yinglong Zhang [1], Xing Huang [3] and Zhangjun Liu [1]

[1] School of Civil Engineering and Architecture, Wuhan Institute of Technology, Wuhan 430205, China; 13235670277@163.com (Y.S.); huxiaodi@wit.edu.cn (X.H.); zhangyl@stu.wit.edu.cn (Y.Z.); liuzhangjun@wit.edu.cn (Z.L.)
[2] State Key Laboratory of Silicate Materials for Architecture, Wuhan University of Technology, Wuhan 430070, China; wusp@whut.edu.cn
[3] Poly Changda Engineering Co., Ltd., Guangzhou 510062, China; huangx202205@163.com
* Correspondence: 19060101@wit.edu.cn

Abstract: Buton Rock Asphalt (BRA) refers to the natural rock asphalt natively produced on the Buton island of Indonesia. It is often used as a modifier to enhance the performance of asphaltpavement. However, the segregation of BRA in BRA-Modified Asphalt (BRA-MA) has restricted its application. This study aims to investigate how the particle size and content of BRA affect the physical properties and storage stability of BRA-MA. Penetration, softening point, viscosity, and viscosity-temperature susceptibility (VTS) were analyzed. The evaluation method of storage stability was discussed and determined. The segregation of BRA in BRA-MA of static storage and transportation process were simulated and tested. The results suggest that the softening point and viscosity were positively correlated to BRA content and inversely determined by particle size. Penetration, VTS, and ductility were reduced due to the decline in particle size and increment of BRA content. The index of segregation value based on viscosity difference showed better statistical and quantitative significances than the softening-point difference in evaluating the storage stability. The particle size and content of BRA are positively correlated to the segregation of BRA-MA. Both the storage temperature and time were positively correlated to the segregation of BRA-MA. We prove that the relationship between specific surface area and segregation are power functional. BRA-MA with BRA whose 50% particle sizes are lower than 13.6 μm showed low segregation in transportation.

Keywords: Buton rock asphalt; storage stability; particle size; physical property

Citation: Su, Y.; Hu, X.; Wan, J.; Wu, S.; Zhang, Y.; Huang, X.; Liu, Z. Physical Properties and Storage Stability of Buton Rock Asphalt Modified Asphalt. *Materials* 2022, 15, 3592. https://doi.org/10.3390/ma15103592

Academic Editor: Francesco Canestrari

Received: 4 April 2022
Accepted: 12 May 2022
Published: 18 May 2022

1. Introduction

Natural rock asphalt refers to the asphalt that exists in rock crevices, which is often used as an additive and modifier for asphalt materials [1–5]. Buton Rock Asphalt (BRA) is a kind of rock asphalt that originates from Buton Island, Sulawesi province, Indonesia [6,7]. It is used as a modifier to prepare Buton Rock Asphalt-Modified Asphalt (BRA-MA) to improve high-temperature and moisture performances of pavement [6,7]. The preparation methods of BRA-modified asphalt mainly include wet processing and dry processing. Wet processing uses BRA as a modifier to prepare BRA-modified asphalt, whose performance is significantly improved compared to base asphalt [8]. Lv et al. [7] reported that the addition of BRA could help to improve the aging resistance of bio-asphalt. The use of BRA can reduce the consumption of petroleum asphalt and cost of materials. It is suggested that BRA is also helpful for the cleaner and cheaper production of asphalt pavement construction, signifying that the application of BRA is beneficial for environment-protection [9] and economic values [10–13]. Consequently, the application prospect of BRA-MA is very extensive.

However, the storage stability of BRA-MA requires further study since the segregation of BRA often occurs in the production plant and transportation of BRA-MA, especially in wet processing. Hence, the storage stability of BRA-MA can be divided into two parts:

static storage [14] in the production plant and the transportation-storage process [15]. Segregation characteristic of the two parts is differential. Thus, the storage stability of BRA-MA during thermal static storage and transportation should be investigated and improved to address the segregation of BRA. Direct observation method, softening-point difference [16], Fourier infrared spectroscopy, and fluorescence microscope were used to evaluate the storage stability of modified asphalt [17]. Gu et al. [18] evaluated the storage stability of waterborne epoxy emulsified asphalt by fluorescence microscope. It was found that the content increase in waterborne-epoxy resin improves the compatibility of waterborne epoxy resin emulsified asphalt. Nevertheless, whether the storage stability evaluation method of polymer-modified asphalt is applicable to BRA-MA is not clear, since BRA is different with polymer modifier. For instance, it is quite difficult to be distinguished under a microscope, thus the micro-observation method is not applicable.

The factors [19] affecting the storage stability of modified asphalt include the content of modifier, storage temperature, storage time, and particle size. T J, Navarro et al. [20] and Liang et al. [21] studied the effects of the storage temperature and sample particle size on the storage stability of tire-rubber powder modified asphalt. It was found that the storage stability of tire-rubber-powder-modified asphalt gradually decreases with the increase in storage temperature and rubber powder particle size. Fang et al. [22] studied the effect of the preparation process on the storage stability of waste polyethylene-modified asphalt by using differential scanning calibration (DSC) and thermal gravimetric analysis (TGA). The preparation process was found to have little effect on the storage stability of modified asphalt. However, how particle size and BRA content determine the storage stability of BRA-MA has not been systematically studied. The effect of storage temperature, storage time, and cooling process on static storage and transportation storage is not yet explicit.

BRA powders of five different sizes were used as modifiers, and BRA-MA with 10%, 20%, 30%, and 40% BRA are introduced in this study. Penetration, softening point, ductility, viscosity, and viscosity-temperature susceptibility are characterized to understand the effect of particle size and BRA content on their physical properties. The softening-point difference test and viscosity difference test are separately conducted and discussed. Therefore, the appropriate evaluation method for the segregation of BRA-MA can be determined. Static storage stabilities by different temperature and time dependencies are characterized to illustrate the static storage process in the production plant before transportation. Storage stability during transportation is also tested in a simulation of the cooling process in a container in the vehicle. It provides a scientific basis for improving the physical properties and storage stability of BRA-MA.

2. Materials and Experimental Methods

2.1. Materials

Part a of Figure 1 illustrates the appearance of BRA, which is brown powder after being crushed. Figure 1b presents the mesoscopic image of BRA through Scanning Electron Microscope analysis, whose magnification is 4500. The image suggests that BRA has a very rough surface and porous structure. Rock asphalt and mineral cannot be clearly distinguished through the mesoscopic image; it proved that rock asphalt was evenly mixed and combined with minerals. The asphalt content of BRA was tested based on the ignition oven method [23], since BRA can be regarded as a mixture of asphalt and mineral. The result shows that the mass percentage of asphalt in BRA is 27.8%, while the minerals account for 71.2%. The moisture content of BRA is 1%. This study also introduced base asphalt of Pen 60~80, which was produced by the Guochuang company, Hubei province, as the material for preparing the BRA-MA. Table 1 illustrates the properties of base asphalt. Penetration, softening point, ductility, and viscosity are characterized according to JTG E20-2011 and JTG F40-2004.

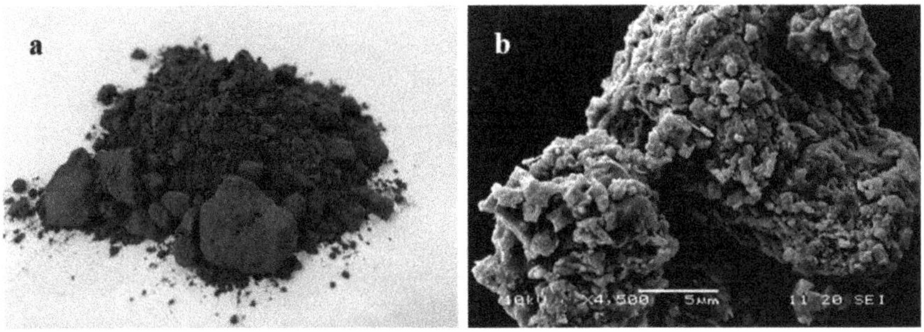

Figure 1. BRA powder (**a**) and its SEM image (**b**).

Table 1. Properties of base asphalt.

Properties	Penetration (25 °C, 100 g, 5 s)	Softening Point	Ductility (5 cm/min, 15 °C)	Brookfield Viscosity (10 r/min, 135 °C)
Unit	0.1 mm	(°C)	cm	Pa·s
Base asphalt	71.2	48.8	≥100	0.142
Standard limits	60~80	≥46.0	≥100	-

2.2. Preparation of BRA-MA

BRA should be pre-treated before preparing various BRA-MA. BRA raw material should be kept in a temperature controlling box at 80 °C for 12 h. Moisture in BRA can therefore be removed. BRA was then ground into five different particle sizes using a planetary ball mill. A Malvern laser particle-size analyzer [21] was used to characterize the particle-size distribution of the five BRA samples. The five BRA samples were named as BRA-1, BRA-2, BRA-3, BRA-4, and BRA-5, respectively. Table 2 presents the particle-size testing results of BRA samples; corresponding d(0.5) particle sizes are 6.26 μm, 9.55 μm, 12.58 μm, 13.60 μm, and 106.22 μm, respectively. The d(0.5) refers to the volume content of particles accounts for 50% whose particle size is smaller than a particle size. It is used to reflects the particle size. Specific surface area was also presented, which were negatively correlated to the particle size.

Table 2. Particle-size testing results of BRA.

BRA Number	d(0.1) (μm)	d(0.5) (μm)	d(0.9) (μm)	Specific Surface Area (m^2/g)
BRA-1	1.20	6.26	21.07	2.02
BRA-2	1.84	9.55	23.87	1.53
BRA-3	2.39	12.58	32.27	1.28
BRA-4	2.26	13.60	37.64	1.26
BRA-5	75.29	106.22	150.55	0.78

Base asphalt should be heated to 155 °C in an agitator for preparing BRA-MA. Rotational speed was set as 2500 r/min for 20 min after half of the BRA was added to the base asphalt. Then, the remaining BRA was added to the base asphalt, while the speed of a high-speed shear agitator was also set as 2500 r/min for another 20 min. Figure 2 shows the high-speed shear agitator for BRA-MA preparation. This special mixing method can ensure the uniformity of BRA in BRA-MA, while the particle size of BRA remained unchanged during shearing.

Figure 2. High-speed shear agitator for BRA-MA preparation.

2.3. Experimental Methods

This study introduced the external-adding method, namely, the mixing proportion is the mass ratio of BRA to base asphalt. The proportions of BRA 1–5 in corresponding BRA-MA were 10%, 20%, 30%, and 40%, respectively. Thus, 20 BRA-MA samples were prepared based on different particle sizes and contents. Figure 3 illustrates the outline of this study. BRA-MA's physical properties, storage-stability evaluation method, static storage stability, and storage stability of transportation process were studied.

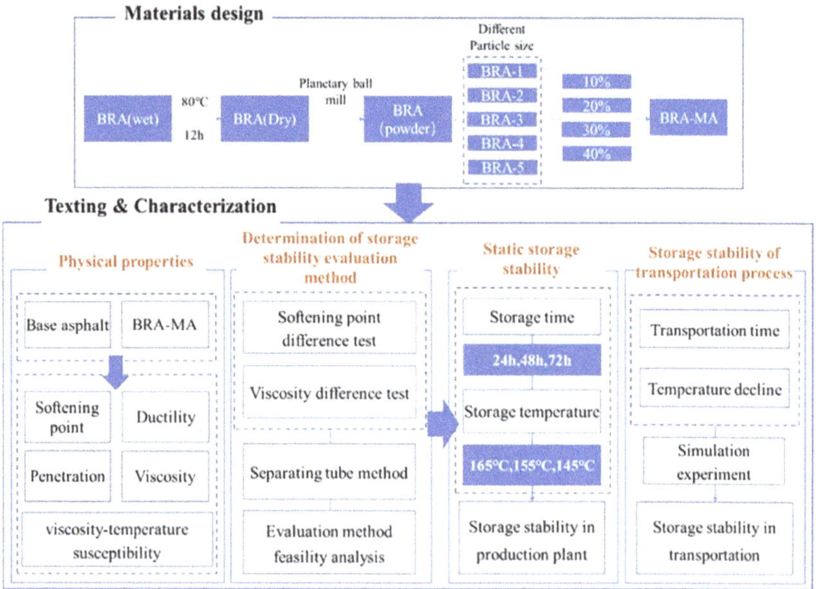

Figure 3. Outline of this study.

Firstly, penetration, softening point, ductility, and viscosity of base asphalt and BRA-MA were characterized. The impact of BRA content and particle size on physical properties can be concluded. Secondly, this study introduced a separating-tube method to indicate the distribution uniformity of BRA in BRA-MA. Two indicators for evaluating storage stability were compared, so that more appropriate evaluation indicators can be obtained. Subsequently, the stability of BRA-MA in factory static storage were investigated. Finally, the storage stability of BRA-MA during the transportation process was simulated according to temperature decline in transportation.

The storage stability of modified asphalt is the key technical requirement for its production and transportation, which may lead to modification failure and certainly affect the service performance of the modified asphalt mixture. In order to study the influence of particle size and BRA content on the storage stability of BRA-MA, the optimum indicator for storage-stability evaluation should be determined. As explained in stage 5 of the outline of this study in Figure 3, the softening-point difference and index of segregation (IS) based on the viscosity difference between the top and bottom part samples were used to characterize the segregation degree of BRA. The two evaluation methods were used separately to analyze the influence of particle size and BRA content on the storage stability of BRA-MA. The feasibility of the two indicators of the methods was discussed. The static significance of the results according to two indicators was analyzed by variance tests. If F is greater than F crit, then the difference between the results is proven. If the p-value is greater than 0.01 and less than 0.05, the difference is significant. If the p-value is less than 0.01, the difference is highly significant.

Figure 4 implies the storage process of BRA-MA in a production plant and during transportation. The produced BRA-MA was contained in a soaking tank with temperature-controlling equipment, which can keep BRA-MA at a relatively stable high temperature. Static storage at a high temperature of BRA-MA can save the time and cost of reheating, especially in short-term construction. Maintaining the high liquidity of BRA-MA facilitates rapid loading onto a transport vehicle. Thus, the storage of BRA-MA in a production plant is static. The static storage of BRA-MA refers to the hot-storage process [22,24,25] in a production plant before transportation. Then, BRA-MA is pumped into the vehicle container for transportation, the transportation time is usually less than 12 h. Both the storage stability of BRA-MA in the production plant and transportation process storage was discussed in this study.

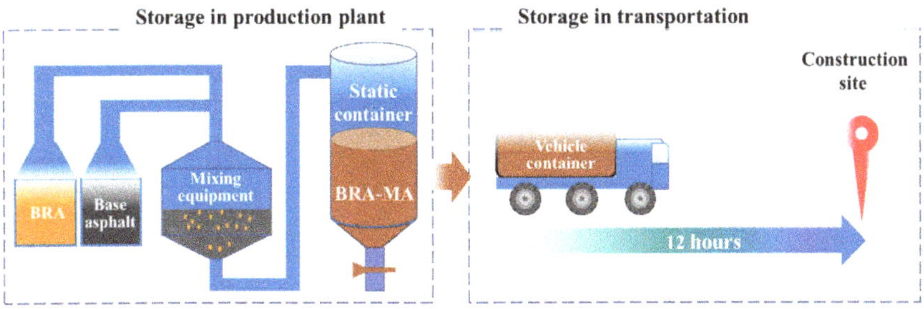

Figure 4. Storage process of BRA-MA in a production plant and during transportation.

The separating tube used in this study was composed of an aluminum sheet with a thickness of 0.4 mm. Therefore, the customized tube was easy to cut and separate. It should be averagely separated and noted as 3 parts by length, which were the top, middle, and bottom parts. Figure 5 expresses how the separating-tube method was conducted to evaluate the storage stability and determination of the evaluation indicator for segregation. At stage 1, the separating tube should be filled with BRA-MA on a specific shelf, exactly when the mixing of BRA and base asphalt was completed. Subsequently, the separating tubes should be kept vertically in a temperature-controlling box according to the established storage conditions, during which the segregation occurs. BRA would gradually settle at the bottom of the tubes due to its gravity. In the next stage, segregated samples were kept in a freezer at −4 °C in 4 h. The segregation rate of BRA would decrease to a very low level due to the low temperature, which can be considered that the segregation of BRA-MA was terminated. As shown in stage 4, separating tubes should be cut and separated into 3 parts by an electric saw. Subsequently, BRA-MA from the top and bottom parts should be characterized by the softening-point and viscosity tests, respectively. Softening point and viscosity of the top and bottom samples were different, as a result of the BRA content difference of the two samples owing to BRA segregation. In the final stage, the softening-point difference and index of segregation based on viscosity difference can be acknowledged, respectively. The optimum evaluation indicator can be therefore be determined. The storage stability of BRA-MA can consequently be calculated and obtained.

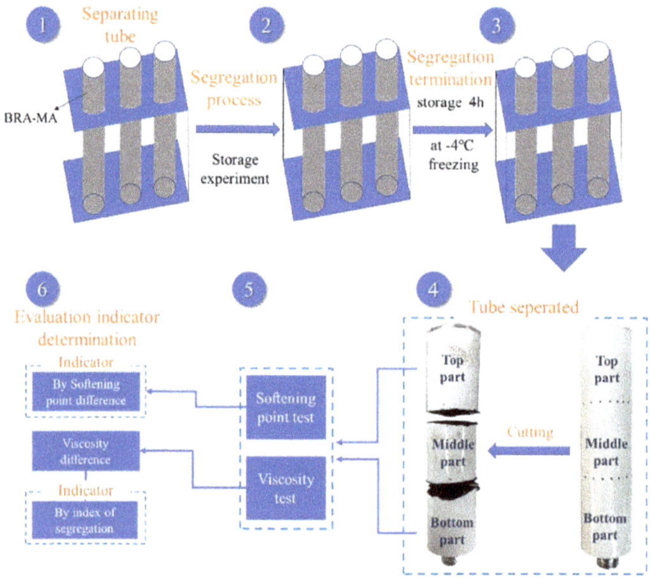

Figure 5. Separating-tube method of index of storage stability.

3. Results and Discussions

3.1. Physical Properties of BRA-MA

3.1.1. Softening Point

Figure 6 presents the results of the softening point of base asphalt and BRA-MA. BRA-MA samples were labeled with BRA type and BRA content. The yellow bars present the data of base asphalt, whose BRA content is 0%. The softening point of BRA-MA showed an obvious rising tendency, along with the increase in BRA content. The softening-point value of BRA-MA with 40% BRA-1 was 54.05 °C, which was 5.3 °C higher than that of base asphalt. On the other hand, the softening point of BRA-MA with the same BRA content decreased from BRA-1 to BRA-5. The results of BRA-MA can meet the standard limits of JTG F40-2004. It proved that the softening point of BRA-MA can be enhanced by a decline in BRA particle size. The results also illustrate that the softening point is positively correlated to BRA content. The high-temperature stability and temperature sensitivity of asphalt is generally evaluated by its softening point [26,27]. The temperature sensitivity of asphalt can be reduced by the addition of BRA. Thus, the increase in BRA content and decline in particle size would help to enhance the high-temperature performance of BRA-MA.

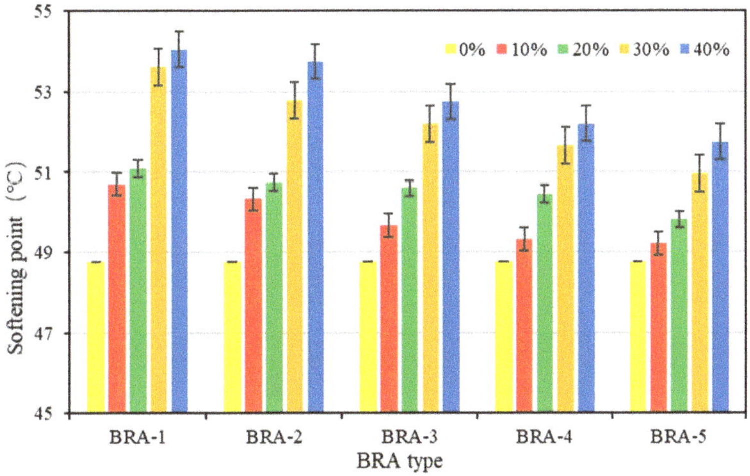

Figure 6. Softening point of base asphalt and BRA-MA.

3.1.2. Penetration

Figure 7 shows the penetration of base asphalt and BRA-MA. Penetration indicates the hardness of asphalt. A higher penetration value means a lower hardness. The yellow bars present the data of base asphalt without BRA. Penetration of BRA-MA decreased as BRA content increased, while base asphalt presented the highest penetration value. Penetration of BRA-MA containing 40% BRA-1 was 4.85 mm, which decreased by 31.9% compared to that of base asphalt. Additionally, the penetration of BRA-MA showed an approximativelyincreasing trend from BRA-1 to BRA-5 at the same BRA content. Therefore, the penetration of BRA-MA increased with the increment in particle size. The results suggest that BRA content negatively affects penetration, while the particle size of BRA shows a contrary effect. Therefore, the addition of BRA can improve the hardness of asphalt.

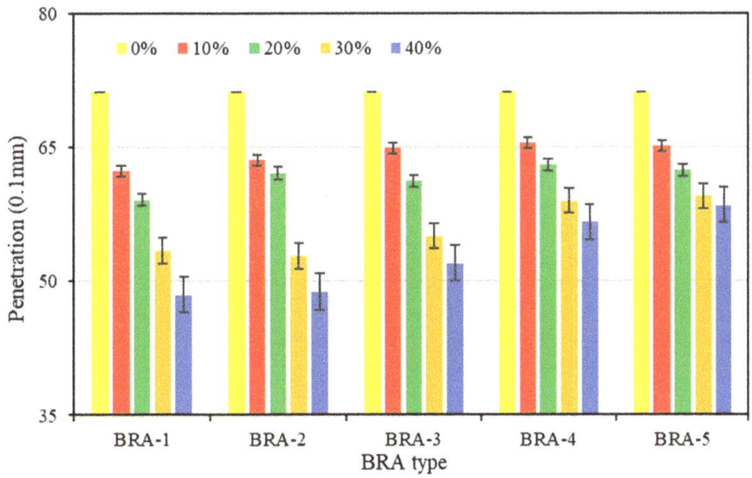

Figure 7. Penetration of base asphalt and BRA-MA.

3.1.3. Ductility

Figure 8 presents the 15 °C ductility of base asphalt and BRA-MA, whose yellow bars are the data of base asphalt. It is obvious that the ductility of base asphalt is about 150 cm, while ductility values of BRA-MA are almost lower than 60 cm. The ductility value of BRA-MA with 40% BRA-5 was 18.5 cm, whose decline rate reached 84.6% compared to base asphalt. This result indicates that the addition of BRA leads to remarkable ductility loss. The ductility of BRA-MA also shows a downward trend as the increment in BRA content and particle size. Thus, BRA content and particle size negatively determine the ductility of BRA-MA. It is believed that it was the inorganic mineral particles in BRA that played a critical role in reducing the ductility of BRA-MA. Although its ductility is much lower than that of base asphalt, whether its low-temperature performance is poor is still hard to define due to the lack of corresponding standard for BRA.

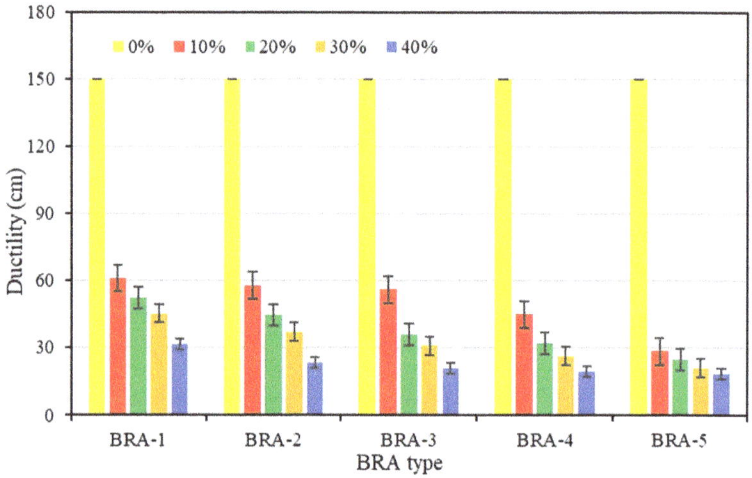

Figure 8. Ductility of base asphalt and BRA-MA.

3.1.4. Viscosity and Viscosity-Temperature Susceptibility (VTS)

Viscosities of 20 BRA-MA at 115 °C, 135 °C, 155 °C, and 175 °C were measured by a Brookfield viscometer. Figures 9–12 illustrate the viscosity of BRA-MA by temperature and particle-size dependency. The viscosity of base asphalt was appended for comparison. The 135 °C viscosity results of BRA-MA can meet the JTG F40-2004 standard limits for modified asphalt. The viscosity of base asphalt was significantly lower than that of BRA-MA. BRA-MA with BRA-1 showed the highest viscosity at the same temperature and BRA content. BRA-MA with 40% BRA-1 showed the highest viscosity, and its growth rate reached 114% compared to that of base asphalt. The viscosity difference between base asphalt and BRA-MA gradually decreased with the increase in temperature. The larger particle size of BRA led to lower viscosity. However, the effect of BRA particle size of viscosity was not very significant when the temperature was over 135 °C. Therefore, the viscosity of asphalt apparently increases with the addition of BRA and a smaller particle size.

Figure 13 explains the effect of BRA content on the viscosity of BRA-MA, which uses a logarithmic longitudinal axis. The viscosity result of BRA-MA with 40% BRA-1 was used for an instance, and corresponding fitting curves of the data were presented. As the BRA-1 content increased from 10% to 40%, the viscosity of BRA-MA showed an obvious rising trend. This result proves that raising BRA content contributes to promote the viscosity of BRA-MA. To conclude, viscosity had a negative correlation with the particle size of BRA, but it was positively affected by BRA content.

VTS [28] refers to the slope of the viscosity–temperature curve with the iterated logarithm of viscosity and logarithm of temperature as the coordinate axes [29]. For instance, Figure 14 demonstrates the straight lines of BRA-MA fitting with BRA-1. According to ASTM D 2493, the smaller absolute value of VTS implies the lower temperature sensitivity of asphalt [30]. The basic equation of VTS is:

$$\mathrm{VTS} = \frac{\lg[\lg(\eta_{T_2})] - \lg[\lg(\eta_{T_1})]}{\lg(T_2) - \lg(T_1)} \quad (1)$$

where the T_1 and T_2 = temperatures (centigrade) of the binder at two known points, and η_{T_1} and η_{T_2} 5 = viscosities of the binder at the same two points. The viscosities of BRA-MA at 115 °C, 135 °C, 155 °C, and 175 °C were used to fit a straight line according to Equation (1).

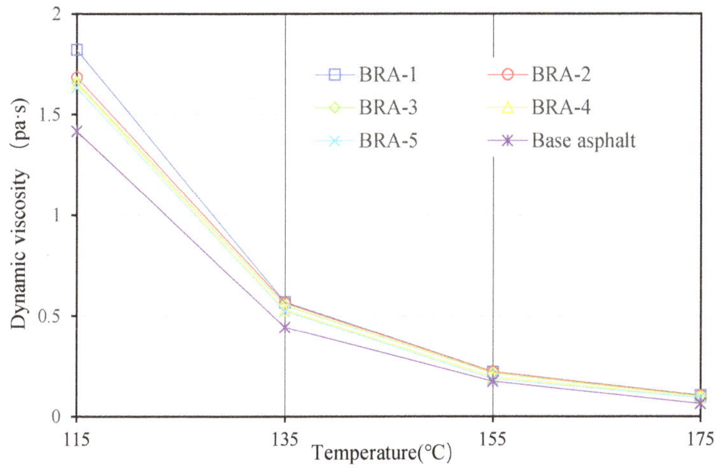

Figure 9. Viscosity of BRA-MA with 10% BRA.

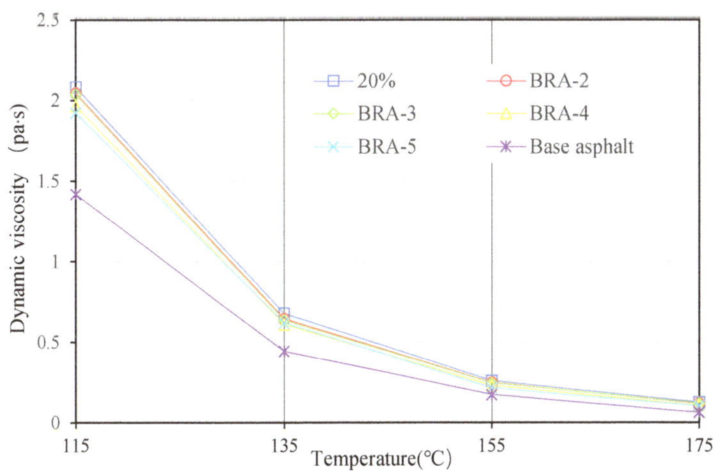

Figure 10. Viscosity of BRA-MA with 20% BRA.

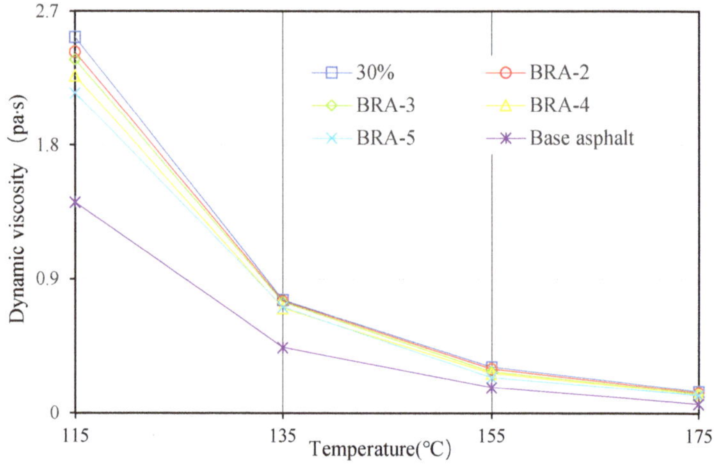

Figure 11. Viscosity of BRA-MA with 30% BRA.

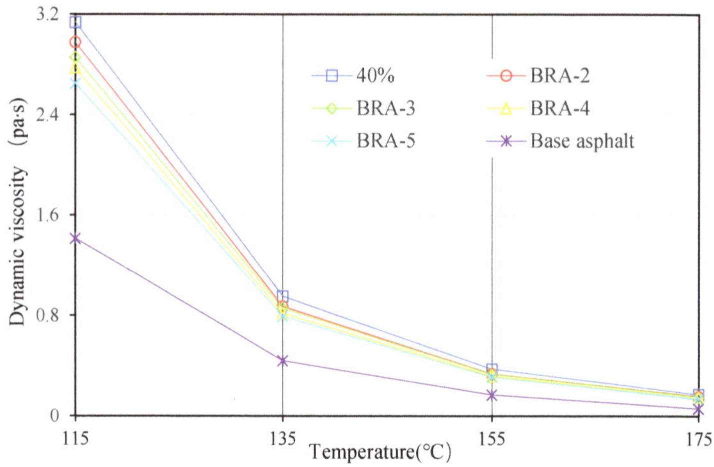

Figure 12. Viscosity of BRA-MA with 40% BRA.

The absolute value of BRA-MA VTS values can be observed in Table 3, which involves the VTS value of base asphalt for comparison. The addition of BRA can reduce the VTS of asphalt. It is clear that the absolute VTS value decreased with the increase in BRA content, which indicates that BRA can reduce the temperature susceptibility of asphalt. Additionally, the smaller particle size also leads to a lower absolute VTS value. Therefore, VTS is negatively correlated to BRA content, but positively determined by particle size. These results indicate that the higher BRA content and smaller particle size reduce BRA-MA's temperature susceptibility, which may improve the BRA-MA mixture's stability at high temperatures.

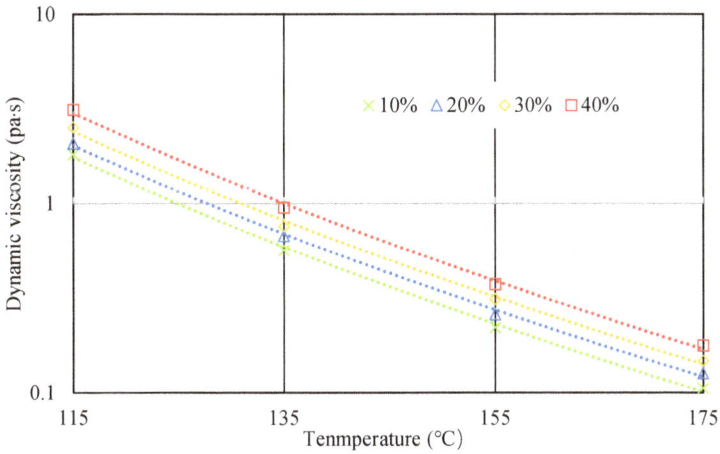

Figure 13. Viscosity of BRA-MA with 40% BRA with logarithmic axis.

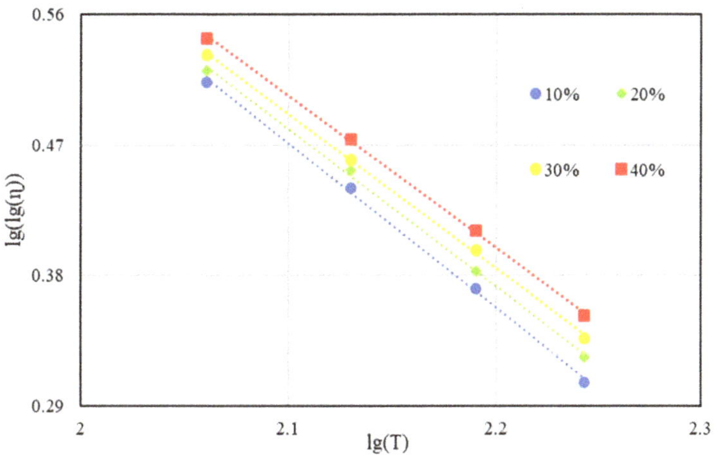

Figure 14. Viscosity–temperature curve of BRA-MA with BRA-1 for VTS.

Table 3. Absolute value of BRA-MA VTS values.

	BRA-1	BRA-2	BRA-3	BRA-4	BRA-5
0%	1.297	1.297	1.297	1.297	1.297
10%	1.135	1.117	1.128	1.164	1.176
20%	1.087	1.112	1.114	1.147	1.166
30%	1.066	1.083	1.104	1.107	1.120
40%	1.048	1.074	1.075	1.081	1.097

3.2. Storage-Stability Evaluation Determination

3.2.1. Softening Point Difference

Softening-point difference [31] is one of the most commonly used indicators in the storage-stability evaluation of polymer-modified asphalt. Figure 6 proves that BRA content is positively correlated to the softening point of BRA-MA; thus, a higher softening point means a higher BRA content. Through the difference of the softening point between the top and bottom samples, the segregation of BRA content between the top and bottom of the separating tube can be determined. If the softening-point difference between the top and bottom is less than 2.2 °C, according to ASTM D 5976, this indicates that the segregation is not serious and the modified asphalt has acceptable storage stability. The equation for calculating the softening-point difference is:

$$\Delta SP = |SP_t - SP_b| \tag{2}$$

where ΔSP refers to the softening-point difference between the top and bottom of separating tube, SP_t and SP_b are the softening points of the top and bottom of the separating tube, respectively. Figure 15 illustrates the softening-point difference of BRA-MA. The softening-point difference of BRA-MA with more than 20% BRA-5 was over 2.2 °C, which indicates their severe segregation. Softening-point difference of BRA-MA with BRA-1~4 was lower than 2.2 °C, suggesting that their segregation is slight. However, it was found that the segregation of BRA-MA with BRA-2, -3, and -4 was also severe by manual inspection, which is contrary to the evaluation standard of the softening-point difference test. Additionally, even if the softening-point difference of two kinds of asphalt is equal, their segregation degrees are thought to be different since their original softening points are different. Therefore, the softening-point difference method may not quantitatively indicate the degree of segregation of BRA-MA.

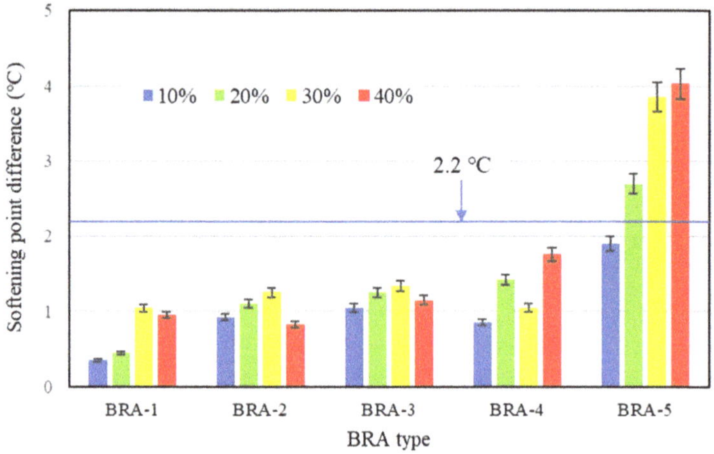

Figure 15. Softening-point difference of BRA-MA.

The variance analysis results show that the F value of the softening-point difference by BRA-content dependency was lower than F crit, proving that BRA content did not show a statistical effect on the softening-point difference. In addition, F-value by particle-size dependency was higher than F crit and the p-value was less than 0.05, which implies that particle size can significantly affect softening-point difference. Therefore, the softening-point difference test failed to reveal how the BRA content influenced the storage stability, nor can it reveal the quantitative segregation of BRA-MA.

3.2.2. IS Based on the Viscosity Difference

IS based on the viscosity difference was used as an indicator for the purpose of developing the susceptibility of detecting the segregation of BRA-MA. As shown in Figure 5, this evaluation method is also based on the separating-tube method. The viscosity of the top and bottom part of separating-tube should be tested at 135 °C. Higher viscosity signifies the corresponding higher BRA content, according to Figure 13. Additionally, IS was used to illustrate the segregation of BRA-MA. The calculation equation is shown in Equations (3) and (4).

$$\Delta\eta = |\eta_t - \eta_b| \tag{3}$$

$$IS = \Delta\eta/\eta_O \tag{4}$$

where $\Delta\eta$ refers to the viscosity difference, η_t and η_b are the viscosities of the top and bottom of the separating tube, η_O is the original viscosity of BRA-MA before segregation, and IS is the index of segregation based on the viscosity difference. A higher IS value means a more serious segregation of BRA. Not only the viscosity difference, but also the original viscosity of the modified asphalt had been considered in the calculation of IS. The segregation of asphalt with different original viscosities can be semi-quantitatively compared through the IS method.

Figure 16 presents the IS of BRA-MA after the segregation process, which kept BRA-MA at 165 °C for 48 h. The value of IS showed an almost rising trend, along with a the BRA content increases. Moreover, the IS value illustrated an increasing tendency with the increment in BRA particle size. The F-value is the statistic of the F-test, which is used to indicate the statistical significance of the two methods. The variance analysis results indicate that both the F-values of IS by BRA content and particle-size dependency are higher than the corresponding F crit. The *p*-values proved that the discrepancy of IS by BRA content and particle-size dependency were significant. Consequently, IS based on the viscosity difference had more static significance in the evaluating influence of BRA content and particle size on segregation compared with the softening difference evaluation. IS was then determined as the method to indicate the storage stability of BRA-MA in the following study.

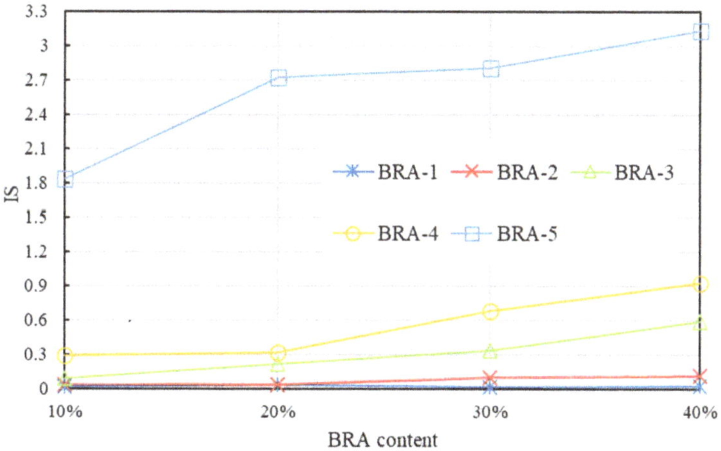

Figure 16. IS of BRA-MA after the segregation process.

3.3. Static Storage Stability in the Production Plant

Storage temperature and time are the key factors affecting the static storage stability of samples. Therefore, how BRA content and particle size affect static storage stability at different temperatures and times was studied. The storage temperatures were 145 °C, 155 °C, and 165 °C, while the storage times were 24, 48, and 72 h. The aging of BRA-MA at high temperatures was negligible in a soaking tank, which cut off the outside air.

Figures 17–19 present the IS results of BRA-MA after the segregation process by storage temperature and time dependency, respectively. The IS values illustrated an obviously rising trend as the BRA content increased, which also showed an increasing tendency along with particle-size increase from BRA-1 to BRA-5. On the other hand, with the extension of storage time, the IS value also showed an increasing tendency It was evident that heightening temperatures would result in higher IS values by comparison of Figures 17–19. Therefore, reducing the temperature, particle size, storage time, and BRA content would help to lower the segregation of BRA-MA in static storage in a static container.

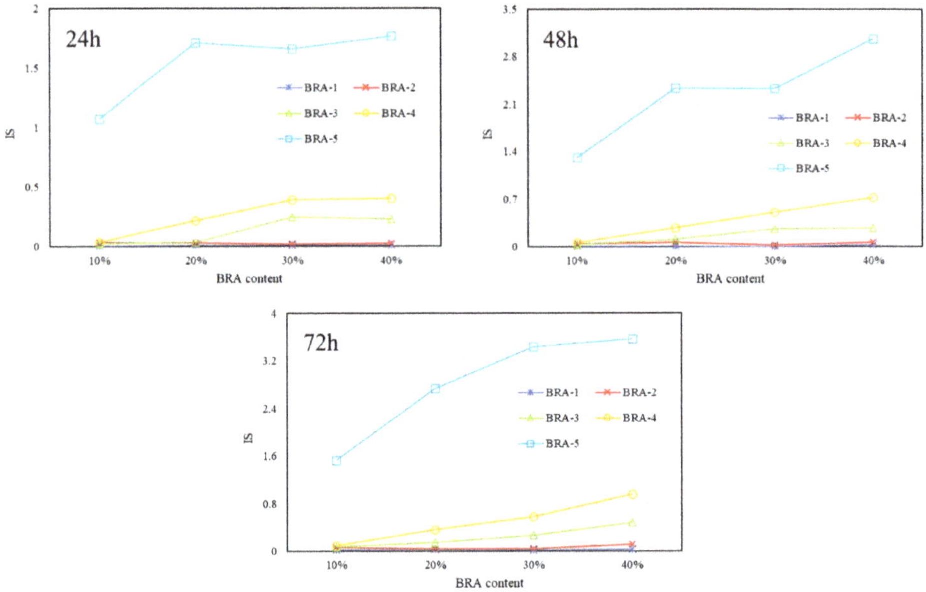

Figure 17. IS of BRA-MA after the segregation process at 145 °C.

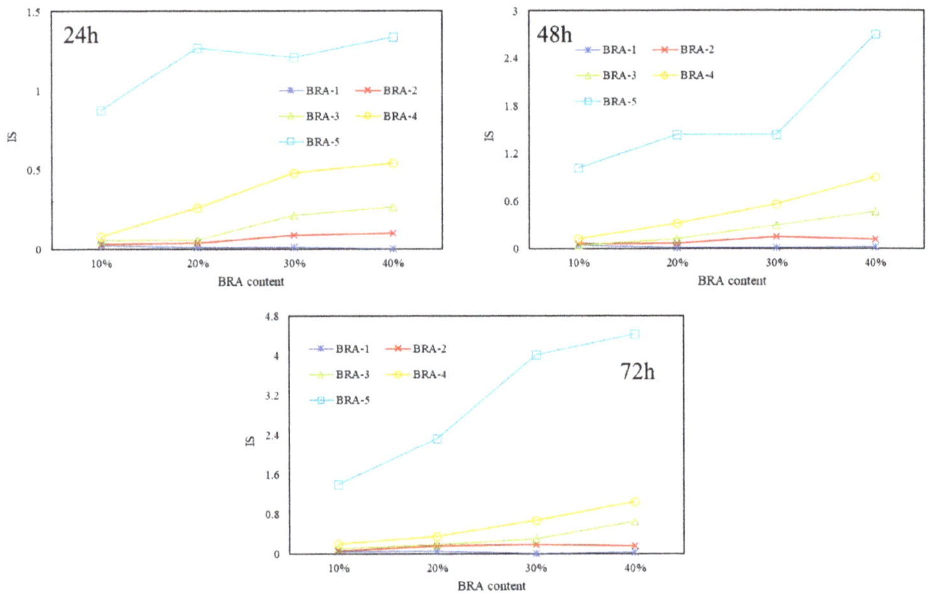

Figure 18. IS of BRA-MA after the segregation process at 155 °C.

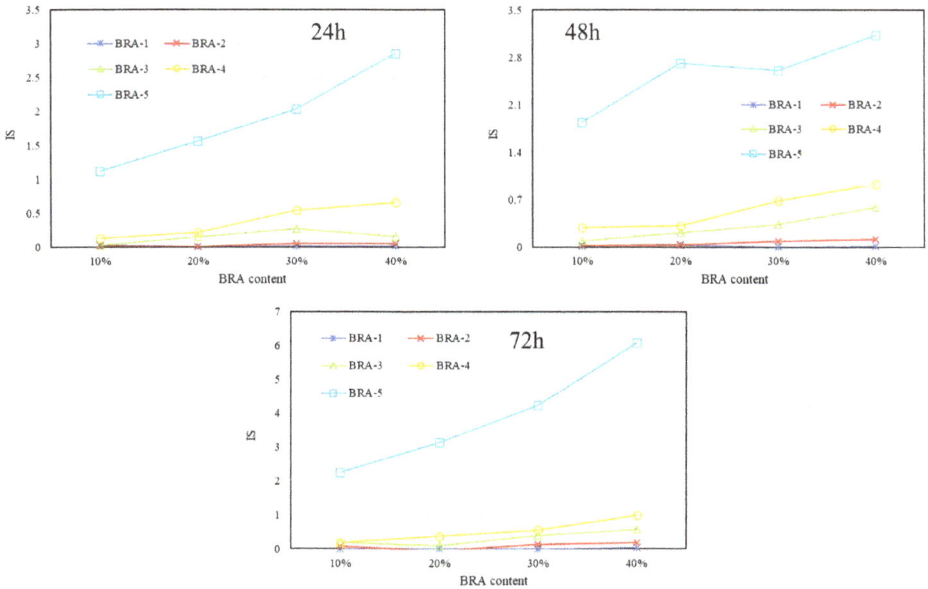

Figure 19. IS of BRA-MA after the segregation process at 165 °C.

The correlation between the specific surface area of BRA and IS was analyzed. Figure 20 presents the IS results by temperature, specific surface area, and BRA content at 145 °C, 155 °C, and 165 °C, respectively, for 24 h' static storage, for instance. The IS value presented a decline as the increment in the specific surface area of BRA, since its particle size was negatively correlated to the specific surface area. Power function curves were used to fit the data point. The result of the fitting equation and R^2 of BRA-MA after the segregation process for 24 h of static storage is shown in Table 3. Corresponding complex correlation coefficients (R^2) were higher than 0.89, suggesting that fitting between the IS value and specific surface area was rational. It revealed that the functional relationship of IS and specific surface area can be fit as Equation (5):

$$\mathrm{IS} = s_{w_i} x^{-a} \quad (5)$$

where s_{w_i} refers to the segregation coefficients, which is positively correlated to the segregation of BRA-MA; x is the variable (specific surface area); and a is the power. This functional relationship of the IS and specific surface area can be used to indicate how particle size that is correlated to the specific surface area determines the segregation of BRA-MA. Table 4 listed the Power function curve-fitting equation and R^2 of BRA-MA. It is obvious that the R^2 is mostly over 0.9, which suggested that the fitting is adequate.

Table 4. Power function curve-fitting equation and R^2 of BRA-MA.

Temperature (°C)	BRA Content (%)	Power Function Curve-Fitting Equation	R^2
145	10	$y = 0.1929x^{-5.216}$	0.9968
	20	$y = 0.3852x^{-5.58}$	0.9930
	30	$y = 0.6138x^{-5.727}$	0.9672
	40	$y = 0.6268x^{-5.505}$	0.9737

Table 4. *Cont.*

Temperature (°C)	BRA Content (%)	Power Function Curve-Fitting Equation	R^2
155	10	$y = 0.2284x^{-3.913}$	0.9935
	20	$y = 0.3813x^{-4.937}$	0.9816
	30	$y = 0.6213x^{-4.967}$	0.9104
	40	$y = 0.7397x^{-5.889}$	0.8968
165	10	$y = 0.2222x^{-4.498}$	0.9928
	20	$y = 0.4174x^{-6.197}$	0.9915
	30	$y = 0.8069x^{-5.256}$	0.9641
	40	$y = 0.8867x^{-5.629}$	0.9713

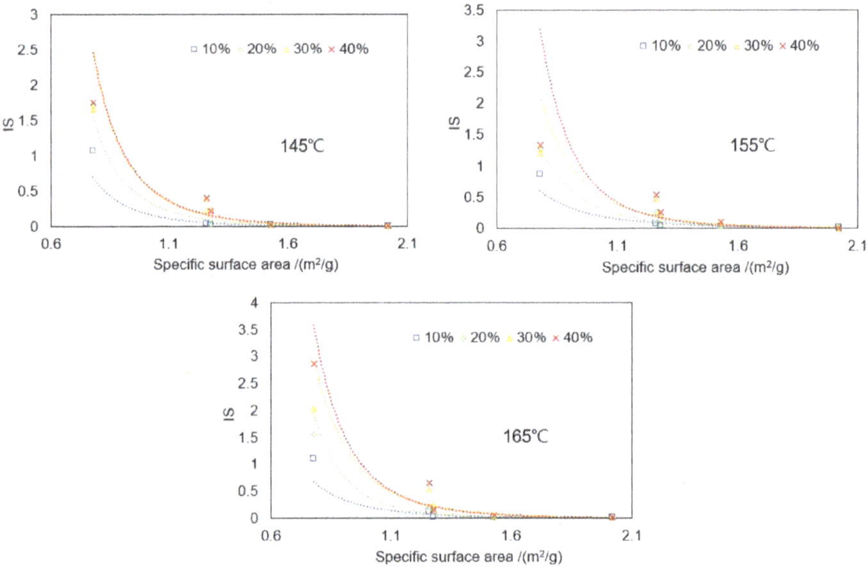

Figure 20. IS of BRA-MA after the segregation process for 24 h static storage at 145 °C, 155 °C, and 165 °C.

3.4. Storage Stability during Transportation

Transportation is a non-negligible process for storage stability and the aging of modified asphalt [32,33]. A simulated experiment was designed to investigate the storage stability variation of BRA-MA during its transportation. BRA-MA was pumped from a static container into a vehicle container that was in a delivery vehicle in a production plant. The original temperature of BRA-MA in the container was found to be around 145 °C due to heat loss in the pumping process. Then, the vehicle should transport BRA-MA to the corresponding construction site (as Figure 3 illustrated). The transportation time is usually less than 12 h, according to the on-the-spot investigation. The heat loss of BRA-MA during the transportation process can be alleviated by the insulation layer and heating equipment of the containers on the vehicles. The temperature decline in BRA-MA of 12 h during the transportation process was 20 °C, which means its temperature decreasing rate is averagely of 5 °C per 3 h. This experiment consequently simulated the segregation of BRA-MA in consideration of its cooling process during transportation.

Figure 21 explains the following stages of this simulated experiment. The original temperature of BRA-MA in the separating tubes was 145 °C. Then, BRA-MA should be placed into a temperature-controlling box, and the initial temperature is 145 °C when the time is marked as 0 h. The temperature of the box should be switched to 140 °C when the storage time is 1.5 h. Then, the temperature of the BRA-MA should be tested for the first time at 3 h to confirm that its temperature has already decreased to 140 °C. Meanwhile, the IS value at 140 °C should be tested according to the viscosity difference test conducted as the temperature test ①. As Figure 21 illustrates, the temperature and IS value of BRA-MA should be tested when its temperature gradually reduces to 135, 130, and 125 °C, as the temperature test ②, ③ and ④, respectively. The temperature of BRA-MA finally decreased to 125 °C after 12 h; the cooling-process simulation of the transportation was terminated.

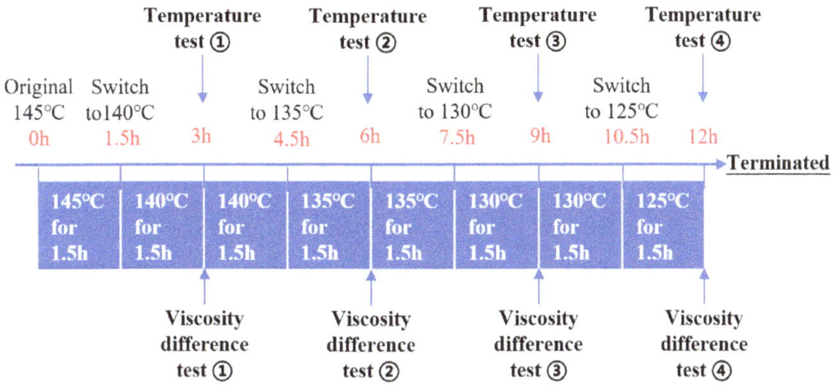

Figure 21. Outline of simulated experiment for the storage stability during transportation.

The IS values of BRA-MA with BRA-1~4 during transportation are provided in Figure 22, while the IS value of BRA-5-based BRA-MA is shown in Figure 23. The abscissa is the temperature during transportation, and its corresponding transportation times are 3, 6, 9, and 12 h. The results show that the IS values of BRA-MA with BRA-1~4 are below 0.2, suggesting that their segregation is very slight. The IS value also showed an increasing trend as the particle size rises. However, BRA-5-based BRA-MA illustrated a high IS value during transportation, proving that its segregation was serious. In addition, the IS values of BRA-MA with BRA-5 presented an increasing tendency in the cooling process. However, IS change of BRA-1~4-based BRA-MA in the cooling process was not explicit, especially when BRA content was below 30%. Therefore, BRA-MA with BRA-1~4 showed low segregation and acceptable storage stability compared to BRA-5-based BRA-MA, although their IS values changed unreasonably. BRA-MA with BRA-1~4 whose d(0.5) particle sizes were lower than 13.6 μm showed low segregation. However, the segregation of BRA-MA with BRA whose d(0.5) particle size was over 106 μm was severe.

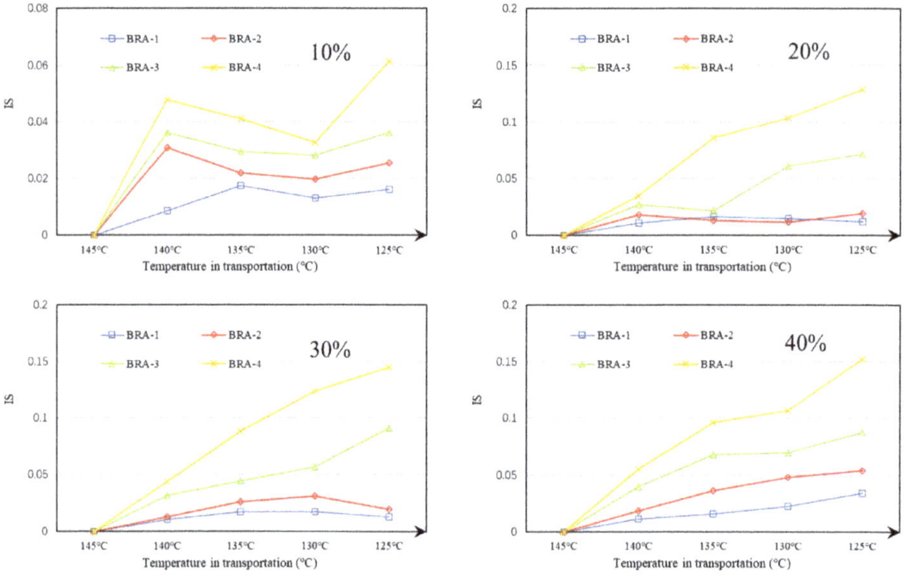

Figure 22. IS of BRA-MA with BRA-1~4 during transportation.

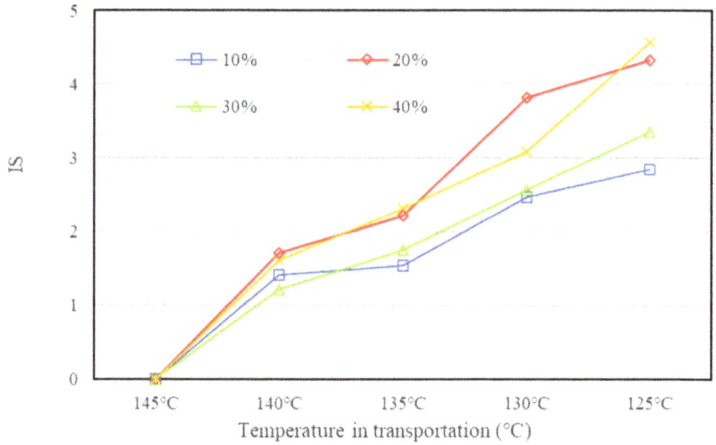

Figure 23. IS of BRA-MA with BRA-5 during transportation.

4. Conclusions

This study introduced BRA of different particle sizes and contents to prepare BRA-MA in order to determine how the factors affect their physical properties and storage stability. Storage-stability evaluation methods were also discussed. Based on the results, the following conclusions can be drawn.

(1) Softening point was positively correlated to BRA content, while the particle size of BRA showed a negative correlation. Penetration of BRA-MA increased as the increment in particle size, while BRA content negatively affected penetration. Both BRA content and particle size negatively determines the ductility of BRA-MA. A larger particle size of BRA resulted in lower viscosity, but a higher BRA content increased the viscosity of BRA-MA. The temperature susceptibility of BRA-MA de-

creased with the increase in BRA content, but the smaller particle size led to lower temperature susceptibility.

(2) Separating-tube method was used to simulate the segregation of BRA-MA in Lab. How BRA-content influence on storage stability could not be significantly reflected by the softening-point difference test, and the segregation degree was not quantitively revealed. The IS value based on the viscosity difference test had a higher statistical significance when evaluating the influence of BRA content and particle size on segregation. The segregation of asphalt with different original viscosity could be quantitatively compared through the IS value based on the viscosity difference test.

(3) Storage stabilities of static and transportation corresponded to storage in a production plant and during the vehicle transportation process. The segregation of BRA-MA illustrated a rising trend as the BRA coffintent and particle size increased. Both storage temperature and time were positively correlated to the segregation of BRA-MA. It was proved that the relationship between the specific surface area and segregation were power functional. This relationship can be used to understand how particle size, which is correlated to specific surface area, determines the segregation of BRA-MA. Based on the simulated experiment of transportation segregation, BRA-MA with BRA-1~4 whose d(0.5) particle sizes ere lower than 13.6 μm, showed low segregation. However, the segregation of BRA-MA with BRA whose d(0.5) particle size was over 106 μm was severe.

Author Contributions: Conceptualization, Y.S., X.H. (Xiaodi Hu), J.W. and X.H. (Xing Huang); methodology, software, Y.S., X.H. (Xiaodi Hu) and J.W.; validation, Y.S., X.H. (Xiaodi Hu), J.W. and S.W.; formal analysis, Y.S., X.H. (Xiaodi Hu), J.W. and S.W.; investigation, Y.S., X.H. (Xiaodi Hu), J.W. and Z.L.; resources, X.H. (Xiaodi Hu), J.W., X.H. (Xing Huang) and S.W.; data curation, J.W. and Y.Z.; writing—original draft preparation, J.W., Y.Z. and Y.S.; writing—review and editing, S.W., Z.L. and X.H. (Xiaodi Hu); visualization, Y.S. and J.W.; supervision, X.H. (Xiaodi Hu) and S.W.; project administration, S.W., Z.L. and J.W.; funding acquisition, S.W., Z.L. and J.W. All authors have read and agreed to the published version of the manuscript.

Funding: This research was funded by Technological Innovation Major Project of Hubei Province grant number (2019AEE023), Key R&D Program of Hubei Province grant number (2020BCB064), Technological Innovation Team in Universities of Hubei Province grant number (Project No. T2020010) And The APC was funded by Scientific Research Starting Foundation of Wuhan Institute of Technology grant number (No. K202021).

Institutional Review Board Statement: Not applicable.

Informed Consent Statement: Not applicable.

Data Availability Statement: Not applicable.

Acknowledgments: The authors are grateful for financial support from the Technological Innovation Major Project of Hubei Province (2019AEE023), Key R&D Program of Hubei Province (2020BCB064), the Plan of Outstanding Young and Middle-aged Scientific and Technological Innovation Team in Universities of Hubei Province (Project No. T2020010), and Scientific Research Starting Foundation of Wuhan Institute of Technology (No. K202021).

Conflicts of Interest: The authors declare no conflict of interest.

References

1. Ky, A.; Wla, B.; Ly, C.; Jo, A.; Man, Z. Evaluation of waste cooling oil and European Rock Asphalt modified asphalt with laboratory tests and economic cost comparison—ScienceDirect. *J. Clean. Prod.* **2021**, *310*, 127364.
2. Suaryana, N. Performance evaluation of stone matrix asphalt using indonesian natural rock asphalt as stabilizer. *Int. J. Pavement Res. Technol.* **2016**, *9*, 387–392. [CrossRef]
3. Liu, C.; Lv, S.; Jin, D.; Qu, F. Laboratory Investigation for the Road Performance of Asphalt Mixtures Modified by Rock Asphalt-Styrene Butadiene Rubber. *J. Mater. Civ. Eng.* **2020**, *33*, 04020504. [CrossRef]
4. Liu, H.; Zhang, Z.; Li, Z.; Li, N. Effects of polyphosphoric acid (PPA), styrene-butadiene: Tyrene (SBS), or rock asphalt on the performance of desulfurized rubber modified asphalt. *J. Appl. Polym. Sci.* **2021**, *138*, 50621. [CrossRef]

5. Wen, Y.; Guo, N.; Wang, L.; Jin, X.; Wen, H. Assessment of various fatigue life indicators and fatigue properties of rock asphalt composite. *Constr. Build. Mater.* **2021**, *289*, 123147. [CrossRef]
6. Fan, X.; Lu, W.; Lv, S.; He, F. Improvement of Low-Temperature Performance of Buton Rock Asphalt Composite Modified Asphalt by Adding Styrene-Butadiene Rubber. *Materials* **2019**, *12*, 2358. [CrossRef]
7. Lv, S.; Peng, X.; Liu, C.; Qu, F.; Zhu, X.; Tian, W.; Zheng, J. Aging resistance evaluation of asphalt modified by Buton-rock asphalt and bio-oil based on the rheological and microscopic characteristics. *J. Clean. Prod.* **2020**, *257*, 120589. [CrossRef]
8. Ming, W.A.; Cx, B. Evaluation of microstructural features of Buton rock asphalt components and rheological properties of pure natural asphalt modified asphalt. *Constr. Build. Mater.* **2020**, *267*, 121132.
9. Yang, C.; Wu, S.; Cui, P.; Amirkhanian, S.; Zhao, Z.; Wang, F.; Zhang, L.; Wei, M.; Zhou, X.; Xie, J. Performance characterization and enhancement mechanism of recycled asphalt mixtures involving high RAP content and steel slag. *J. Clean. Prod.* **2022**, *336*, 130484. [CrossRef]
10. Wan, J.; Wu, S.; Xiao, Y.; Fang, M.; Song, W.; Pan, P.; Zhang, D. Enhanced ice and snow melting efficiency of steel slag based ultra-thin friction courses with steel fiber. *J. Clean. Prod.* **2019**, *236*, 117613. [CrossRef]
11. Xue, Y.; Wei, X.; Zhao, H.; Wang, T.; Xiao, Y. Interaction of spent FCC catalyst and asphalt binder: Rheological properties, emission of VOCs and immobilization of metals. *J. Clean. Prod.* **2020**, *259*, 120830. [CrossRef]
12. Cui, P.; Wu, S.; Xiao, Y.; Hu, R.; Yang, T. Environmental performance and functional analysis of chip seals with recycled basic oxygen furnace slag as aggregate. *J. Hazard. Mater.* **2020**, *405*, 124441. [CrossRef] [PubMed]
13. Wang, F.; Xie, J.; Wu, S.; Li, J.; Zhang, L. Life cycle energy consumption by roads and associated interpretative analysis of sustainable policies. *Renew. Sustain. Energy Rev.* **2021**, *141*, 110823. [CrossRef]
14. Leviton, A.; Pallansch, M.J. High-Temperature Short-Time Sterilized Evaporated Milk. III. The Influence of the Lipid Phase on Heat and Storage Stability. *J. Dairy Sci.* **1961**, *44*, 633–643. [CrossRef]
15. Yang, Z.; Kumar, A.; Huhnke, R.L. Review of recent developments to improve storage and transportation stability of bio-oil. *Renew. Sustain. Energy Rev.* **2015**, *50*, 859–870. [CrossRef]
16. Galooyak, S.S.; Dabir, B.; Nazarbeygi, A.E.; Moeini, A. Rheological properties and storage stability of bitumen/SBS/montmorillonite composites. *Constr. Build. Mater.* **2010**, *24*, 300–307. [CrossRef]
17. Yoshida, H.; Zhang, G.Z.; Kitamura, T.; Kawai, T. Compatibility of Polymer Blends Evaluated by Crystallization Dynamics. Simultaneous DSC-FTIR method. *J. Therm. Anal. Calorim.* **2001**, *64*, 577–583. [CrossRef]
18. Gu, Y.; Tang, B.; He, L.; Yang, F.; Wang, H.; Ling, J. Compatibility of cured phase-inversion waterborne epoxy resin emulsified asphalt. *Constr. Build. Mater.* **2019**, *229*, 116942. [CrossRef]
19. Xiaolong, Z.; Aimin, S.; Biao, D.; Yuqiao, T.; Xiaonan, H. Evaluation and Analysis of Variance of Storage Stability of Asphalt Binder Modified by Nanotitanium Dioxide. *Adv. Mater. Sci. Eng.* **2017**, *2017*, 6319697.
20. Navarro, F.J.; Partal, P.; Martınez-Boza, F.; Gallegos, C. Thermo-rheological behaviour and storage stability of ground tire rubber-modified bitumens. *Fuel* **2004**, *83*, 2041–2049. [CrossRef]
21. Ming, L.; Xue, X.; Fan, W.; Sun, H.; Yan, Y.; Xing, B. Viscous properties, storage stability and their relationships with microstructure of tire scrap rubber modified asphalt. *Constr. Build. Mater.* **2015**, *74*, 124–131.
22. Fang, C.; Pei, L.; Yu, R.; Liu, X. Preparation process to affect stability in waste polyethylene-modified bitumen. *Constr. Build. Mater.* **2014**, *54*, 320–325. [CrossRef]
23. Ym, A.; Pp, A.; Wei, H.A.; Mz, A.; Bh, B. Quantifying the effective mobilized RAP content during hot in-place recycling techniques. *J. Clean. Prod.* **2021**, *314*, 127953.
24. Wang, R.; Xiong, Y.; Yue, M.; Hao, M.; Yue, J. Investigating the effectiveness of carbon nanomaterials on asphalt binders from hot storage stability, thermodynamics, and mechanism perspectives. *J. Clean. Prod.* **2020**, *276*, 124180. [CrossRef]
25. Zhang, R.; Wang, H.; Jiang, X.; You, Z.; Yang, X.; Ye, M. Thermal Storage Stability of Bio-Oil Modified Asphalt. *J. Mater. Civ. Eng.* **2018**, *30*, 04018054. [CrossRef]
26. Feih, S.; Manatpon, K.; Mathys, Z.; Gibson, A.G.; Mouritz, A.P. Strength degradation of glass fibers at high temperatures. *J. Mater. Ence* **2009**, *44*, 392–400. [CrossRef]
27. Liu, W.; Li, H.; Lin, H.; Du, X.; Liu, S. Effects of a Fuel-Resistant Modifier on the High-Temperature Characteristics of Asphalt Binders. *Sustainability* **2021**, *13*, 7924. [CrossRef]
28. Rasmussen, R.O.; Lytton, R.L.; Chang, G.K. Method to Predict Temperature Susceptibility of an Asphalt Binder. *J. Mater. Civ. Eng.* **2002**, *14*, 246–252. [CrossRef]
29. Wang, P.; Yong, W.; Zhao, K.; Dan, C.; Wong, A. Evolution and locational variation of asphalt binder aging in long-life hot-mix asphalt pavements. *Constr. Build. Mater.* **2014**, *68*, 172–182. [CrossRef]
30. Zhao, X.; Yan, K.; Yang, S.; Peng, H. Laboratory Research on the Properties of Warm Amorphous Poly Alpha Olefin–Modified Asphalt Mixture Using Sasobit and Deurex. *J. Mater. Civ. Eng.* **2018**, *30*, 04018076. [CrossRef]
31. Zhang, C.; Yu, J.; Xue, L.; Sun, Y. Investigation of γ-(2,3-Epoxypropoxy)propyltrimethoxy Silane Surface Modified Layered Double Hydroxides Improving UV Ageing Resistance of Asphalt. *Materials* **2017**, *10*, 78. [CrossRef] [PubMed]
32. Hallmark-Haack, B.L.; Hernandez, N.B.; Williams, R.C.; Cochran, E. Ground Tire Rubber Modification for Improved Asphalt Storage Stability. *Energy Fuels* **2019**, *33*, 2659–2664. [CrossRef]
33. Zhu, Y.; Zhang, J.; Si, C.; Yan, T.; Li, Y. Laboratory Evaluation on Performance of Recycled Asphalt Binder and Mixtures under Short-Term Aging Conditions. *Sustainability* **2021**, *13*, 3404. [CrossRef]

MDPI
St. Alban-Anlage 66
4052 Basel
Switzerland
Tel. +41 61 683 77 34
Fax +41 61 302 89 18
www.mdpi.com

Materials Editorial Office
E-mail: materials@mdpi.com
www.mdpi.com/journal/materials